Advances in Experimental Medicine and Biology

# Proteomics, Metabolomics, Interactomics and Systems Biology

Volume 1511

*Advances in Experimental Medicine and Biology* provides a platform for scientific contributions in the main disciplines of the biomedicine and the life sciences. This series publishes thematic volumes on contemporary research in the areas of microbiology, immunology, neurosciences, biochemistry, biomedical engineering, genetics, physiology, and cancer research. Covering emerging topics and techniques in basic and clinical science, it brings together clinicians and researchers from various fields.

*Advances in Experimental Medicine and Biology* has been publishing exceptional works in the field for over 40 years, and is indexed in SCOPUS, Medline (PubMed), EMBASE, Reaxys, EMBiology, the Chemical Abstracts Service (CAS), and Pathway Studio.

2023 CiteScore: 5.9

Timothy Daniel Veenstra

# Proteomic Applications in Virology

Springer

Timothy Daniel Veenstra
College of Medicine
Lake Erie College of Osteopathic Medicine
Bradenton, FL, USA

ISSN 0065-2598 ISSN 2214-8019 (electronic)
Advances in Experimental Medicine and Biology
ISSN 2730-6216 ISSN 2730-6224 (electronic)
Proteomics, Metabolomics, Interactomics and Systems Biology
ISBN 978-3-032-22339-5 ISBN 978-3-032-22340-1 (eBook)
https://doi.org/10.1007/978-3-032-22340-1



This Springer imprint is published by the registered company Springer Nature Switzerland AG
The registered company address is: Gewerbestrasse 11, 6330 Cham, Switzerland

*To my beautiful wife, you are everything I could ever desire. Thank you for filling my life with love and joy*

# Preface

I have lived through several historical events during my lifetime. Although I was too young to remember it, the first of these was Neil Armstrong taking the first steps on the moon. I clearly remember where I was when I heard about the explosion of the Space Shuttle Challenger in 1986. Although I watched the Tiananmen Square protests in 1989, I honestly didn't understand their political implications. The attack on the World Trade Center in New York City (i.e., 9/11) happened the morning my family and I were preparing to move across the United States. Although a lot of negativities came from these attacks, both domestically and foreign, patriotism across the United States was on full display as people had flags adorning their yards and cars for several weeks. It was a time that brought the country together.

The most recent historical event that we have lived through as a global community is the coronavirus disease 2019 (COVID-19) pandemic. This pandemic has been unlike any that almost everyone alive today has lived through. With an estimate of over 7 million deaths from the COVID-19 pandemic, it dwarfed the number killed during the 2009 H1N1 pandemic. With the mandate for social distancing, the COVID-19 pandemic brought with it the closure of the world's economy and social fabric. While I hoped that it would bring with it a spirit of perseverance and community as everyone struggled through this pandemic, the result was division and disagreement that seems to continue within other sectors of our global society.

I have also lived during one of the greatest times to be a scientist. Over the past four decades the polymerase chain reaction was developed, the human genome was sequenced, immunotherapy breakthroughs helped in the battle against cancers, and proteomics technology rapidly advanced. Although I didn't know it at the time, my first introduction to proteomics technology was my fourth year honors project as an undergraduate student. I was working on a method to purify a protein from *Campylobacter jejuni* with the purpose of testing the ability of fast-atom bombardment mass spectrometry (FAB-MS) to measure its molecular weight. This project was in the early 1990s before the use of MS in analyzing proteins was commonplace. Fast forward a few years and my post-doctoral supervisor discussed pursuing a two-dimensional polyacrylamide gel electrophoresis (2D-PAGE) and MS strategy to identify proteins that were regulated by vitamin D. While we didn't pursue this project, we did use a yeast two-hybrid (Y2H) assay to identify proteins that interact with the calcium-binding protein calbindin $D_{28K}$.

As proteomics began to mature in the late 1990s, I had gained enough experience to be named the director of a proteomics laboratory that the National Cancer Institute was building. Some of the lab's earliest efforts were characterizing the plasma proteome. By combining our results with those of three other labs, we came up with a list of 1175 unique plasma proteins, which was considered a major accomplishment 20 years ago. Today, it is commonplace for proteomic labs to not simply identify thousands of proteins within a proteome, but quantitate them, measure their phosphorylation status, and determine protein-protein interactions (PPIs). Beyond the technology, what has truly advanced proteomics is the cooperation with other scientific disciplines. During its infancy, MS labs were isolated as basic research labs did not yet understand how proteomics could further their research. Nowadays, proteomics is widely used thoughout basic research not only resulting in greater insight into how cells and organisms function but enabling higher quality hypotheses to be generated.

These developments in technology and cooperation are evident throughout the chapters of this book. None of the projects described in this book simply identify proteins present within a virus or its host cell. Rather, they focus on challenges such as identifying how viral infection effect PPIs within the host cell, finding circulating prognostic biomarkers, and using proteomic data to identify drugs that can be repurposed to combat viral infection or identify druggable targets within the host that can be used to treat infection by several different types of viruses. My hope is that this book continues to advance this trend of scientific cooperation by showing proteomic experts how their technology can impact viral research and virus experts how they can utilize proteomic technology to learn more about how viruses impact life on our planet.

One thought that kept occurring to me during the writing of this book is not how far proteomics has come in the past two decades, but where it will be two decades from now. I often remind young scientists and medical professionals that their grandparents were alive before the structure of DNA was known. Now it is possible to sequence an entire human genome in less than a day. Soon every newborn will have the opportunity to have its genome sequenced. I am now envisioning the day in which proteomic profiling of serum or plasma becomes part of a routine physical examination. This capability will enable the identification of changes in an individual's proteome that can diagnose diseases that may be present at preclinical stages. There is still a lot to do to achieve this goal, but the progress in getting there has been exceptional.

Bradenton, FL, USA Timothy Daniel Veenstra

# Acknowledgments

Writing a book requires many things from different people. The writer must be knowledgeable and persistent; two attributes that they must supply themselves. What the writer needs more than anything that is supplied by others is encouragement. I can remember speaking with my brother and telling him I should not have accepted this task as I started to feel I was doing it to satisfy my own ego. Having known me for several decades, he told me that I was really doing it because of the challenge. Those words went a long way in keeping me going. The greatest encouragement I received, however, was from my amazing wife Christine. She could not count the number of times I discussed this book with her. I talked about things I had learned and how much more work I had left to finish. More than anything, however, she listened to my frustration in the progress I was making. She would sit patiently at night and on weekends as I edited chapters. She always responded with words that kept me moving forward. To Jacob, Prairie, and Benjamin, I want you to know that I am prouder of you than anything I have ever accomplished in my career. Ultimately, what made me complete this book was that I gave Springer my word, and my conscience would not let me go back on that promise. On that note, I want to thank Springer for their patience. As I often do, I overestimated my efficiency and the amount of time I would be able to dedicate to researching and writing this book, so thank you for sticking with me.

**Competing Interests** The author has no competing interests to declare that are relevant to the content of this manuscript.

# Contents


**1 Proteomic Technology** . . . . . 1
1.1 Introduction . . . . . 2
1.2 How Has Proteomics Changed Protein Science? . . . . . 3
1.2.1 Hypothesis- to Discovery-Driven . . . . . 4
1.3 Proteomic Technology . . . . . 6
1.3.1 Mass Spectrometry . . . . . 6
1.3.2 Bottom-Up Versus Top-Down . . . . . 8
1.4 The Critical Need for Protein Fractionation . . . . . 9
1.4.1 Polyacrylamide Gel Electrophoresis . . . . . 10
1.4.2 Liquid Chromatography . . . . . 11
1.4.3 Protein Complex Isolation . . . . . 14
1.4.4 Immunoprecipitation . . . . . 14
1.4.5 Pull-Down Assays . . . . . 15
1.4.6 Co-fractionation Mass Spectrometry . . . . . 15
1.4.7 Cross-Linking and Proximity-Linking Assays . . . . . 16
1.4.8 Yeast Two-Hybrid Assays . . . . . 17
1.5 Protein Identification . . . . . 17
1.5.1 Tandem Mass Spectrometry . . . . . 18
1.5.2 Electron Capture and Electron Transfer Dissociation . . . . . 19
1.5.3 A Typical MS Proteomic Analysis . . . . . 20
1.6 Protein Arrays . . . . . 20
1.6.1 Types of Protein Arrays . . . . . 22
1.6.2 Target Protein Arrays . . . . . 22
1.6.3 Antibody Arrays . . . . . 23
1.6.4 Reverse-Phase Protein Arrays . . . . . 23
1.6.5 Tissue Arrays . . . . . 24
1.7 Conclusions . . . . . 24
References . . . . . 25

**2 Basic Virology** . . . . . 29
2.1 Introduction . . . . . 29
2.2 Are Viruses Alive? . . . . . 31
2.3 How Are Viruses Named? . . . . . 31
2.4 Size of Viruses . . . . . 32
2.5 Virus Structure . . . . . 33
2.5.1 Enveloped Virus . . . . . 34
2.5.2 Non-enveloped Viruses . . . . . 35
2.5.3 Viral Shape . . . . . 35
2.6 Viral Genomes . . . . . 36
2.7 Genomic Classification of Viruses . . . . . 37
2.7.1 Size of Viral Genomes . . . . . 39
2.7.2 What's Encoded in a Viral Genome? . . . . . 39
2.8 Viral Infection . . . . . 40
2.8.1 Viral Egress . . . . . 41
2.9 Viral Replication . . . . . 41
2.9.1 HIV Replication . . . . . 41
2.9.2 Influenza A Replication . . . . . 44
2.10 Conclusions . . . . . 46
References . . . . . 47

**3 Proteomics in the Diagnosis of Viral and Pathogenic Infections** . . . . 51
3.1 Introduction . . . . . 51
3.2 Clinical Virology . . . . . 52
3.2.1 Indirect Detection of Viral Cytopathic Effects . . . . . 53
3.2.2 Indirect Detection of Viral Antibodies . . . . . 54
3.2.3 Direct Detection Methods . . . . . 56
3.2.4 Immunological-Based Direct Detection . . . . . 56
3.2.5 Nucleic Acid-Based Direct Detection . . . . . 57
3.2.6 Direct Detection Using Microarrays . . . . . 58
3.2.7 Future Nucleic Acid-Based Detection Methods . . . . . 59
3.3 Proteomic Contributions to Clinical Virology . . . . . 60
3.3.1 Proteomics in Clinical Bacteriology . . . . . 60
3.3.2 Mass Spectrometry in Clinical Virology . . . . . 61
3.4 Clinical Diagnosis of SARS-CoV-2 . . . . . 64
3.4.1 Clinical Detection of SARS-CoV-2 Using Proteomic Patterns . . . . . 65
3.4.2 Clinical Diagnosis of SARS-CoV-2 Using Targeted Proteomics . . . . . 66
3.5 Limitations of Targeted Proteomic Clinical Virology Assays . . . . 70
3.6 Conclusions . . . . . 71
References . . . . . 72

**4 Viral Prognosis Using Proteomics** . . . . . 75
4.1 Introduction . . . . . 75
4.2 Viral Characteristics That Affect Severity of Disease . . . . . 76

4.3 Host Characteristics That Affect Severity of Disease .......... 77
4.4 The Value of Prognosis .......................... 80
4.4.1 Prognosis of Virus Infections........................ 80
4.4.2 Prognosis Based on Modified World Health Organization Score.............................. 82
4.5 Plasma Analysis to Identify Prognostic Markers of COVID-19 ................................ 83
4.5.1 Stratification of Patients with COVID-19 ............ 84
4.5.2 Protein Clusters Identified in Different COVID-19 Trajectory Groups .......................... 86
4.5.3 Biological Pathways Associated with Negative Outcomes.................................. 86
4.5.4 Biomarkers Associated with Death in Trajectory Group 4 Patients.................................. 89
4.5.5 Biological Pathways Associated with Positive Outcomes .................................. 89
4.5.6 Specific Proteins for COVID-19 Prognosis in Trajectory Group 4 .............................. 90
4.5.7 Prognostic Severity Biomarker Outcome Across All COVID-19 Cohorts .......................... 91
4.6 Search for Biomarkers That Indicate the Need for Intensive Care.................................... 92
4.7 Long COVID ................................. 93
4.7.1 Identification of Biomarkers Prognostic of Long COVID ................................. 94
4.8 Conclusion ................................. 97
References.................................... 98

**5 Identification of Virus-Host Interactions** ................ 103
5.1 Introduction ................................ 104
5.2 Virus-Host Cell Interactions ...................... 104
5.2.1 Viral Entry ............................. 105
5.3 Proteomic Methods for Identifying Virus-Host Interactions ..... 106
5.3.1 Viral/Host Cell Receptor Interaction Identification Using Mass Spectrometry ..................... 107
5.3.2 Sample Preparation Techniques.................... 110
5.3.3 Yeast Two-Hybrid Identification of Protein-Protein Interactions.............................. 111
5.3.4 Pull-Down Assays ......................... 114
5.4 Virus-Host Protein-Protein Interaction Databases............ 117
5.4.1 Analysis of PPIs Within Databases ................ 118
5.4.2 Using Databases to Explain Symptoms of Viral Infections ............................... 120
5.4.3 Using PPI Databases to Identify Potential New Drugs ... 121
5.5 Conclusions ................................ 122
References.................................... 123

**6 The Pathology of Viral Infections** . . . 127
6.1 Introduction . . . 127
6.2 Pathogenesis of Viruses . . . 129
6.2.1 Identifying Host Proteins Involved in Viral Entry . . . 129
6.2.2 Identifying Host Cell Receptors Via Immunoprecipitation . . . 131
6.2.3 Identifying Host Cell Receptors Via Proximity Labeling . . . 133
6.3 Identifying Proteins Involved in Viral Replication . . . 136
6.3.1 Host Proteins Within Virions . . . 136
6.4 Characterizing Changes in the Proteome of Virus-Infected Cells . . . 137
6.4.1 Dynamic Changes in the Proteome of Adenovirus-Infected Cells . . . 138
6.4.2 Validation of Host Proteins Affected by Adenovirus Infection . . . 143
6.5 Proteomic Analysis of Virus Infection Using Protein Isotopic Labeling . . . 144
6.5.1 Proteomic Analysis of H1N1- and H5N1-Infected Mast Cells . . . 145
6.6 Viral Egress . . . 148
6.6.1 Proteomic Analysis of Viral Egress . . . 148
6.6.2 Proteomic Analysis of Non-lytic Release . . . 149
6.6.3 Validation of Quasi-Enveloped Hepatitis A Exosome Proteins . . . 152
6.7 Conclusions . . . 153
References . . . 154

**7 Proteomic Analysis of COVID-19 Infection** . . . 159
7.1 Introduction . . . 160
7.2 The SARS-CoV-2 Life Cycle . . . 161
7.3 SARS-CoV-2 Infects Proteomic Research . . . 163
7.4 Novel Proteomic Tests to Diagnose COVID-19 . . . 165
7.5 Host Response to COVID-19 . . . 167
7.5.1 Proteomic Analysis of Blood Proteins . . . 167
7.5.2 Determining the Tissue Origin of SARS-CoV-2-Affected Proteins . . . 169
7.5.3 Serum Proteins Traceable to SARS-CoV-2 Infection Site . . . 170
7.5.4 Tissue Origin of Blood Proteins Identified in SARS-CoV-2-Infected Patients . . . 171
7.6 Value in Identifying Serum Proteins Affected by SARS-CoV-2 . . . 172
7.7 Proteomic Analysis of SARS-CoV-2 Cell Surface Interactions . . . 174

7.8 A SARS-CoV-2/Human Cell Interaction Map. . . . . . . . . . . . . . . . 175
7.8.1 Global Mapping of SARS-CoV-2/Human Protein-Protein Interactions . . . . . . . . . . . . . . . . . . . . . . . . 176
7.9 Intelligent Drug Repurposing Through Protein-Protein Interaction Data . . . . . . . . . . . . . . . . . . . . . . . . . . . . . . 180
7.10 Proteomic Databases of COVID-19 Studies . . . . . . . . . . . . . . . . . 181
7.10.1 CoVProt . . . . . . . . . . . . . . . . . . . . . . . . . . . . . . . . 181
7.10.2 COVIDpro . . . . . . . . . . . . . . . . . . . . . . . . . . . . . . 182
7.10.3 CoViMAPP . . . . . . . . . . . . . . . . . . . . . . . . . . . . . 183
7.10.4 CoV3D. . . . . . . . . . . . . . . . . . . . . . . . . . . . . . . . . 183
7.11 Conclusions . . . . . . . . . . . . . . . . . . . . . . . . . . . . . . . . . . . . . . 184
References. . . . . . . . . . . . . . . . . . . . . . . . . . . . . . . . . . . . . . . . . . . . . 184

**8 Proteomic Characterization of HIV Infection** . . . . . . . . . . . . . . . . . . . 189
8.1 Introduction . . . . . . . . . . . . . . . . . . . . . . . . . . . . . . . . . . . . . . . 189
8.2 Human Immunodeficiency Virus . . . . . . . . . . . . . . . . . . . . . . . . . . 190
8.2.1 The Structure of HIV . . . . . . . . . . . . . . . . . . . . . . . . . . . . 192
8.2.2 Replication and Life Cycle of HIV . . . . . . . . . . . . . . . . . . 194
8.3 Proteomics and HIV. . . . . . . . . . . . . . . . . . . . . . . . . . . . . . . . . . 196
8.3.1 Circulating Biomarkers of HIV Infection . . . . . . . . . . . . . 197
8.3.2 Analysis of Plasma Proteome at Different Stages of HIV Infection. . . . . . . . . . . . . . . . . . . . . . . . . . . . . . . . 197
8.3.3 The Effect of HIV Infection on Functional Pathways . . . . 199
8.3.4 Biomarkers Associated with Different Stages of HIV Infection . . . . . . . . . . . . . . . . . . . . . . . . . . . . . . . . . . . . 202
8.3.5 Plasma Biomarkers for Monitoring Viral Load . . . . . . . . . 202
8.3.6 Plasma Biomarkers of HIV Disease Progression. . . . . . . . 204
8.4 Identifying Biomarkers of Comorbidities in People Living with HIV . . . . . . . . . . . . . . . . . . . . . . . . . . . . . . . . . . . . . . . . . . . 206
8.4.1 Plasma Proteomic Analysis Using Proximity Extension Assays . . . . . . . . . . . . . . . . . . . . . . . . . . . . . . . . . . . . . . 207
8.4.2 Identification of Comorbidity Biomarkers. . . . . . . . . . . . . 208
8.4.3 Tissue and Cell Origin of Comorbidity Biomarkers . . . . . 211
8.4.4 Association of Putative Biomarkers with Cardiovascular Disease . . . . . . . . . . . . . . . . . . . . . . . . . . . . . . . . . . . . . . 211
8.4.5 What Was Found . . . . . . . . . . . . . . . . . . . . . . . . . . . . . . . 211
8.4.6 Correlation Between Putative HIV Biomarkers and the Microbiome. . . . . . . . . . . . . . . . . . . . . . . . . . . . . 212
8.5 Challenges in Investigating Viral Infections . . . . . . . . . . . . . . . . . 214
8.6 The Need for Canonical Proteins in Proteomics . . . . . . . . . . . . . . 214
8.6.1 Normalization Using Canonical Proteins. . . . . . . . . . . . . . 215
8.7 Conclusions . . . . . . . . . . . . . . . . . . . . . . . . . . . . . . . . . . . . . . . 216
References. . . . . . . . . . . . . . . . . . . . . . . . . . . . . . . . . . . . . . . . . . . . . 217

**9 Proteomic Analysis of Influenza** . . . . . . . . . . . . . . . . . . . . . . . . . . . . . 223
9.1 Introduction . . . . . . . . . . . . . . . . . . . . . . . . . . . . . . . . . . . . . . . . 223
9.2 The Influenza Viruses. . . . . . . . . . . . . . . . . . . . . . . . . . . . . . . . . 225
9.2.1 The Structure of Influenza Viruses. . . . . . . . . . . . . . . . . . . 226
9.2.2 Genomic Structure of Influenza. . . . . . . . . . . . . . . . . . . . . 227
9.3 The Influenza Virus Life Cycle . . . . . . . . . . . . . . . . . . . . . . . . . . . 228
9.3.1 Virus Attachment and Entry. . . . . . . . . . . . . . . . . . . . . . . . 228
9.3.2 Synthesis of Viral RNA . . . . . . . . . . . . . . . . . . . . . . . . . . . 230
9.3.3 Synthesis of Viral Proteins. . . . . . . . . . . . . . . . . . . . . . . . . 230
9.3.4 Packaging of RNA and Assembly of Virus . . . . . . . . . . . . 230
9.3.5 Virus Budding and Release . . . . . . . . . . . . . . . . . . . . . . . . 231
9.4 Proteomic Analysis to Identify Pan-Viral Therapies to Treat Influenza A Infections . . . . . . . . . . . . . . . . . . . . . . . . . . . . . . . . . 231
9.4.1 Proteomic Identification of Host-Directed Therapy Targets . . . . . . . . . . . . . . . . . . . . . . . . . . . . . . . . . . . . . . . . . . 232
9.4.2 Identification of Host Kinase Activity Affected by Influenza Infection . . . . . . . . . . . . . . . . . . . . . . . . . . . . 235
9.4.3 Recognition of Host-Directed Therapies Against Influenza. . . . . . . . . . . . . . . . . . . . . . . . . . . . . . . . . . . . . . . . 237
9.4.4 Identifying a “Universal” Host-Directed Therapeutic . . . . 239
9.4.5 Connecting Therapy to Mechanism. . . . . . . . . . . . . . . . . . 240
9.4.6 The Job Is Not Yet Finished. . . . . . . . . . . . . . . . . . . . . . . . 240
9.5 The Role of Adjuvants in Vaccination . . . . . . . . . . . . . . . . . . . . . 241
9.5.1 Tissue Proteomics to Understand Adjuvant Mechanism. . . . . . . . . . . . . . . . . . . . . . . . . . . . . . . . . . . . . 243
9.5.2 Mechanism of Action of Adjuvants . . . . . . . . . . . . . . . . . . 244
9.5.3 Proteins Unique to Radiofrequency Adjuvant Treatment . . . . . . . . . . . . . . . . . . . . . . . . . . . . . . . . . . . . . . 249
9.5.4 Impact on Understanding the Mechanism of Action of Adjuvants. . . . . . . . . . . . . . . . . . . . . . . . . . . . . . . . . . . . 250
9.6 Conclusions . . . . . . . . . . . . . . . . . . . . . . . . . . . . . . . . . . . . . . . . 251
References. . . . . . . . . . . . . . . . . . . . . . . . . . . . . . . . . . . . . . . . . . . . . 253

**10 Proteomic Applications in Vaccine Development**. . . . . . . . . . . . . . . . 259
10.1 Introduction . . . . . . . . . . . . . . . . . . . . . . . . . . . . . . . . . . . . . . . 260
10.2 Characteristics of a Good Vaccine . . . . . . . . . . . . . . . . . . . . . . . . 261
10.3 Identification of Optimal Antigens and Epitopes. . . . . . . . . . . . . . 262
10.3.1 Epitope Mapping Using Peptide Arrays . . . . . . . . . . . . . . 264
10.3.2 Construction of Synthetic Peptide Arrays for Broad Proteome Coverage . . . . . . . . . . . . . . . . . . . . . . . . . . . . . . 264
10.3.3 Polyclonal Antibody Testing Using a Protein Array . . . . . 265
10.4 On- and Off-Target Binding. . . . . . . . . . . . . . . . . . . . . . . . . . . . . 266
10.4.1 Physiological Effect of Off-Target Antibody Binding . . . . 267
10.5 Advantages and Disadvantages of Peptide Array Antibody Validation. . . . . . . . . . . . . . . . . . . . . . . . . . . . . . . . . . . . . . . . . . 269

10.6 Immunopeptidomics .................................... 270
10.6.1 Using MHC-I as a Prefilter for Antigen Recognition .... 271
10.6.2 Identification of Immunopeptides .................. 272
10.6.3 Application of Immunopeptidomics to Develop Tuberculosis Vaccines .......................... 273
10.7 Identification of Potential Vaccine Candidates Using Existing Proteomic Datasets .................................... 275
10.8 *Wuchereria bancrofti* Vaccine Development via Proteomics and Immunoinformatics ................................ 276
10.8.1 Design and Characterization of a Multi-epitope Vaccine Construct ....................................... 278
10.8.2 Immune Simulation of Predicted Vaccine Construct ..... 279
10.9 Vaccine Candidate Identification Using a Proteomics/Bioinformatics Strategy ...................... 280
10.9.1 Analysis of Biological Samples to Identify Vaccine Candidates ...................................... 281
10.9.2 Antigen Identification Using Subtractive Proteomics .... 281
10.9.3 Bioinformatic Analysis of *Acinetobacter baumannii* Vaccine Candidates ............................ 282
10.10 Conclusions ........................................ 283
References .............................................. 284

**Index** ................................................ 289

# Chapter 1
# Proteomic Technology

**Abstract** The proteome is defined as the total protein complement of a genome—whether that genome is associated with an individual cell, tissue, or organism (Wasinger et al., Electrophoresis 16:1090–1094, 1995). Proteomics refers to the application of technologies to characterize the proteome. The term "proteome" was coined by researchers at the University of Sydney, Australia, in a paper published in 1995 (Wasinger et al., Electrophoresis 16:1090–1094, 1995). But did proteomics really begin in 1995 or were researchers doing proteomic research long before? As a PhD student, I would tell people that my field of expertise was protein science, as I was involved in determining the three-dimensional structures of proteins using nuclear magnetic resonance (NMR) spectroscopy. As my career evolved to using mass spectrometry (MS) to analyze complex mixtures of proteins, my field of expertise was proteomics. While still studying proteins, what had changed was the number of proteins I was analyzing per experiment. In a nutshell, proteomics is simply the large-scale study of proteins. While protein scientists have been studying proteins for almost 200 years (Mulder, Bulletin des Sciences Physiques et Naturelles en Néerlande 104, 1838), proteomics has increased the throughput and the ability to characterize proteins in the context of other proteins within their environment. While well-known for its high-throughput ability to identify and quantitate proteins, proteomics encompasses several different facets of protein characterization including such things as structure, function, protein-protein interactions, posttranslational modifications, and protein quantitation. Data acquired under any of these broad categories can be applied to greater understand something part of the cell system.

**Keywords** Proteomics • Discovery-driven • Mass spectrometry • Liquid chromatography • Protein complex • Pull-down assay • Yeast two-hybrid assay • Protein array



T. D. Veenstra, *Proteomic Applications in Virology*, Advances in Experimental Medicine and Biology 1511, https://doi.org/10.1007/978-3-032-22340-1_1

## 1.1 Introduction

The origins of modern-day proteomics can be traced to the use of two-dimensional (2D) gel electrophoresis to separate proteins extracted from guinea pig, mouse, and *Escherichia coli* (Scheele 1975; Klose 1975; O'Farrell 1975). The ability to identify proteins separated on these gels was added a decade later with the addition of amino acid sequencing to the process (Aebersold et al. 1987; Kennedy et al. 1988). It took almost another decade to develop the technology that would allow proteins separated on gels to be identified using MS (Wilm et al. 1996). In one of these early studies, ten proteins from yeast that had been separated on a one-dimensional (1D) gel were identified (Shevchenko et al. 1996). While gel electrophoresis was a critical technology in the development of proteomics, investigators are currently identifying thousands of proteins per study using fractionation techniques that obviate the use of gel electrophoresis (Fig. 1.1).

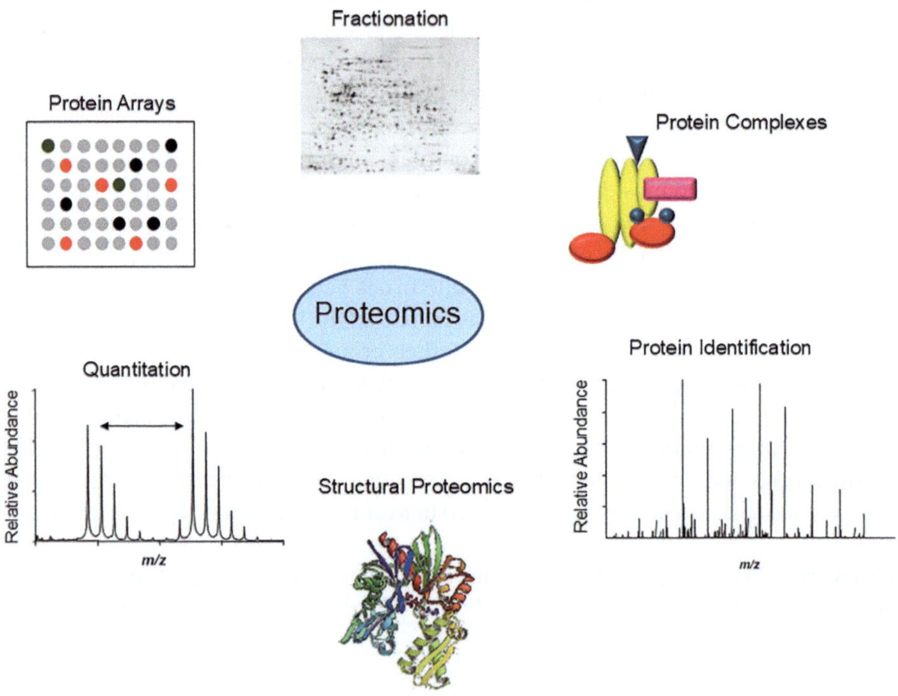


**Fig. 1.1** The various types of experiments conducted in proteomics. While it has expanded over the years to include such fields as structural proteomics and protein arrays, the major areas in proteomics are centered around protein identification, quantitation, and characterization of protein complexes

## 1.2 How Has Proteomics Changed Protein Science?

Up until a few decades ago, the proteome was studied using a reductionist, hypothesis-driven strategy. A reductionist strategy assumes that a complex system can be explained using a "divide and conquer" approach (Van Regenmortel 2004). The system is broken down into simpler units that are analyzed, and results obtained from these individual studies are collated to explain how the system functions. The reductionist strategy has been effective in explaining many important processes that occur within living systems, especially in the areas of metabolism, developmental biology, immunology, and human disease. Reductionism is the primary strategy used to diagnose diseases and monitor their treatment (Ahn et al. 2006a, b). Many complex biological systems, however, cannot be explained through studying their individual parts. Not only do reductionist methods underestimate their complexity, but they generally use simple laws to explain the behavior of the constituents within these systems (Van Regenmortel 2004). A purely reductionist strategy can limit the progress of many areas of medical research including disease progression, drug discovery, vaccine development, and transplant medicine.

Owing to its ability to concurrently analyze multiple proteins, proteomics has enabled a systems biology strategy to be applied to the proteome (Veenstra 2021). In its purest form, systems biology doesn't divide a complex system into its simplest units. Rather, it uses a holistic approach so that the interconnectedness between proteins can be measured (Aderem 2005). By applying computational and mathematical tools to the copious amount of data, both the function of the entire system can be explained, and predictions can be made as to how the entire system would be affected if any component within it was perturbed. Since biological systems are extremely complex and possess evolving properties, only a systems strategy would be capable of explaining how they function and adapt. Their complexity also requires that scientists from a variety of disciplines, including molecular biology, computational science, physics, engineering, chemistry, and mathematics, cooperate in developing models that explain how biological systems function.

What is required to gain a systems biology view of a cell or organism, which is constantly changing due to internal, external, and temporal influences? Three major factors that need to be considered are context, time, and space (Bardini et al. 2017). Context measures all the proteins that are specifically involved in a process, time measures temporal changes that occur to these proteins, and space measures the location within the system that the proteins occupy. Measuring these factors requires technology that can acquire data within a reasonable time and computational tools that can determine the best-fit model that accommodates the data.

### 1.2.1 Hypothesis- to Discovery-Driven

Another major advance catalyzed by the development of proteomics is the movement from hypothesis- to discovery-driven studies (Anderson 2008). Biology has been dominated by hypothesis-driven studies in which experiments are designed to test the validity of a hypothesis proposed to explain an observed phenomenon. Often, the hypothesis is focused on a very specific, detailed question (see Fig. 1.2). An example of a hypothesis-driven experiment would be to design an experiment to measure the phosphorylation state of a protein responsible for catalyzing a metabolic reaction. Experiments would focus on the phosphorylation state of the protein as a whole or specific amino acids within the protein. If the protein's (or specific amino acids') phosphorylation state differs when the metabolic reaction occurs, the hypothesis that the phosphorylation state of the protein is responsible for catalyzing the metabolic reaction would be accepted as true. If the phosphorylation state of the protein does not change, the hypothesis would be rejected, and a new hypothesis would be proposed and tested. Partially due to technological limits, a vast majority of biological research has been performed using a hypothesis-driven strategy.

In contrast, discovery-driven studies are designed to collect as much data as possible within a single experiment. Unlike a hypothesis-driven study, the question posed in a discovery-driven study is very broad, for example, contrasting the

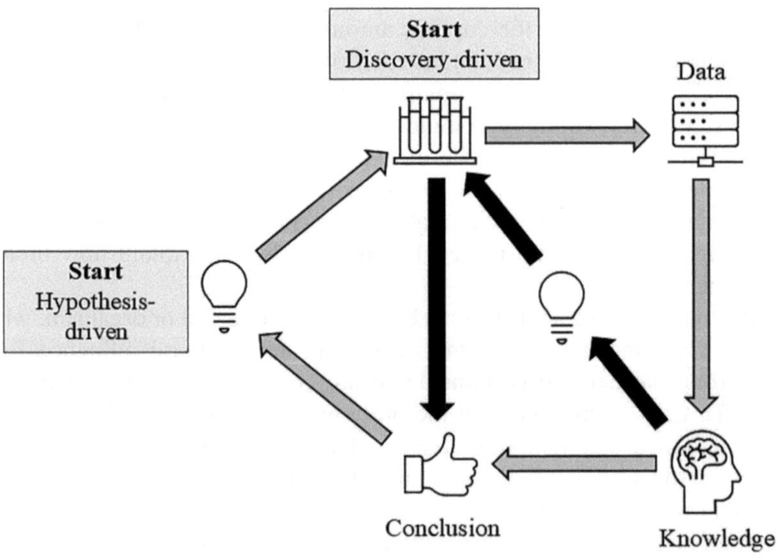


**Fig. 1.2** Comparison of hypothesis- and discovery-driven scientific methods. A hypothesis-driven study starts with a hypothesis that is tested via experimentation. The data from the experiment(s) is used to generate knowledge that leads to a conclusion that supports or refutes the original hypothesis. A discovery-driven study begins with experimentation. The data from the experiment(s) is used to generate knowledge that is used to develop a hypothesis, which is tested using an additional experiment(s). The results of these focused experiments are used to generate conclusions

example of the hypothesis-driven experiment previously described, a discovery-driven experiment would attempt to collect phosphorylation data on as many proteins as possible. The data would be analyzed in an attempt to recognize any associations between the metabolic reaction and proteins whose phosphorylation state changed. The data would generate a list of possible phosphorylated candidates that would need to be tested using a hypothesis-driven study to further delineate the specific phosphorylation event(s) associated with the metabolic reaction. Hypothesis- and discovery-driven studies have something in common—both strategies generate further testable hypotheses. In contrast, hypothesis-driven studies *explain*, while discovery-driven studies *describe* some aspects of the system being analyzed. From a practical point of view, the biggest difference between the two strategies is the amount of data acquired. Hypothesis-driven studies obtain a lot of specific data on a small number of proteins, while discovery-driven studies obtain sparse amounts of data on many proteins. Since so many proteins are analyzed concurrently, a discovery-driven study generally requires sophisticated software programs to analyze the data. Data generated in hypothesis-driven studies, on the other hand, can often be visually examined or analyzed using simple software algorithms.

While both strategies have been conducted for as long as science has existed, the past three decades have seen an increase in the number of labs dedicated to discovery-driven studies. This increase was initiated by the Human Genome Project (HGP), which focused on determining the complete chemical makeup of the human genetic code (International Human Genome Consortium 2001; Venter et al. 2001). The HGP has been a tremendous boost to the research and treatment of diseases both in terms of speed and depth. The technologies and computational tools that have been developed during the HGP and subsequent genomics era have reduced the time to sequence an entire human genome from 13 years to less than 8 h (Goenka et al. 2022).

The genome does not, however, provide all the information needed to develop a working model of how a cell functions. While the genome provides the blueprint for the cell, proteins carry out its functions. In addition, the existence of any DNA sequence does not guarantee the presence of its corresponding protein at any given time point. The sequences within the genome do not directly describe the structure, function, or location of the proteins they encode. Only by analyzing proteins directly can the types and positions of functionally important posttranslational modifications be identified. At its basic level, the genome provides a list of the proteins that can exist but does not give any information which of these proteins actually exist at any given time and how they interact with each other to form pathways and networks. The biggest discordance between the genome and proteome is that the genome changes very little during an organism's lifetime (except for epigenetic changes) while its proteome is in a constant state of flux. This flux includes all the different characteristics of a protein: its abundance level, modification states, subcellular location, interacting partners, etc. In addition, the proteome of a cell is significantly more sensitive to its environment than its genome.

Owing to the success of the HGP, the expectations for proteomics were huge (Cho 2007). Unfortunately, technology challenges along with the complexity and dynamic nature of the proteome led to several growing pains. In retrospect, the goal of characterizing a complete proteome was not a singular goal. Proteomics is like calculus; it is the rate of change that is important, not simply a protein's presence. However, to determine something's rate of change, it is necessary to verify its presence. Proteomics has necessarily become segregated into different goals including identifying proteins present within a proteome, measuring their abundance, characterizing their posttranslational modifications, determining their location within the cell, and recognizing proteins that interact to form functional complexes (Ning et al. 2011). All these characteristics, along with their change over time between different samples, are ultimately the goal of proteomics.

## 1.3 Proteomic Technology

Proteomics has had several bumps along the road, but like all worthwhile science endeavors it will continue to suffer growing pains as it further matures. A historical perspective of progress to date, however, shows that proteomics has made tremendous progress since its inception. In 2004, a total of 1175 gene products were reported in a manuscript that collated the results from four separate proteomic studies of the human plasma proteome (Anderson et al. 2004). Fast forward ahead 15 years, a proteomic study identified more than 6000 proteins within erythrocytes, platelets, platelet-rich plasma, platelet-free plasma, and whole blood taken from individuals (Geyer et al. 2019). This increased proteomic coverage is driven almost solely by developments in analytical instrumentation and computer hardware and software. These technological improvements have benefited proteomics as well as basic research studies such as identifying protein interactions, posttranslationally modified proteins, and mutations. Standard techniques used to measure changes in protein expression levels or modifications (Western blotting, etc.) have become increasingly replaced by proteomic technologies such as MS owing to the accuracy, speed, and direct identification afforded using this instrumentation.

### *1.3.1 Mass Spectrometry*

While MS continues to be the most used technology for gathering proteomic data, it is not the only one. Even MS relies on complementary technologies, such as sample preparation, protein fractionation, and bioinformatics to achieve its goals. A mass spectrometer is made up of four components (see Fig. 1.3): an ionization source, a mass analyzer, a detector, and a computer workstation to control the instrument (Cooks et al. 1983). The ionization source is responsible for converting solution-phase molecules into gas-phase ions. I don't want the reader to get confused by the use of the term "ion." Mass spectrometers measure *ions*, not neutral molecules. In the context of proteomics, the ions being measured are simply

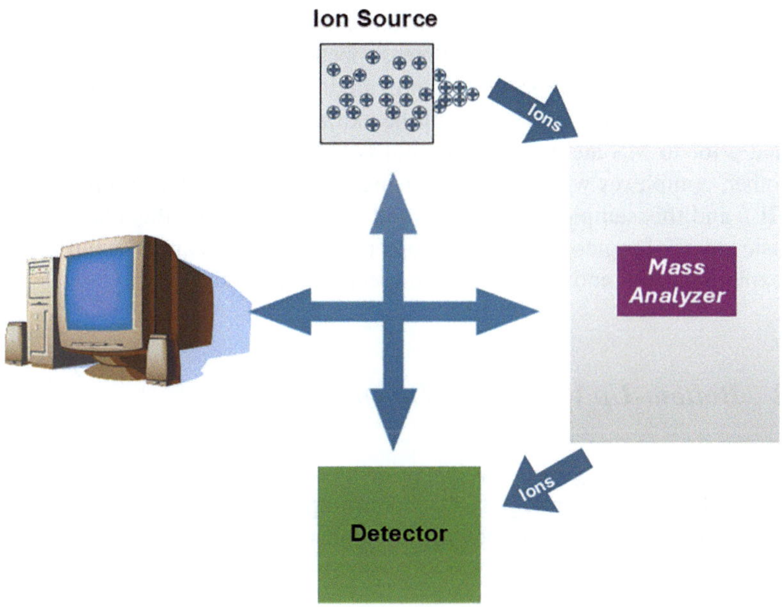


**Fig. 1.3** The basic components of a mass spectrometer for conducting proteomic analysis. The basic components include an ionization source (which ionizes the molecules that transform them from a liquid to a gaseous state), a mass analyzer (which moves and manipulates the ions), a detector (which records the mass-to-charge ratios of the ions), and a computer workstation (which controls and synchronizes the mass spectrometer)

proteins or peptides that are ionized owing to the solvent conditions and the ion source. While there are several different types, most proteomic studies use either electrospray ionization (ESI) or matrix-assisted laser desorption/ionization (MALDI) sources (Fenn et al. 1989; Tanaka et al. 1988). Of these two, ESI is used much more than MALDI owing to its ability to ionize molecules that are directly eluted from a liquid chromatography (LC) column (Huang and Henion 1990). The mass analyzer manipulates the ions created in the source prior to their detection. This manipulation can include several different features from simply allowing all ions to move to the detector to isolating specific ions to fragmenting them for the purpose of determining some specific characteristic of the ion. For proteomic analysis, the two most valuable manipulations performed within the mass analyzer are the isolation and fragmentation (also known as dissociation) of the ions. There are many different types of mass analyzers available today including ion traps, linear ion traps, Fourier transform ion cyclotron resonance spectrometers, Orbitraps, triple quadrupoles, triple-quadrupole time of flight (TOF), TOF/TOF, etc. While the type of instrument used depends on the research question being asked, Orbitrap-based mass analyzers have become the workhorse instruments in proteomics today based on their speed (i.e., ability to rapidly identify molecules), sensitivity, and resolution (Scheltema et al. 2014).

While proteomics is dominated by MS, the results rely heavily on how the sample is prepared prior to it entering the mass spectrometer. The goal of any proteomic study will dictate how the sample is prepared. If the goal is to identify as many proteins as possible, the proteome will be extracted from cells or tissues and fractionated prior to MS analysis. If the goal is to identify proteins that interact with each other, complexes will be isolated using a technique such as immunoprecipitation (IP), and this sample will be analyzed using MS. While this chapter provides the basic tenets of proteomics, other chapters will describe experimental details that help explain the aim and results of each specific viral study.

## *1.3.2 Bottom-Up Versus Top-Down*

The two primary strategies by which proteins are identified using MS need to be discussed. These two strategies are bottom-up and top-down proteomics (see Fig. 1.4). Bottom-up and top-down proteomics differ in the form in which the

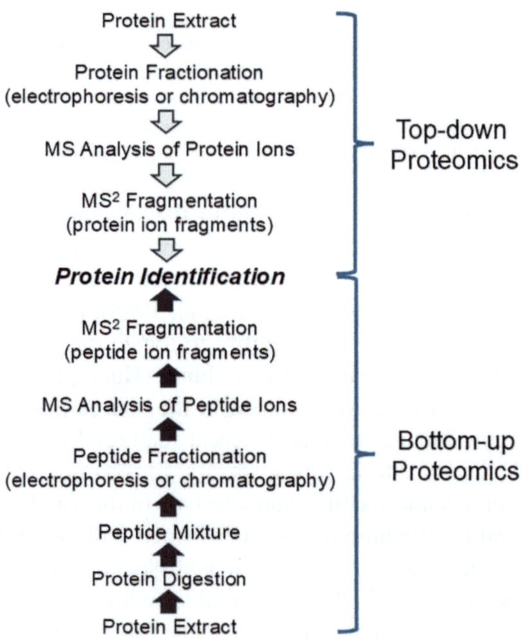


**Fig. 1.4** Top-down versus bottom-up proteomics. In top-down proteomics, intact proteins are fractionated prior to being introduced into the mass spectrometer and being fragmented using tandem mass spectrometry ($MS^2$). In bottom-up proteomics, proteins extracted from the sample are enzymatically digested into peptides. The peptides are fractionated prior to being introduced into the mass spectrometer and then fragmented using $MS^2$. While great strides have been made in top--down proteomics over the past few years, bottom-up proteomics remains the dominant strategy for identifying proteins in complex mixtures

proteins are introduced into the mass spectrometer. In bottom-up proteomics (Miller and Smith 2023), the proteins are digested into peptides using a protease. While trypsin is the most common choice, other proteases are also used (e.g., Lys-C, Glu--C, etC.). The peptides are identified using MS and tandem MS ($MS^2$ or MS/MS, described later), and this identification is used to assume the presence of a protein in the original sample. In top-down proteomics, the intact protein is introduced into the mass spectrometer (Kelleher 2004). Whether the protein is introduced intact or as peptides, the mass spectrometer fragments the ions into smaller pieces that are detected and software analyzed to reveal the protein's/peptide's identity. Owing to many factors, especially greater chromatography separation and MS fragmentation efficiency, bottom-up remains the dominant method for MS-based proteomics. Since it is the dominant method, the reader should assume that any MS-based result discussed in this book was acquired using bottom-up methods unless otherwise indicated.

The continued development of mass spectrometers with greater sensitivity has enabled an increasingly greater number of proteins to be identified in complex proteome samples. An important feature is duty cycle (i.e., the ratio of ions that are produced in the source that are analyzed). Mass spectrometers with higher duty cycles can measure signals from a greater number of ions since they do not require as much time to produce a readable signal (McLuckey et al. 1994). Resolution—the ability to distinguish signals from unique ions—enables the mass spectrometer to select a greater number of ions for identification in a process known as data--dependent analysis (DDA) (Deschamps et al. 2023). Higher resolution also enables multiple ions that are selected and fragmented simultaneously to be individually identified using a process known as data-independent analysis (DIA) (Canterbury et al. 2014). The number of proteins that can be identified with these improvements still pales in comparison to the complexity of the human proteome. As of 2021, the Human Proteome Organization (HUPO) had collated empirical evidence for 18,357 human proteins (Omenn et al. 2023). This number basically indicates that at least some evidence exists—even as minimal as a single peptide—that a protein is being produced from an annotated human gene. This number does not include proteoforms resulting from splice variants, posttranslational modifications (PTMs), and sequence variations, which are estimated to be as high as several million (Smith et al. 2021).

## 1.4 The Critical Need for Protein Fractionation

Let's do a thought experiment. There are 17 different sports balls used in Olympic events ranging in size and weight from a table tennis ball to a men's basketball. I hand all 17 balls to an individual knowledgeable in Olympic sports and ask them to identify as many as they can in 5 s, but their eyes must be closed. They drop the balls and just keep one in their hands. After attempting to identify which ball they kept, they reach down and grab another. Most people knowledgeable in sports would be able to identify three balls during the allotted time. Now, let's repeat the experiment, but this time I am going to hand over only one ball at a time, and the individual still has 5 s to identify it. After this time, I hand them another ball, and they have another 5 s to identify the ball. This sequence is repeated until all 17 balls have been handed to the individual. Most people knowledgeable in sports would correctly identify approximately 12 of the balls (blindly distinguishing men's and women's water polo balls and men's and women's handballs would be difficult).

This thought experiment is a simplified illustration of the value of fractionation to proteomics. In most proteomic experiments, mass spectrometers isolate and fragment ions into smaller pieces that can be used to identify them. The enormous number of proteins in a human proteome make it impossible to simply infuse the entire sample into the instrument and expect to identify anything except the highest abundant proteins. Fractionation provides the "divide and conquer" strategy that enables mass spectrometers to isolate and fragment ions that enter the instrument over an extended time frame. Even with high-resolution separation, many ions still enter the mass spectrometer at the same time; however, the instrument is given the opportunity to isolate and identify a greater number of ions over the course of the experiment. While not given as much fanfare, fractionation is as critical as the mass spectrometer itself in obtaining insightful results. Over the years, there have been several different fractionation strategies utilized depending on the complexity of the proteome sample and the goal of the study. I will describe some of them below.

### *1.4.1 Polyacrylamide Gel Electrophoresis*

Marc Wilkins coined the term "proteomics" in 1995 with his study of the simple proteome of *Mycoplasma genitalium* (Wasinger et al. 1995). This study used two--dimensional polyacrylamide gel electrophoresis (2D-PAGE) to separate the proteins based on their isoelectric points (pI) and molecular masses ($M_r$). Visualized protein spots were individually cored from the gel and identified using a combination of MALDI-MS and amino acid analysis. Although 2D-PAGE has developed to the point at which it can resolve thousands of proteins, each protein must be individually processed and identified (Harper and Speicher 2019). This process makes 2D-PAGE very laborious and time-consuming. There are, however, some advantages to using 2D-PAGE. When quantitatively comparing proteomes, only visualized proteins that show a difference in intensity between the compared gels need to be analyzed. Since individual protein spots are analyzed, the number of peptides identified for each protein (i.e., protein coverage) is usually higher than when using online fractionation methods. In addition, identified peptides can be reliably assigned to specific proteins. Posttranslationally modified forms of proteins are readily identified using 2D-PAGE as they often form distinct lines of spots on the gel (Bergström Lind et al. 2013). The use of 2D-PAGE for proteomics continued to grow between 1995 and 2010. Since 2010, the number of proteomic studies using 2D-PAGE fractionation has continued to decline, largely due to its replacement by higher-throughput methods (that will be described later). Nevertheless, 2D-PAGE was (and continues to be) instrumental in the field of proteomics.

While 2D-PAGE is mostly associated with proteomics, 1D-PAGE is also commonly used. Its use grew as mass spectrometers with greater performance characteristics were developed. One-dimensional PAGE is generally used for fractionating protein complexes extracted from cells via affinity isolation to identify protein-protein interactions (PPIs). Instead of using a hypothesis-driven method such as Western blotting, PPIs are being characterized using MS to identify proteins separated using 1D-PAGE (see Fig. 1.5). In this method, fractionated proteins are cut from the gel and digested into peptides that are identified using MS (Heck and Van Den Heuvel 2004). Not only does this method identify proteins within the complex, but it can determine the PTM states of various amino acids within proteins.

### *1.4.2 Liquid Chromatography*

The combination of LC with MS was a revolutionary step forward in proteomics (Huang and Henion 1990). The direct coupling of reversed-phase (RP) LC and MS dominates proteomics because of the acidity and volatility of mobile phases used in RPLC (e.g., 0.1% formic acid in acetonitrile, etc.), make it highly compatible with ESI. The ability to operate RPLC at very low flow rates (i.e., in the nL/min range) also makes it ideal for ESI. Fractions can be eluted directly from an RPLC column into the ESI source of a mass spectrometer, enabling the automation of peptide and protein identification. This coupling severely decreases the labor required to analyze complex proteomes. The coupling of RPLC with mass spectrometers is so ubiquitous that the term LC/MS is generally understood to refer to RPLC/MS.

Many scientists are familiar with RPLC. There are many differences, however, between RPLC used for proteomics and RPLC for other applications such as protein purification. Reversed-phase chromatography columns coupled to mass spectrometers are usually less than 100 μm in diameter and operate at nL/min flow rates (Lenčo et al. 2022). Small diameter columns operating at low flow rates provide very high resolution and allow the mass spectrometer's ion source to efficiently promote the solution-phase ions into the gas-phase (i.e., desolvation). The higher the separation resolution, the fewer the number of molecules enter the mass spectrometer per unit time. Minimizing their number provides the mass spectrometer with the opportunity to isolate and identify a greater number of molecules. On the practical side, low flow rates enable small amounts of sample to be analyzed, something that is important when utilizing precious clinical samples.

The combination of high-resolution RPLC separations and the speed of modern mass spectrometers still do not come close to matching the number of species in a complex proteome sample. For this reason, prefractionation is often performed so that a single proteome sample is analyzed using many RPLC/MS runs (El Rassi and Puangpila 2017). While there are many options, the most popular prefractionation method is strong cation exchange (SCX). It is so commonly used and provided such

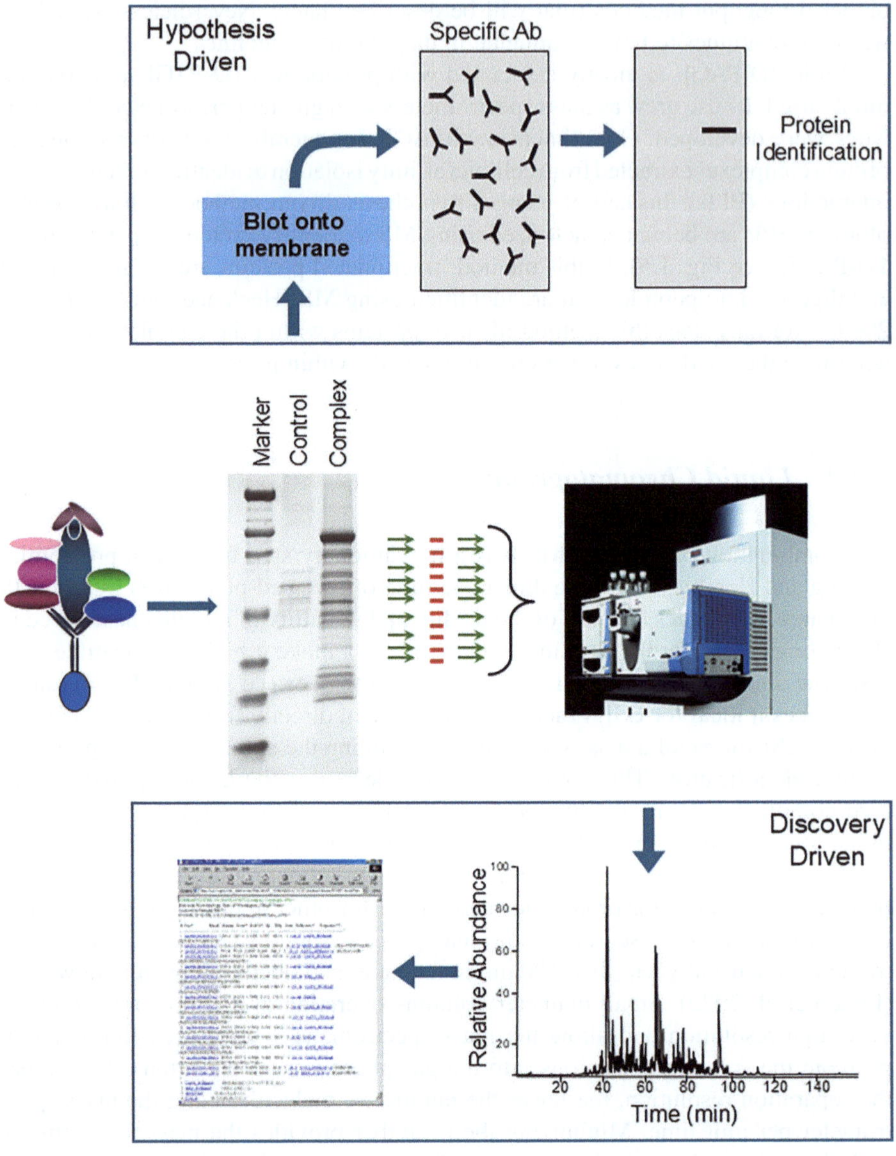


**Fig. 1.5** Identification of protein-protein interactions using discovery- and hypothesis-driven methods. In a discovery-driven method, protein complexes are extracted from cells using an affinity molecule (e.g., antibody) that recognizes a target protein. The proteins within the complex are fractionation and identified using tandem mass spectrometry ($MS^2$). In the figure, the complex is separated using gel electrophoresis, but the complex can be fractionated using a variety of different methods, including both electrophoresis and chromatography. In a hypothesis-driven method, the extracted protein complex is separated using gel electrophoresis and then blotted onto a nitrocellulose membrane. The blot is probed using a specific antibody to determine if the protein that binds to the antibody was part of the original protein complex

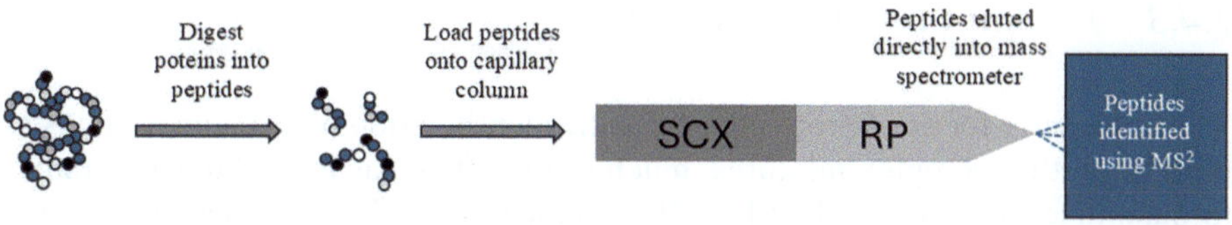


**Fig. 1.6** Multidimensional protein identification technology (MudPIT) for identification of proteome samples. In MudPIT, the proteome sample is digested into peptides that are loaded onto a biphasic capillary column comprised of strong cation exchange (SCX) and reversed-phase (RP) components. The peptides are fractionated using these two orthogonal chromatography methods directly into a mass spectrometer, which identifies them using tandem mass spectrometry ($MS^2$)

a technological advance in proteomics that it deserves to be highlighted in this chapter. When using SCX, the proteome sample is digested into peptides and generally separated into 10–20 SCX fractions. Each fraction is individually analyzed using RPLC/MS. This SCX-RPLC fractionation scheme was originally developed in the lab of Dr. John Yates (see Fig. 1.6). It was formally called MudPIT (multidimensional protein identification technology) but is commonly referred to as shotgun proteomics (Washburn et al. 2001). Shotgun proteomics has been used extensively to identify and compare the abundances of proteins in samples such as tissue and blood for many purposes including the discovery of diagnostic biomarkers, identifying PPIs, and changes in protein abundances due to external perturbations (Tirumalai et al. 2003; Tsamourgelis et al. 2023; Yates et al. 2009).

While combining separation methods allows greater proteome coverage, the number of separation methods and fractions collected must be carefully weighed against analysis time. While current MS-based proteomic studies are high-throughput compared to historical counterparts, they still require much more time than typical genome sequencing or microarray analysis. In addition, increasing the number of fractions that require MS analysis reaches a point of diminishing returns as the number of new proteins identified no longer directly correlates with the number of fractions analyzed.

In top-down proteomics, RPLC is the primary fractionation method for separating intact proteins prior to their infusion into the ion source of the mass spectrometer (Drown et al. 2022). Other fractionation techniques, such as capillary electrophoresis (Johnson et al. 2022), have been used, but RPLC has a greater loading capacity allowing greater sensitivity for the overall analysis. Top-down proteomics has some advantages over bottom-up proteomics. These include minimizing the complexity of the sample by keeping proteins intact rather than digesting them into peptides, proteins are identified directly and not through peptide surrogates that are scattered throughout the LC/MS chromatogram, and PTMs are directly assigned to their protein of origin. While it is making great strides, top-down proteomics still lags far behind bottom-up proteomics since the technologies necessary for effective intact protein separation and $MS^2$ fragmentation are not as efficient as those available for peptide analysis.

### 1.4.3 Protein Complex Isolation

While the early focus of proteomics was the identification of as many proteins as possible within complex mixtures, much of the effort has moved toward studies to quantitate proteins or identify PPIs. Characterizing PPIs has become such a major field in proteomics that it is worth a short primer on some of the techniques used to isolate protein complexes prior to reading the future chapters. While proteins within isolated complexes are usually identified using LC/MS, isolating the complex requires different purification and fractionation techniques than simple protein extraction and SCX. The major difference is cell lysis conditions. When extracting a proteome for the purpose of identifying proteins or quantifying their abundance, harsh conditions (i.e., high concentration of salt and detergent) can be used to extract as much protein as possible. When the focus is on PPIs, cell lysis and protein extraction conditions need to be gentle enough (i.e., low salt and detergent concentration) to preserve non-covalent interactions between proteins.

### 1.4.4 Immunoprecipitation

Immunoprecipitation is the most common method used to isolate protein complexes for identifying PPIs using MS (Bjorck and Kronvall 1984). Immunoprecipitation is like co-immunoprecipitation (co-IP), except IP refers to the isolation of a single protein, whereas co-IP refers to the isolation of multiple proteins that are within a complex. In either case, an antibody is used to capture a protein or protein complex from a mixture. In many cases, the antibody is coupled to a solid support such as agarose or magnetic particles (Wang et al. 2022; Carneiro et al. 2016). Agarose beads are highly porous, which allows antibodies to bind to the internal matrix of the beads, giving them a high binding capacity. Agarose beads, however, can also result in a greater number of nonspecific interactions than magnetic beads (Zeng et al. 2022). A major advantage of magnetic particles is the ability to isolate antibody-bound particles by placing magnets against the tube in which the complex is being isolated. This method allows the cell lysate to be aspirated without the risk of removing the magnetic beadbound protein complexes (Stévenin et al. 2021). Magnetic bead extraction does not require centrifugation, eliminating a step that often results in the loss of low-affinity PPIs. Magnetic beads are more amenable to high-throughput automation than agarose beads. A third option is the use of magnetic agarose beads that are comprised of cross-linked agarose beads surrounding a ferrimagnetic core (Brechmann et al. 2022). Since they are larger than conventional magnetic beads, they possess greater antibody-binding capacity than magnetic beads while retaining all the advantages of magnetic beads.

### 1.4.5 Pull-Down Assays

Pull-down assays are like IPs except they use molecular tags attached to recombinant proteins that are isolated from the mixture via a specific ligand that is attached to a solid resin (Marino et al. 2013; Wang et al. 2022). The tagged proteins are expressed within cells where they act as the target for the ligand "bait" that is bound to the resin. Many diverse types of ligands can be used; however, there are three main types used today. The first are small molecule ligands. Examples of this class are glutathione or immobilized metals (e.g., $Ni^{2+}$, $Cu^{2+}$, and $Zn^{2+}$) that bind to proteins tagged with either glutathione S-transferase (Honjoh et al. 2022) or hexahistidine tags (i.e., $His_6$) (Chang 2006), respectively. The second type uses immobilized protein ligands to capture recombinant proteins expressing a peptide tag. This type includes calmodulin-binding peptide tags that bind calmodulin. The third type captures recombinant proteins coupled to a small peptide tag using an antibody coupled to a resin. Some commonly used examples include FLAG (DYKDDDDK), c-myc (CEQKLISEEDL), and hemagglutinin peptides (YPYDVPDYA) (Zhao et al. 2013). Often, the recombinant protein is designed with multiple copies of the peptide tag at either the amino or carboxy termini.

Each of these methods has their advantages and disadvantages. The major advantage of expressing a tagged protein within cells is that it increases the amount of that protein within the mixture. This increase allows a greater amount of protein complex to be isolated, which is critical for proteins such as kinases and phosphatases that are usually only present in cells at low levels. The flip side to this advantage is that overexpressing a protein within a cell can result in nonspecific interactions that may not be readily recognized as nonspecific in the final data. In addition, adding a tag to a protein can interfere with its ability to interact with its normal binding partners via steric hindrance or cause some change in the wild-type protein's structure. These factors need to be considered when using any of these pull-down methods, which is why results obtained from PPI studies require careful validation using orthogonal techniques.

### 1.4.6 Co-fractionation Mass Spectrometry

Co-fractionation mass spectrometry (CF-MS) is a recently developed method for identifying PPIs (see Fig. 1.7). The method is dependent on using conditions that maintain protein complexes in their native state throughout the proteome extraction and prefractionation steps prior to MS analysis. It precludes the use of organic solvents, high heat, strong acids, as well as high detergent and high salt concentrations during proteome extraction and fractionation. Many different types of fractionation methods such as SCX, anion exchange chromatography, isoelectric focusing, glycerol gradient separation, etc. can be used to fractionate the protein complexes if the gradient conditions do not disrupt the PPIs (Low et al. 2021). Fractions eluting from

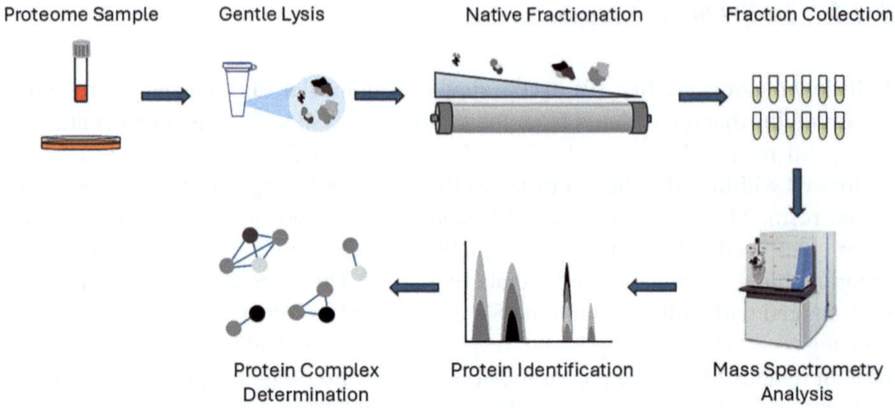


**Fig. 1.7** Co-fractionation mass spectrometry (CF-MS) for identifying protein-protein interactions. In CF-MS, proteome samples are extracted from biological samples using gently lysis conditions to maintain non-covalent interactions between proteins. The proteins are fractionated under native condition to maintain protein complexes in their native state. Proteins that are identified in the same fractions as they elute into the mass spectrometer are hypothesized to interact with one another

the column are collected, digested into peptides, and analyzed using RPLC coupled directly online with MS. Protein-protein interactions are hypothesized based on the identification of proteins within the same fraction and previous knowledge of their function and location within the cell. Co-fractionation MS obviates the need for recombinant proteins or affinity tags coupled to solid resins. Careful validation needs to be conducted when using CF-MS as the wide range of affinities and non-specific interactions that occur between proteins can lead to false-positive results. Also, CF-MS does not detect many low-affinity PPIs as it is difficult to replicate native conditions during the extraction and fractionation steps.

### *1.4.7 Cross-Linking and Proximity-Linking Assays*

Two additional methods used in proteomics to identify PPIs are cross-linking mass spectrometry (CL-MS) (Chen and Rappsilber 2023) and proximity-labeling mass spectrometry (PL-MS) (Pino and Schilling 2021). In CL-MS, a chemically reactive cross-linking reagent is added to a proteome sample. The CL reagent forms covalent bonds with specific amino acids within proteins. Most CL reagents have a minimum of two functional groups, so they will react with specific amino acids in close proximity: resulting in the formation of intra- and inter-protein cross-links. After enzymatically digesting the cross-linked protein mixture, the resulting peptides are identified using RPLC/MS. Cross-linked peptides are identified using $MS^2$ to determine their amino acid sequence and modifications that result from the attached cross-linker. The MS data is then analyzed to identify PPIs since proteins in close proximity will be cross-linked by the CL reagent.

Proximity-labeling MS, which is like CL-MS, uses an engineered protein that has an enzyme fused to it. This engineered protein serves as the bait for other proteins that interact with it. The most commonly used enzyme that is attached to the bait protein is a mutant form of biotin ligase A, known as BioID. The BioID enzyme covalently modifies proteins (or other biomolecules) that are physically adjacent to the bait protein with biotin, for example, proteins within a 10–20-nm radius of a bait protein that has been labeled with BioID will be labeled with biotin (Liu et al. 2020). Labeled proteins are extracted from the cell mixture using streptavidin, which has a strong affinity for biotin. Mass spectrometry identification of biotin-labeled proteins suggest they are all proximal to each other. Besides PPIs, PL-MS has been used for identifying protein-RNA interactions (Weissinger et al. 2021). Both CL-MS and PL-MS methods are designed to identify low- to medium-affinity PPIs, whereas IPs and pull-down methods are more adept at identifying medium- to high-affinity PPIs since low-affinity non-covalent interactions may not survive the sample preparation steps.

### *1.4.8 Yeast Two-Hybrid Assays*

While identifying PPIs using MS has become a dominant technology, yeast two-hybrid (Y2H) screening has (and continues to) contributed much to proteomics (Fields and Song 1989). Unlike MS methods, Y2H screening identifies PPIs in vivo (see Fig. 1.8). In Y2H screening, a plasmid containing the DNA-binding domain (BD) of a transcription factor is fused to a bait protein and this construct is transfected into yeast cells expressing a library of plasmids containing target proteins fused to the activation domain (AD) of the same transcription factor. Yeast colonies are multiplied in liquid culture and then spread on solid media plates to generate individual colonies expressing bait and specific target chimeric proteins. Cells in which the bait and target proteins interact result in the expression of the gene regulated by the transcription factor formed by the DNA-BD and AD indicating a positive PPI. The plasmid is extracted from these colonies and sequenced to identify the interacting proteins. The Y2H system has the advantage of detecting PPIs in vivo; however, the technique is only capable of identifying binary interactions, but yeast three-hybrid (Y3H) systems are available to measure interactions between three different molecules.

## 1.5 Protein Identification

Up to this point, the fractionation and introduction of the sample into the mass spectrometer have been described. This section deals with how proteins are identified. Remember, most proteomic experiments are bottom-up, meaning that *peptides* are

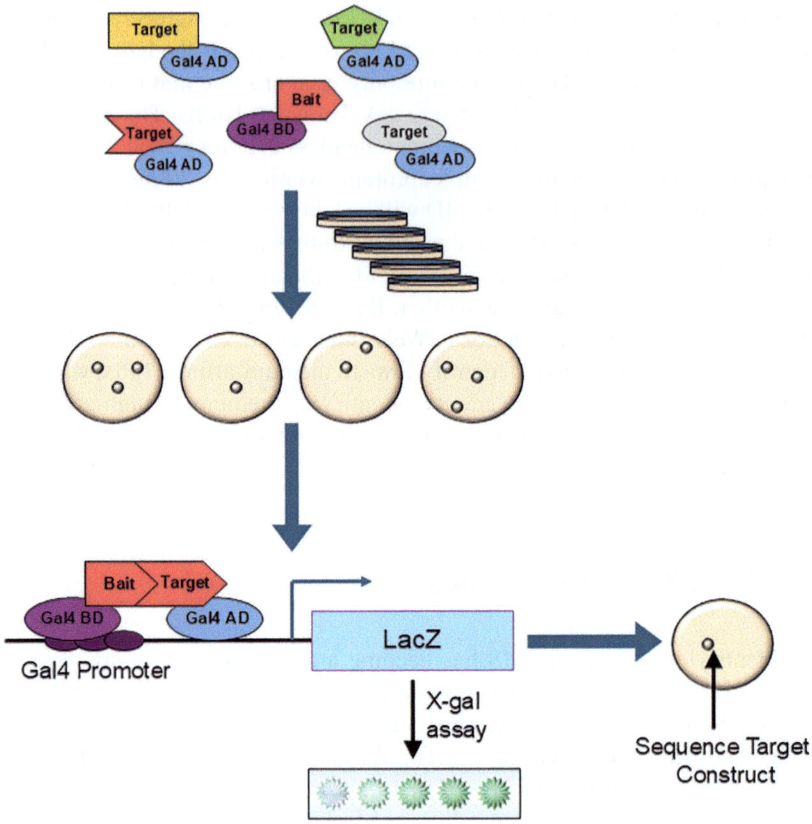


**Fig. 1.8** Yeast two-hybrid method for identification of protein-protein interactions. A plasmid containing the bait protein is fused to the DNA-binding domain (BD) of a transcription factor (e.g., Gal4). The DNA-BD plasmid is transformed into yeast cells that contain a library of plasmids containing target proteins fused to the activation domain (AD) of the same transcription factor. If the bait and target proteins bind to one another in the yeast cells cultured on solid media, the BD and AD of the transcription factor will come into close proximity of one another resulting in the expression of a reporter gene (e.g., LacZ). Positive yeast colonies are extracted from the media, and their plasmids are sequenced to identify the interacting bait and target proteins

being identified and this identification is being used as a surrogate for the presence of the protein in the sample.

## *1.5.1 Tandem Mass Spectrometry*

The most common method for identifying peptides in an MS-based experiment is $MS^2$ in which collision-induced dissociation (CID), also known as collisionally activated dissociation (CAD), is used to fragment the peptide into smaller fragments

(McLuckey et al. 1994). In CID, the kinetic energy of a selected ion is increased by applying an electric potential. The accelerated ion is then collided with an inert gas (e.g., helium, nitrogen, or argon) causing the ion to fragment into smaller pieces, which are then detected. The *m/z* ratios of these fragments are then used to identify the peptide that entered the mass spectrometer. The reader should not let *m/z* ratio confuse them. It is simply a unit of measurement that represents the MW of the molecule divided by its charge state (which is usually +1, +2, or +3 in the MS analysis of peptides). Software algorithms have been developed that take the raw $MS^2$ data and compare it to predicted $MS^2$ spectra generated from protein sequences contained within a database. Since the most prominent sites of fragmentation are the peptide bonds between amino acids (which have defined MWs), these algorithms can predict what an $MS^2$ spectrum of a peptide should resemble. Another factor that makes $MS^2$ spectra predictable is the fact that not all peptide bonds within a single peptide will break during an $MS^2$ experiment. What happens instead is populations of the peptide will be formed that contain various lengths of its amino acids joined together.

Since it is a critical point to understand how peptides are identified, I am going to risk being verbose in ensuring that the reader has a basic understanding of how peptides are identified using $MS^2$. Let's use the following word to represent a 13-amino acid peptide: VIRALPROTEOME. If this peptide was analyzed using $MS^2$, it would produce data containing peaks representing populations of letters of various lengths including the following:

| | | | |
|---|---|---|---|
| VIRALPR | IRALPRO | ALPROTEO | PROT |
| IRAL | ALPRO | OTEOME | TEOM |
| IRALPROTE | LPROTEOME | ROTEOM | PROTE |
| PROTEO | PROTEOME | VIRAL | RAL |

While other sequences could be produced, the key thing to notice is how the letters remain in the order they exist in the intact peptide. In addition, the peptide's MW is determined by the mass spectrometer prior to it being fragmented. Using fragmentation information along with its MW makes it possible to readily identify this peptide. Tandem MS is critical since it is impossible to accurately identify a peptide based solely on its MW owing to the degeneracy of MWs within a human proteome. The most distinct characteristic of a peptide is its amino acid sequence, the precise information that is provided using $MS^2$.

### *1.5.2 Electron Capture and Electron Transfer Dissociation*

While CID is the dominant method for identifying peptide ions, electron transfer dissociation (ETD) and electron capture dissociation (ECD) methods are particularly efficient for top-down proteomics in which intact proteins must be fragmented

(Zhurov et al. 2013). Both methods are effective at identifying PTMs as they leave the modification attached to the amino acid. In ECD, energetic electrons bind to multiply protonated ions generated by the mass spectrometer. The energy released from the electrons causes the ion to fragment. In comparison to CID spectra of peptide ions, ECD spectra show a greater extent of peptide bond cleavages; however, both amino acid sequence and PTM information are preserved. Like ECD, ETD produces high energy electrons obtained through a gas-phase reaction between the positively charged protein ions generated in the mass spectrometer and anions (e.g., anthracene or fluoranthene). These electrons react with the protein ion to cause fragmentation. Both amino acid sequence and PTM information are preserved in ETD.

### *1.5.3 A Typical MS Proteomic Analysis*

To illustrate how peptides in a complex proteome sample are identified using a bottom-up approach, I am going to use the elements presented in Fig. 1.9 as an illustration which describes the $MS^2$ portion using an ion trap mass spectrometer. In the first stage, the proteins are digested into tryptic peptides. This collection of peptides is injected onto an RPLC column, which is coupled directly into the ion source of the mass spectrometer. At any point in time, several peptide ions enter the mass spectrometer. These ions are guided into the ion trap portion of the instrument, where they are temporarily stored. A peptide ion with a specific *m/z* value (usually the one with the highest abundance) is isolated in the trap. The energy of the peptide ion is increased, and an inert gas is added to the trap. The peptide ion collides with the gas, causing its fragmentation. The fragments are ejected from the trap and sent to the detector. The mass spectrometer repeats this process of isolation, fragmentation, and detection on the next most abundant peptide ion. When analyzing very complex mixtures, the instrument is operated in a data-dependent acquisition (DDA) mode where each full MS scan is followed by a specific number (usually ten or more) $MS^2$ scans where the most abundant peptide molecular ions are sequentially isolated in the ion trap and subjected to CID. State-of-the-art mass spectrometers can perform this cycle thousands of times per hour affording the identification of thousands of peptides in a single LC/$MS^2$ experiment. The key for the reader going forward is not to get lost in the technical details but realize that this process allows thousands of proteins in complex samples to be identified. While there are subtle differences, this method is also used to identify such features as PTMs within proteins.

## 1.6 Protein Arrays

While MS has been the dominant proteome technology, protein arrays have also had a major impact in proteomics. Protein arrays (also known as microarrays) are high-throughput and highly specific, can simultaneously measure thousands of proteins,

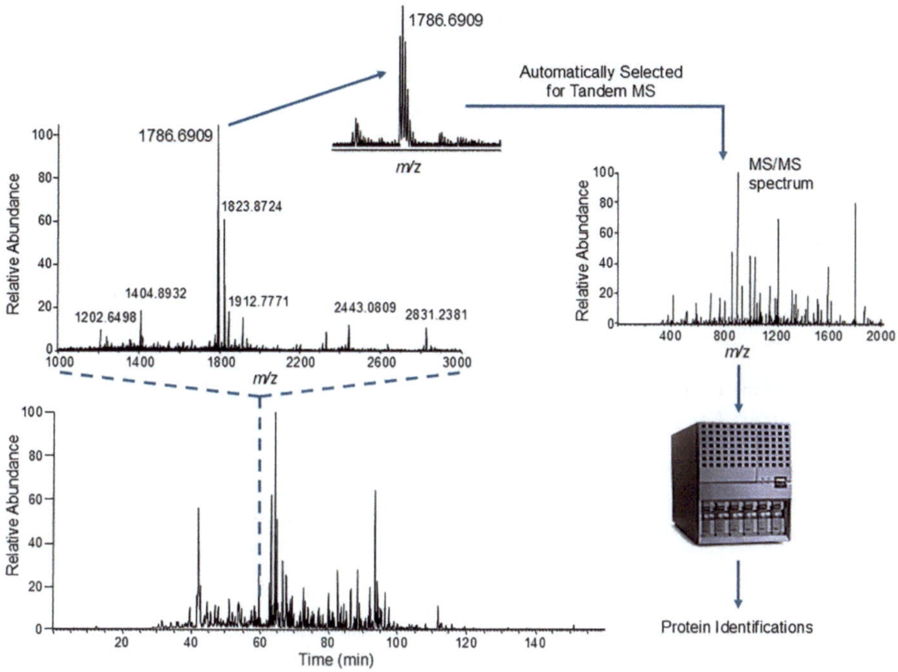


**Fig. 1.9** Identification of proteins using data-dependent tandem mass spectrometry (MS[2]). The mass spectrometer measures the mass-to-charge (m/z) ratio of peptides that enter the instrument at a specific time point. Individual peptides are isolated based on their intensity measured in the preceding MS scan. The isolated peptide (in this case m/z 1786.6909) is fragmented via collision--induced dissociation (CID), which collides the peptide with an inert gas within the collision cell of the mass spectrometer. The fragment ions are recorded to generate the MS[2] spectrum, which is analyzed against a database of in silico predicted fragment pattern of peptides within a database. If a peptide is identified, it is correlated to a specific protein

and require only a small amount of sample. Protein arrays evolved from the development of DNA arrays, which were designed to characterize nucleic acids (Schena et al. 1995). The concept of protein arrays was originally conceived by Dr. Roger Ekins who initially described arrayed ELISAs (Ekins 1980). He recognized how time and money could be saved by using antibody-based arrays to simultaneously measure multiple analytes. What truly accelerated the development of protein arrays were advances in protein (and specifically antibody) expression, which provide the "baits" that are used to capture ligands. As we will see in later chapters that describe applications, protein arrays have been extensively used to detect and characterize viruses as well as determine their effect on host cells.

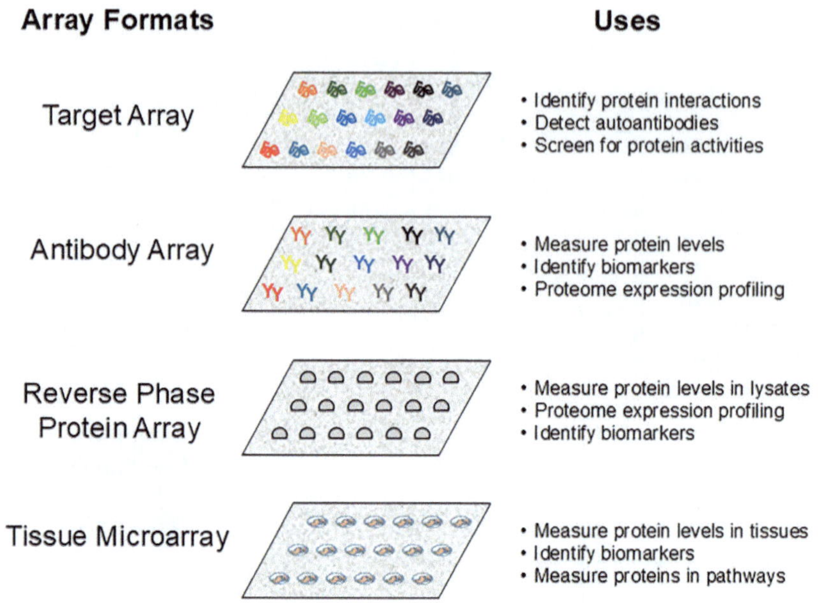


**Fig. 1.10** Four major types of arrays used in proteomics and their uses

## *1.6.1 Types of Protein Arrays*

In a typical protein array, many different "baits" are coupled to specific locations on a solid surface that is then probed with a sample of interest. Solid surfaces such as polyvinylidene fluoride membranes, glass, microtiter wells, beads, and other particles (e.g., gold nanoparticles) have all been used to couple baits. There are four major types of arrays: target protein arrays, antibody arrays, reverse-phase protein arrays (RPPAs), and tissue arrays (see Fig. 1.10).

## *1.6.2 Target Protein Arrays*

Target protein arrays consist of purified proteins or peptides immobilized on a solid surface (Melton 2004). These arrays can identify various interactions between proteins and other molecules such as other proteins, antibodies, DNA, metabolites, etc. In specialized formats, they can also be used to measure enzyme activity. The greatest technological barrier to preparing a target protein array is producing the bait proteins. The bait proteins are usually expressed in cells or in vitro translation systems. They are individually purified and attached to the solid surface. While it sounds simple, expressing and purifying proteins for target arrays is anything but. Creating the required infrastructure requires a lot of investment in labor and technology. On the technical side, recombinant proteins do not always efficiently express in prokaryotic or eukaryotic systems, and if it does express, there is no guarantee it will fold properly or retain its native conformation during purification or printing on the solid surface.

Cell-free protein arrays were developed to overcome the challenges of expressing, purifying, and printing (Yu et al. 2018). In a cell-free system, RNA or DNA is deposited on the array slide, and the encoded proteins are translated by applying a cell-free expression system (e.g., rabbit reticulocyte lysate) to each spot. Since cell-free protein arrays require less sample handling and the expression lysates contain chaperone proteins to assist with correct folding, they often produce arrays with a greater percentage of active proteins. In addition, printing nucleic acids instead of purified proteins makes array production simpler since the technology for producing synthetic DNA or RNA strands is more robust.

### *1.6.3 Antibody Arrays*

Owing to their ability to bind proteins, it is not surprising that the first types of protein arrays used antibodies as baits (Aparna and Tetala 2023). Once they are attached to the surface, using an antibody array is quite straightforward. A cell or tissue lysate containing proteins that are fluorescently labeled are introduced onto the array. The array is washed to eliminate nonspecifically bound proteins, and the captured proteins are detected using a fluorescent scanner. Not surprisingly, the biggest issue related to these arrays is nonspecific binding, which occurs with any array-based technique. The three formats of antibody arrays are forward-phase, reverse-phase, and sandwich array. In a forward-phase array, antibodies are printed on the solid surface and probed with a complex mixture of proteins, such as a cell or tissue lysate. In a reverse-phase array, proteins are printed on the surface and probed with a specific antibody to determine if a specific protein of interest is in the sample. A fluorescently labeled secondary antibody is then used to detect the protein/antibody interaction. In a sandwich array, antibodies are printed on the surface that is then probed with a complex mixture of proteins (like a forward-phase array). A second antibody that also binds to the protein(s) of interest is added. Because of issues related to nonspecificity and cross-reactivity, many antibody arrays utilize a sandwich format since two antibodies that bind to different epitopes of the same protein must bind simultaneously to generate a positive signal.

### *1.6.4 Reverse-Phase Protein Arrays*

The second type of protein array that utilizes antibodies are RPPAs (Wulfkuhle et al. 2003). Unlike reverse-phase arrays that print proteins on surfaces, RPPAs immobilize cell lysates onto solid surfaces and probe the spots using specific antibodies. In an RPPA, two antibodies are used. The primary antibody is for detecting the protein of interest within the spotted lysate, while a fluorescently labeled secondary antibody binds to the primary antibody. An RPPA can test a single lysate against several different antibodies or just a single antibody depending on the aim of the study. Since they require very small volumes of lysate per spot, most RPPAs can be created using the lysate from only a few thousand cells.

### *1.6.5 Tissue Arrays*

Tissue arrays are essentially RPPAs except that instead of spotting lysates on a slide, they use sections from fresh frozen or formalin-fixed paraffin-embedded tissues (Dragunow 2020). In many ways, tissue arrays are like high-throughput, multiplexed immunohistochemistry (IHC). Regions of interest are cored from the tissue and sectioned into thin slices. These slices are placed onto glass slides, and like IHC, they are interrogated using specific antibodies. There are two popular modes of tissue arrays. In one mode, a single antibody is used to assay the same tissue region acquired from many different subjects or many different tissue regions acquired from a single patient. In the other mode, a single tissue type is spotted at multiple locations, which are then screened using many different antibodies.

## 1.7 Conclusions

A cell's proteome is a very complex mixture of proteins. As with any complex mixture, there are a myriad of different characteristics that can be analyzed. For instance, early studies in proteomics were mostly interested in simply identifying as many proteins as possible. The next steps in proteomic technology focused on quantifying proteins within proteomes. Other key characteristics that are routinely investigated include PTMs, PPIs, subcellular locations, etc. With such a variety of different characteristics, proteomics has developed many different technologies to address these issues. Proteomics was initially, and continues to be, dominated by MS-based techniques. Along with advanced mass spectrometers, sample preparation and separation techniques have been developed to focus on specific proteomic characteristics.

While not as commonly used for proteomics, protein array technologies have exponentially increased in capability over the past couple of decades. While both protein arrays and MS are often used to quantify proteins, these techniques are generally considered to complement each other rather than directly compete. Currently, MS methods are better suited for discovery-driven studies, while protein arrays are better suited for hypothesis-driven studies. For example, characterizing thousands of proteins in samples taken from cancer and healthy patients using MS may lead to the finding of a novel biomarker. While MS was the discovery tool, it may not be the best technique to measure the biomarker in many clinical samples for diagnostic purposes, since a mass spectrometer can only analyze a single sample per experiment. A protein array would be a higher-throughput technology for measuring this biomarker in clinical samples since it can measure hundreds or thousands of samples concurrently. In the future, MS and protein array techniques will serve to complement each other so that proteomic results are quickly turned into clinical applications.

The purpose of this introductory chapter is to educate those who are more interested in the results provided by proteomics rather than how proteomic technology works. To truly appreciate and have confidence in the results, however, a basic understanding of how the data is acquired is necessary. While a single chapter cannot fully describe all the experimental details, it should provide the reader with a basic understanding of some of the more basic proteomic methods. Other chapters within this book will provide more specific details of procedures used to characterize specific aspects of the viral proteome, especially those focused on how viruses impact their hosts.

## References


Aderem A (2005) Systems biology: its practice and challenges. Cell 121:511–513

Aebersold RH, Leavitt J, Saavedra RA, Hood LE, Kent SB (1987) Internal amino acid sequence analysis of proteins separated by one- or two-dimensional gel electrophoresis after in situ protease digestions on nitrocellulose. Proc Natl Acad Sci USA 84:6970–6974

Ahn AC, Tewari M, Poon CS, Phillips RS (2006a) The limits of reductionism in medicine: could systems biology offer an alternative? PLoS Med 3:e208

Ahn AC, Tewari M, Poon CS, Phillips RS (2006b) The clinical applications of a systems approach. PLoS Med 3:e209

Anderson C (2008) The end of theory: the data deluge makes the scientific method obsolete. Wired

Anderson NL, Polanski M, Pieper R, Gatlin T, Tirumalai RS, Veenstra TD et al (2004) The human plasma proteome: a nonredundant list developed by combination of four separate sources. Mol Cell Proteomics 3:311–326

Aparna GM, Tetala KKR (2023) Recent progress in development and application of DNA, protein, peptide, glycan, antibody, and aptamer microarrays. Biomolecules 13:602

Bardini R, Politano G, Benso A, Di Carlo S (2017) Multi-level and hybrid modelling approaches for systems biology. Comput Struct Biotechnol J 15:396–402

Bergström Lind S, Artemenko KA, Elfineh L, Zhao Y, Bergquist J, Pettersson U (2013) Post translational modifications in adenovirus type 2. Virology 447:104–111

Bjorck L, Kronvall G (1984) Purification and some properties of streptococcal protein G, a novel IgG-binding reagent. J Immunol 133:969–974

Brechmann NA, Jansson M, Hagg A, Hicks R, Hyllner J, Eriksson K et al (2022) Proof-of-concept of a novel cell separation technology using magnetic agarose-base beads. Magnetochemistry 8:34–48

Canterbury JD, Merrihew GE, MacCoss MJ, Goodlett DR, Shaffer SA (2014) Comparison of data acquisition strategies on quadrupole ion trap instrumentation for shotgun proteomics. J Am Soc Mass Spectrom 25:2048–2059

Carneiro DG, Clarke T, Davies CC, Bailey D (2016) Identifying novel protein interactions: proteomic methods, optimization approaches and data analysis pipelines. Methods 95:46–54

Chang IF (2006) Mass spectrometry-based proteomic analysis of the epitope-tag affinity protein complexes in eukaryotes. Proteomics 6:6158–6166

Chen ZA, Rappsilber J (2023) Protein structure dynamics by crosslinking mass spectrometry. Curr Opin Struct Biol 80:102599

Cho WCS (2007) Proteomics technologies and challenges. Genomics Proteomics Bioinformatics 5:77–85

Cooks RG, Busch KL, Glish GL (1983) Mass spectrometry: analytical capabilities and potentials. Science 222:273–291

Deschamps E, Calabrese V, Schmitz I, Hubert-Roux M, Castagnos D, Afonso C (2023) Advances in ultra-high-resolution mass spectrometry for pharmaceutical analysis. Molecules 28:2061

Dragunow M (2020) Human brain neuropharmacology: a platform for translational neuroscience. Trends Pharmacol Sci 41:777–792

Drown BS, Jooß K, Melani RD, Lloyd-Jones C, Camarillo JM, Kelleher NL (2022) Mapping the proteoform landscape of five human tissues. J Proteome Res 21:1299–1310

Ekins R (1980) More sensitive immunoassays. Nature 284:14–15

El Rassi Z, Puangpila C (2017) Liquid-phase based separation systems for depletion, prefractionation, and enrichment of proteins in biological fluids and matrices for in-depth proteomics analysis – an update covering the period 2014-2016. Electrophoresis 38:150–161

Fenn JB, Mann M, Meng CK, Wong SF, Whitehouse CM (1989) Electrospray ionization for mass spectrometry of large biomolecules. Science 246:64–71

Fields S, Song O (1989) A novel genetic system to detect protein-protein interactions. Nature 340:245–246

Geyer PE, Voytik E, Treit PV, Doll S, Kleinhempel A, Niu L et al (2019) Plasma profiling to detect and avoid sample-related biases in biomarker studies. EMBO Mol Med 11:e10427

Goenka SD, Gorzynski JE, Shafin K, Fisk DG, Pesout T, Jensen TD et al (2022) Accelerated identification of disease-causing variants with ultra-rapid nanopore genome sequencing. Nat Biotechnol 40:1035–1041

Harper S, Speicher WD (2019) Comparing complex protein samples using two-dimensional polyacrylamide gels. Curr Protoc Protein Sci 96:e87

Heck AJ, Van Den Heuvel RH (2004) Investigation of intact protein complexes by mass spectrometry. Mass Spectrom Rev 23:368–389

Honjoh H, Tanikawa M, Wada-Hiraike O, Oda K, Inaba H, Kukita A et al (2022) MED1, a novel binding partner of BRCA1, regulates homologous recombination and R-loop processing. Sci Rep 12:17140

Huang EC, Henion JD (1990) LC/MS and LC/MS/MS determination of protein tryptic digests. J Am Soc Mass Spectrom 1:158–165

International Human Genome Consortium (2001) Initial sequencing and analysis of the human genome. Nature 409:860–921

Johnson KR, Gao Y, Greguš M, Ivanov AR (2022) On-capillary cell lysis enables top-down proteomic analysis of single mammalian cells by CE-MS/MS. Anal Chem 94:14358–14367

Kelleher NL (2004) Top-down proteomics. Anal Chem 76:197A–203A

Kennedy TE, Gawinowicz MA, Barzilai A, Kandel ER, Sweatt JD (1988) Sequencing of proteins from two-dimensional gels by using in situ digestion and transfer of peptides to polyvinylidene difluoride membranes: application to proteins associated with sensitization in Aplysia. Proc Natl Acad Sci USA 85:7008–7012

Klose J (1975) Protein mapping by combined isoelectric focusing and electrophoresis or mouse tissues. A novel approach to testing for induced point mutations in mammals. Humangenetik 26:231–243

Lenčo J, Jadeja S, Naplekov DK, Krokhin OV, Khalikova MA, Chocholouš P et al (2022) Reversed-phase liquid chromatography of peptides for bottom-up proteomics: a tutorial. J Proteome Res 21:2846–2892

Liu CH, Chien MJ, Chang YC, Cheng YH, Li FA, Mou KY (2020) Combining proximity labeling and cross-linking mass spectrometry for proteomic dissection of nuclear envelope interactome. J Proteome Res 19:1109–1118

Low TY, Syafruddin SE, Mohtar MA, Vellaichamy A, Rahman ANS, Pung YF et al (2021) Recent progress in mass spectrometry-based strategies for elucidating protein-protein interactions. Cell Mol Life Sci 78:5325–5339

Marino N, Marshall JC, Collins JW, Zhou M, Qian Y, Veenstra T et al (2013) Nm23-h1 binds to gelsolin and inactivates its actin-severing capacity to promote tumor cell motility and metastasis. Cancer Res 73:5949–5962

McLuckey SA, Van Berkel GJ, Goeringer DE, Glish GL (1994) Ion trap mass spectrometry. Using high-pressure ionization. Anal Chem 66:737A–743A

Melton L (2004) Protein arrays: proteomics in multiplex. Nature 429:101–107

Miller RM, Smith LM (2023) Overview and consideration in bottom-up proteomics. Analyst 148:475–486

Mulder GJ (1838) Sur la composition de guelgues substance animales. Bulletin des Sciences Physiques et Naturelles en Néerlande 104

Ning Z, Zhou H, Wang F, Abu-Farha M, Figeys D (2011) Analytical aspects of proteomics: 2009-2010. Anal Chem 83:4407–4426

O'Farrell PH (1975) High resolution two-dimensional electrophoresis of proteins. J Biol Chem 250:4007–4021

Omenn GS, Lane L, Overall CM, Pineau C, Packer NH, Cristea IM et al (2023) The 2022 report on the human proteome from the HUPO human proteome project. J Proteome Res 22:1024–1042

Pino L, Schilling B (2021) Proximity labeling and other novel mass spectrometric approaches for spatiotemporal protein dynamics. Expert Rev Proteomics 18:757–765

Scheele GA (1975) Two-dimensional gel analysis of soluble proteins. Characterization of Guinea pig exocrine pancreatic proteins. J Biol Chem 250:5375–5385

Scheltema RA, Hauschild JP, Lange O, Hornburg D, Denisov E, Damoc E et al (2014) The Q Exactive HF, a benchtop mass spectrometer with a pre-filter, high-performance quadrupole and an ultra-high-field Orbitrap analyzer. Mol Cell Proteomics 13:3698–3708

Schena M, Shalon D, Davis RW, Brown PO (1995) Quantitative monitoring of gene expression patterns with a complementary DNA microarray. Science 270:467–470

Shevchenko A, Wilm M, Vorm O, Mann M (1996) Mass spectrometric sequencing of proteins silver-stained polyacrylamide gels. Anal Chem 68:850–858

Smith LM, Agar JN, Chamot-Rooke J, Danis PO, Ge Y, Loo JA et al (2021) The human proteoform project: defining the human proteome. Sci Adv 7:eabk0734

Stévenin V, Giai Gianetto Q, Duchateau M, Matondo M, Enninga J, Chang YY (2021) Purification of infection-associated macropinosomes by magnetic isolation for proteomic characterization. Nat Protoc 16:5220–5249

Tanaka K, Waki H, Ido Y, Akita S, Yoshida Y, Yoshida T (1988) Protein and polymer analyses up to m/z 100 000 by laser ionization time-of-flight mass spectrometry. Rapid Commun Mass Spectrom 2:151–153

Tirumalai RS, Chan KC, Prieto DA, Issaq HJ, Conrads TP, Veenstra TD (2003) Characterization of the low molecular weight human serum proteome. Mol Cell Proteomics 2:1096–1103

Tsamourgelis A, Swann P, Chouliaras L, O'Brien JT (2023) From protein biomarkers to proteomics in dementia with Lewy bodies. Ageing Res Rev 83:101771

Van Regenmortel MH (2004) Reductionism and complexity in molecular biology. Scientists now have the tools to unravel biological and overcome the limitation of reductionism. EMBO Rep 5:1016–1020

Veenstra TD (2021) Systems biology and multi-omics. Proteomics 21:e2000306

Venter JC, Adams MD, Myers EW, Li PW, Mural RJ et al (2001) The sequence of the human genome. Science 291:1304–1351

Wang S, Wu R, Lu J, Jiang Y, Huang T, Cai YD (2022) Protein-protein interaction networks as miners of biological discovery. Proteomics 22:e2100190

Washburn MP, Wolters D, Yates JR (2001) Large-scale analysis of the yeast proteome by multidimensional protein identification technology. Nat Biotechnol 19:242–247

Wasinger VC, Cordwell SJ, Cerpa-Poljak A, Yan JX, Gooley AA, Wilkins MR et al (1995) Progress with gene-product mapping of the Mollicutes: Mycoplasma genitalium. Electrophoresis 16:1090–1094

Weissinger R, Heinold L, Akram S, Jansen RP, Hermesh O (2021) RNA proximity labeling: a new detection tool for RNA-protein interactions. Molecules 26:2270

Wilm M, Shevchenko A, Houthaeve T, Breit S, Schweigerer L, Fotsis T et al (1996) Femtomole sequencing of proteins from polyacrylamide gels by nano-electrospray mass spectrometry. Nature 379:466–469

Wulfkuhle JD, Aquino JA, Calvert VS, Fishman DA, Coukos G, Liotta LA et al (2003) Signal pathway profiling of ovarian cancer from human tissue specimens using reverse-phase protein microarrays. Proteomics 3:2085–2090

Yates JR, Ruse CI, Nakorchevsky A (2009) Proteomics by mass spectrometry: approaches, advances, and applications. Annu Rev Biomed Eng 11:49–79

Yu X, Petritis B, Duan H, Xu D, LaBaer J (2018) Advances in cell-free protein array methods. Expert Rev Proteomics 15:1–11

Zeng X, Zeng WH, Zhou J, Liu XM, Huang G, Zhu H et al (2022) Removal of nonspecific binding proteins is required in co-immunoprecipitation with nuclear proteins. BioTechniques 73:289–296

Zhao X, Li G, Liang S (2013) Several affinity tags commonly used in chromatographic purification. J Anal Methods Chem 2013:581093

Zhurov KO, Fornelli L, Wodrich MD, Laskay ÜA, Tsybin YO (2013) Principles of electron capture and transfer dissociation mass spectrometry applied to peptide and protein structure analysis. Chem Soc Rev 42:5014–5030

# Chapter 2
# Basic Virology

**Abstract** Current generations are living through a historical event in virology: the severe acute respiratory syndrome coronavirus 2 (SARS-CoV-2) pandemic. If you asked people when the last major pandemic was, many would probably say the Spanish flu of 1918–1920. Pandemics happen more frequently than most people think (Sampath et al., Cureus 13:e18136–18144, 2021). There have been at least five recorded pandemics between the Spanish flu and the recent SARS-CoV-2 pandemic. While the SARS-CoV-2 pandemic has brought with it indescribable sorrow and pain, it has also increased people's awareness of the impact viruses can have on life on Earth. Terms such as hygiene, modes of infection, community spread, quarantine, vaccines, etc. have become part of our daily conversations. As this book's first chapter provided a basic review of proteomic technologies for virologists, this chapter provides a basic review of virology for proteomic scientists. While virologist may find this chapter unsatisfying, hopefully proteomic scientists will learn enough to be able to understand how proteomic technology can be used to further detail the structure, function, and life cycle of viruses. This chapter will provide both generic and specific examples of viral structures, how they infect cells, and their effects on both the infected cell and the entire organism.

**Keywords** Virus • Virus structure • Enveloped virus • Non-enveloped virus • Virus shape • Viral genome • Classification of viruses • Infection • Viral egress • Virus replication

## 2.1 Introduction

As Chap. 1 was a basic primer in proteomic technology for those readers whose expertise is virology, this chapter is a basic primer of virology for those readers whose expertise is proteomics. Before getting into specific details describing various characteristics of viruses, it is important to have a 30,000-foot view of the basic life cycle of a virus. Viruses are obligate parasites—meaning they are reliant on a

 
T. D. Veenstra, *Proteomic Applications in Virology*, Advances in Experimental Medicine and Biology 1511, https://doi.org/10.1007/978-3-032-22340-1_2

host cell for replication and survival. To access the tools necessary for its replication and survival, the virus must cross the host cell's plasma membrane. To enter, the virus needs to bind to a receptor (and often more than one) to cross the membrane and enter the cell's cytoplasm. There are several different methods by which the virus penetrates the membrane, but the goal is the same—deliver its genome so that it can be replicated and used as a template to produce the necessary proteins to replicate the virus. Once its genome is replicated and all the necessary proteins are translated, the viral progeny are assembled. The final step is the release, or budding, of the viral progeny from the host cell. During this process, some viruses "envelope" themselves with lipids from the cell's membrane. In summary, the classical viral life cycle can be divided into five stages (see Fig. 2.1): (1) binding to a host cell

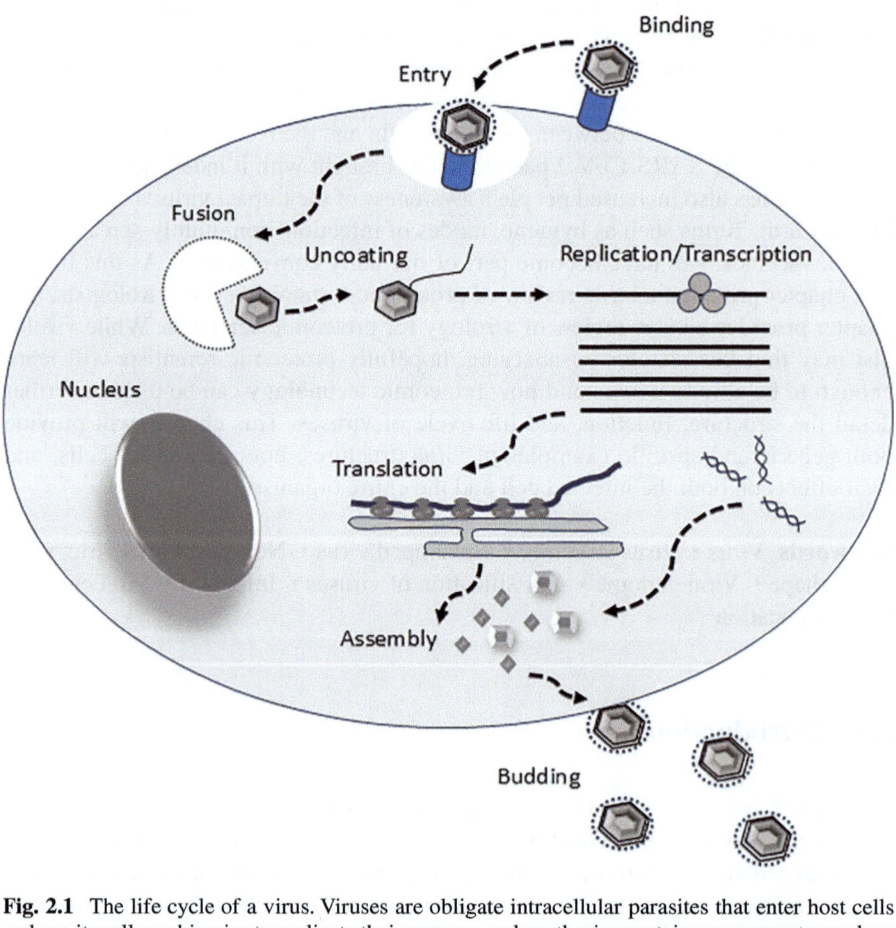


**Fig. 2.1** The life cycle of a virus. Viruses are obligate intracellular parasites that enter host cells and use its cell machineries to replicate their genome and synthesize proteins necessary to produce progeny. The progeny are released from the host cell to invade other nearby hosts. (Reproduced by permission from Villanueva et al. (2005), https://creativecommons.org/licenses/by-sa/4.0/)

receptor, (2) entry into the cytoplasm, (3) replication of its genome and proteins, (4) assembly of the progeny viruses, and (5) budding of the virus out of the host cell (Villanueva et al. 2005). More details on viral structures, processes, and functions will be described in this chapter, but the important thing is to remember is its basic life cycle.

## 2.2 Are Viruses Alive?

One of the most intriguing questions about viruses is whether they are alive or not. Most virologists do not consider viruses as living organisms as they do not meet many of the definitions of life that exist today. For example, viruses do not have cells, are incapable of producing energy through metabolism, and cannot grow or reproduce independently. Others will argue, however, that they do reproduce (although with the help of living cells) and evolve just like living cells (Villarreal 2004). They may point to the obligate intracellular bacterium *Rickettsia* that also requires other cells to reproduce yet is considered a living cell (Rymaszewska and Piotrowski 2024). Recently discovered nucleocytoplasmic large DNA viruses are similar in size to some bacteria and have genomes of approximately 2.5 million nucleotides (Xian and Xiao 2020). Within their genomes are genes that encode metabolic genes involved in nitrogen metabolism, fermentation, sphingolipid biosynthesis, and carbon metabolism (i.e., glycolysis, gluconeogenesis, Kreb's cycle, etc.). Current evidence suggests that these genes affect the metabolism of the host cell rather than act independently within the virus; however, their presence strengthens the argument of those who consider viruses as living.

Complicating the debate, Dr. Patrick Forterre of the Pasteur Institute suggested that viruses can be both living and nonliving (Forterre 2016). His argument describes viruses as alternating between an inactive, nonliving, state outside of a host cell and a metabolically active, living state within a host cell. Dr. Forterre termed the intracellular, metabolically active state of a virus a virocell. To him, viruses are akin to spores, whose potential is only realized when the proper conditions are met. Regardless of whether you consider viruses as living or not, they continue to play a major role in the existence of life on our planet.

## 2.3 How Are Viruses Named?

The International Committee on Taxonomy of Viruses (ICTV) established guidelines for naming newly discovered viruses (https://ictv.global/about/taxonomy). Viruses do not use the naming convention of listing the genus and species together for living organisms, such as *Homo sapiens*, *Canis lupus familiaris*, or *Saccharomyces cerevisiae*. Virus names should also not include any person's name, although historically a few viruses such as Epstein-Barr and Rous sarcoma viruses were named

**Table 2.1** Name origins of selected viruses

| Name | Origin |
|---|---|
| Viruses named after the condition they cause | |
| Human immunodeficiency virus (HIV) | Leads to immunodeficiency |
| Human papillomavirus (HPV) | Causes benign epithelial tumors (i.e., papillomas) |
| Rabies virus | Causes madness (from Latin *rabies*) |
| Hepatitis virus | Cause liver inflammation |
| Viruses named after the location of discovery | |
| Ebola virus | Named after Ebola River in the Democratic Republic of the Congo |
| Marburg virus | Named after Marburg, Germany |
| West Nile virus | Named after West Nile district in Uganda |
| Coxsackievirus | Named after Coxsackie, New York |
| Viruses named after individual(s) who discovered it | |
| Epstein-Barr virus | Named after Michael Anthony Epstein and Yvonne Barr |
| Rous sarcoma virus | Named after Peyton Rous |
| JC virus | Named after patient, John Cunningham |
| Viruses named after their properties | |
| Herpesvirus | Named for lesions that "creep" (from Latin *herpein*) across the skin |
| Poliovirus | Named for "gray" (from Latin polios) lesions that infect the spinal cord |
| Influenza virus | Named because the fifteenth-century epidemic was believed to be "influenced by the stars" |
| Coronavirus | Names after crown-like appearance (from Latin *corona*) |
| Picornavirus | Named because they are small (from Latin *pico*) |

after people. A selected name for a virus should be meaningful and easy to use or pronounce. Guidelines issued by the World Health Organization in 2015 called on scientists, authorities, and the media to avoid using names of places, people, or animals when naming viruses and the infectious diseases that they cause. Unfortunately, the guidelines don't apply retroactively to established diseases whose names are well known and commonly used worldwide. Therefore, conditions like Middle East respiratory syndrome will continue to be associated with this region of the world. Historically, viruses have been named after the clinical condition they cause, their location of discovery, the individual(s) who discovered it, or one of its properties (see Table 2.1).

## 2.4 Size of Viruses

The most obvious feature of viruses is their tiny size. Most viruses (e.g., influenza, adenovirus) are on the order of 100 nm (i.e., 1000 Å) in diameter; however, the smallest are only about 20–30 nm in diameter (e.g., parvovirus, poliovirus) (Louten

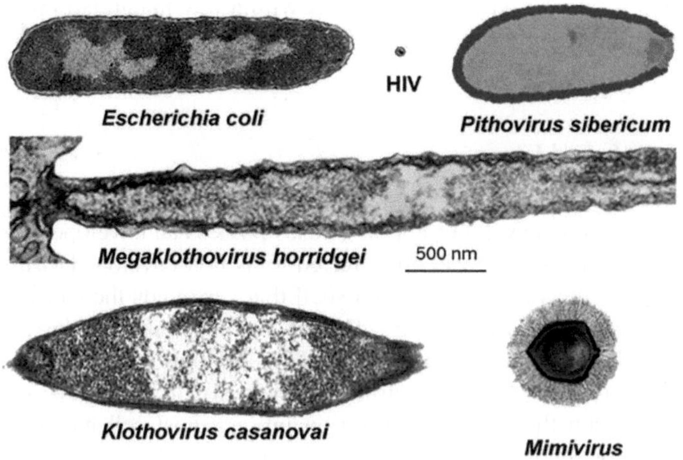


**Fig. 2.2** Comparison of the size of giant viruses *Mimivirus*, *Megaklothovirus horridgei*, *Klothovirus casanovai*, and *Pithovirus sibericum* with that of human immunodeficiency virus (HIV) and the *Escherichia coli*. (Reproduced by permission from Barthelemy et al. (2019), https://creativecommons.org/licenses/by-sa/4.0/)

2016). Considering that a covalent chemical bond is 1–2 Å in length, most viruses are about 1000 atoms in diameter. Giant viruses with diameters in the range of 400–450 nm have recently been isolated (Bosmon et al. 2025). This class includes amoeba-infecting viruses such as *Mimivirus* and *Megavirus chilensis* (i.e., *Megavirus*). The giant viruses *Mimivirus*, *Pithovirus sibericum*, *Klothovirus casanovai*, and *Megaklothovirus horridgei* (currently the largest virus on record with a length of 4 micrometers) are all comparable in size to that of *E. coli* (see Fig. 2.2) (Barthelemy et al. 2019). Other viruses, such as Ebola and Marburg, may have diameters typical for viruses but exist as long threads approaching lengths near 1000 nm (Dupuy et al. 2023). An average human cell is about 10,000 nm (i.e., 10–30 μm) in diameter, which means they are about 100 times larger than the viruses that are infecting them. While bacteria are generally 20–30 times larger than most viruses, mycobacteria are typically only 200–300 nm in diameter making them similar in size to giant viruses.

## 2.5 Virus Structure

For all the physiological responses they generate, viruses have very simple molecular structures. They are essentially nucleic acids in a shell. They contain no cytoplasm or functional organelles. Their nucleic acid component, referred to as their genome, is some form of DNA or RNA. Their genome is protected within a protein shell known as a capsid. Some viruses have an additional envelope surrounding the capsid, which is comprised mainly of lipids with a few integrated proteins. In the following sections, I will discuss these features in more details.

### 2.5.1 *Enveloped Virus*

All viruses possess a capsid, but not all viruses possess an envelope (see Fig. 2.3). Those that only possess a capsid are referred to as non-enveloped viruses (Louten 2016). The envelope is primarily a lipid shell that surrounds the capsid, but it also contains proteins and carbohydrates. The lipids that make up the envelope are obtained from the host cell as the virus exits (Ono 2010). If it contains an outer envelope, it is a strong indication that the virus exits the host cells via exocytosis. However, some viruses obtain their envelope from an internal host cell membrane such as nuclear, Golgi, or rough and smooth endoplasmic reticulum membranes (Welsch et al. 2007). Proteins within the viral envelope can be encoded by either the viral or the host genome, but the lipids and carbohydrates originate strictly from the host cell. Enveloped viruses have some advantages compared to capsid-only viruses. They are better protected from the host's immune system and enzymes (Wisskirchen et al. 2014). Envelope-embedded glycoproteins often serve as ligands that bind to receptors on the host cell enabling them to be taken up more readily by the host. Entry into the host cell is usually accomplished via fusion of the virus envelope with the membrane of the host cell. Entry can also be gained, however, by the engulfing of the entire virus by the host cell membrane, a process known as endocytosis (Klasse et al. 1998). Endocytosis of the virus creates an endosome within the host cell, and the virus envelope fuses with the endosome's membrane to release its contents.

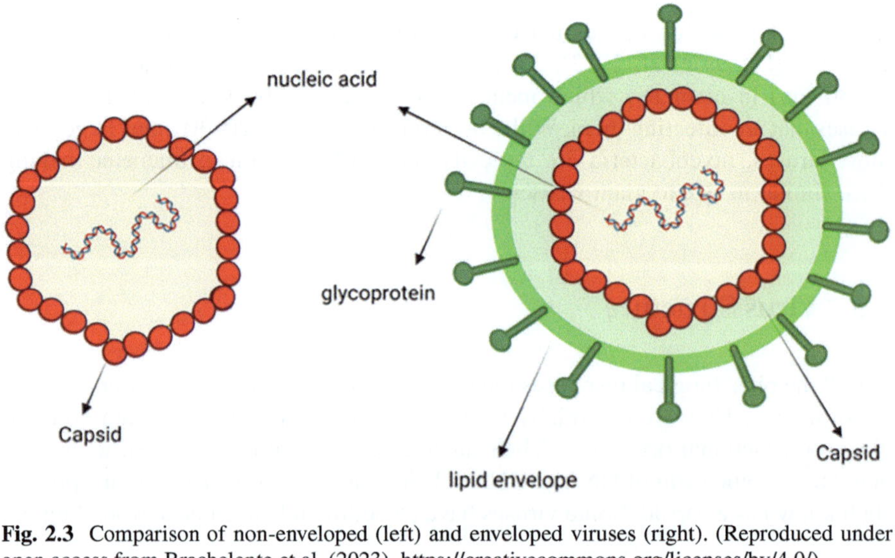


**Fig. 2.3** Comparison of non-enveloped (left) and enveloped viruses (right). (Reproduced under open access from Brachelente et al. (2023), https://creativecommons.org/licenses/by/4.0/)

### 2.5.2 Non-enveloped Viruses

As mentioned above, all viruses contain a capsid (Louten 2016; Krupovic and Koonin 2017; Brachelente et al. 2023). The proteins that form the capsid are encoded within the nucleic acid of the virus. Viruses that only have a capsid surrounding their genome are known as non-enveloped, or naked, viruses (see Fig. 2.3). While it may not seem intuitive, there are advantages to being a non-enveloped virus (Zeng et al. 2023). Only having a capsid means that the virus does not need to encode for envelope proteins and can survive with a smaller genome allowing the virus to expend less energy and more efficiently reproduce. Collectively, these advantages increase the virus's ability to thrive since it can replicate faster, which leads to potentially higher mutation rates that enable it to adapt to changing conditions or infect additional hosts. While it may not seem intuitive, capsid viruses are generally more resistant to environmental extremes such as heat, pH, dryness, and various chemicals (especially detergents) (Tarannum and Ahmed 2023). The lipid components of enveloped viruses are all susceptible to these environmental extremes. Resistance to extreme conditions allows capsid viruses to survive outside the host longer, which increases their chances of transmission through both increased longevity and additional routes of infection beyond direct host contact since they can survive on contaminated surfaces. Many capsid viruses have specialized proteins that attach to host cell receptors and enter the host via endocytosis. This process eliminates the need for membrane fusion, which is required for enveloped virus to enter their host (Más and Melero 2013).

### 2.5.3 Viral Shape

One way that viruses are classified is by the shape of their capsid. While there are different shape classifications reported in literature, one main classifier uses the categories helical, polyhedral, spherical, and complex (see Fig. 2.4) (Parvez 2020). Helical capsids are comprised of multiple subunits of a single protein, called capsomeres, arranged around a central axis to form a helix (Gelderblom 1996). The helix may arrange itself to create a hollow "tubelike" structure, which houses the virus's genome. This helical arrangement gives these viruses rod-shaped or filamentous structures. The size of a helical virus depends on the length of its genome along with how it is packaged and the number of capsomeres that comprise the capsid. Helical viruses can bend and contort under different conditions, allowing the virus to adopt different shapes. Examples of viruses with helical capsids include tobacco mosaic, Ebola, and rabies virus.

Polyhedral viruses have capsids that are many-sided in the form of an icosahedron (Louten 2016). An icosahedron is a geometric solid with 20 triangular faces that somewhat resembles a soccer ball. This highly efficient shape requires the fewest possible capsomeres to encapsulate the virus's genome. Both poliovirus and

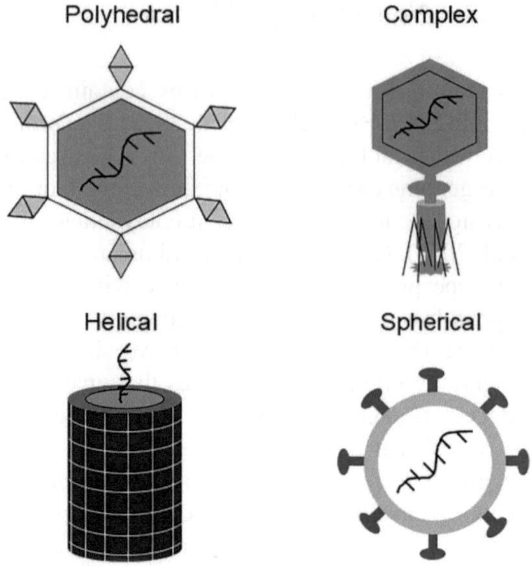


**Fig. 2.4** Basic shapes of viruses

rhinovirus have polyhedral shapes (Parvez 2020; Sherman et al. 2020). Spherical viruses are often considered a subclass of polyhedral viruses because when they are examined closely they are actually icosahedral in shape (Gelderblom 1996). The triangular faces are arranged to make the virus appear spherical; however, their icosahedral shape can be discerned when viewed under an electron microscope. The capsids of complex viruses are neither purely helical nor purely icosahedral. They generally have additional structures such as protein tails or complex outer walls (San Martín 2013). The most recognizable complex viruses are bacteriophages, which have an icosahedral head (which contains the genome) and a helical tail (Kikuchi and King 1975). Fibers in the tail region help attach the virus to the host cell through interactions with surface receptors. The tail can then form a tube which serves as a "molecular syringe" to inject the viral genome into the host (Schwemmlein et al. 2018). All the various viral shapes described can exist in either enveloped or non-enveloped forms.

## 2.6 Viral Genomes

All viruses must contain nucleic acids that serve as its genome. Just like any cellular organism, a virus's genome provides a blueprint for generating the biomolecules necessary for it to thrive. Unlike cells, in which their genomes are encoded using double-stranded DNA (dsDNA), viral genomes are highly variable. While some viruses utilize dsDNA as their genomic material, others contain single-stranded DNA (ssDNA) or single- or double-stranded RNA (ssRNA or dsRNA) (Mahmoudabadi and Phillips 2018). Single-stranded RNA can either be positive- (+ssRNA) or negative-sense (-ssRNA). Positive-sense RNA can be transcribed into mRNA, while -ssRNA is complementary to mRNA and must be converted to +ssRNA by an RNA polymerase prior to its message being translated. These genomes can be packaged as a single piece or multiple segments. Genomes can be in circular forms or linear, regardless of whether they are single- or double-stranded.

An interesting feature of viral genomics is the large amount of genomic diversity they collectively possess. While an individual virus does not contain a lot of genetic material compared to higher-order eukaryotes, collectively, they possess an enormous amount of genetic variation. In fact, viruses comprise a large portion of the total genomic diversity on Earth owing to their number of taxonomic units, rapid infectious cycle, error-prone replication, and large burst size (i.e., the number of virions released from an infected cell) (Sanjuán and Domingo-Calap 2021). Recent metagenomic studies have estimated the number of viral taxonomic groups at over 800 million, with only about 3% of the viral genetic diversity having been discovered (Lu et al. 2025). This large genomic diversity coupled with a lack of characterization makes it difficult to trace the evolutionary history of viruses as they have no shared universal markers such as 16S ribosomal sequences that are useful in classifying and placing organisms in their proper evolutionary lineage.

## 2.7 Genomic Classification of Viruses

Since they have no universal genomic markers, viruses have historically been classified based on notable morphological characteristics, as proposed by the ICTV in 1962, or based on the way they produce mRNA (i.e., Baltimore classification as proposed in 1971) (Fauquet and Fargette 2005; Baltimore 1971). The ICTV classifies viruses into seven orders: *Herpesvirales*, large eukaryotic dsDNA viruses; *Caudovirales*, tailed dsDNA viruses that typically infect bacteria; *Ligamenvirales*, linear dsDNA archaea-infecting viruses; *Mononegavirales*, nonsegmented -ssRNA viruses that infect plants and animals; *Nidovirales*, +ssRNA viruses that infect vertebrates; *Picornavirales*, small +ssRNA viruses that infect plants, insects, and animals; and *Tymovirales*, monopartite +ssRNA viruses composed of a single nucleic acid that infect plants (Fauquet and Fargette 2005). The Baltimore classification scheme also groups viruses into seven categories ultimately based on the process of mRNA production. These classifications include both the genomic structure of the virus and any intermediates that are formed on the route to producing mRNA (see Fig. 2.5a) (Baltimore 1971). These categories are dsDNA viruses (Group I), ssDNA viruses (Group II), dsRNA viruses (Group III), +ssRNA viruses (Group IV), -ssRNA viruses (Group V), +ssRNA-RT viruses with DNA intermediates (Group VI, commonly known as retroviruses), and dsDNA-RT retroviruses (Group VII). Groups VI and VII replicate using a reverse transcriptase. A simpler version of the Baltimore classification system, known as nucleotide type, has also been proposed (Gelderblom 1996). It is like the Baltimore classification; however, it groups viruses into only

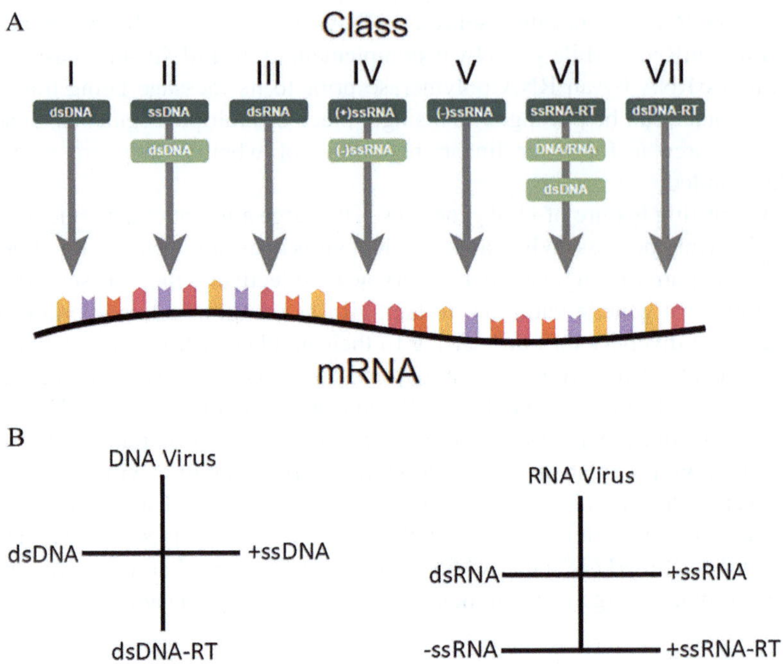


**Fig. 2.5** The nucleotide-type (above) and Baltimore classification (below) system of viruses. The nucleotide-type classification system groups viruses into the type of genetic material they possess and whether it is single- or double-stranded or utilizes reverse transcriptase (RT) enzyme to generate transcriptional-ready genetic material. The Baltimore classification system also categorizes viruses based on their nucleic acid type (DNA or RNA), number of nucleic acid strands (single- or double-stranded) but includes how they generate mRNA during their replication cycle. ("The Baltimore classification of viruses" by Thomas Splettstoesser is licensed under CC BY 3.0)

two groups based on whether their nucleotide type is DNA or RNA (see Fig. 2.5b). The DNA type includes dsDNA, ssDNA, and dsDNA that replicate through an RNA intermediate using a reverse transcriptase enzyme (dsDNA-RT). The RNA group includes dsRNA, +ssRNA, -ssRNA, and +ssRNA-RT. The final classification system is the host-domain classification based on the host domain that the virus infects. This classification system is comprised of eukaryotic, bacterial, and archaeal viruses (Mahmoudabadi and Phillips 2018). The host-domain classification system is based on the hypothesis that viruses evolved along with their specific host domain. Evidence for this hypothesis is bolstered by the observation that there are no known instances of a host from one domain being infected by viruses that infect another domain. An analysis of almost 2400 viruses that could be confidently associated with a specific host showed that approximately 58% infected eukaryotes, 40% infected bacteria, and less than 2% infected archaea.

### 2.7.1 *Size of Viral Genomes*

Since Frederick Sanger and his team sequenced the ssDNA virus phiX174, high-throughput sequencing methods have enabled the rapid deciphering of the entire genome sequences of countless viruses (Sanger et al. 1977). As of early 2025, there were almost 35,000 complete viral genome sequences deposited within the virus-host DB (Mihara et al. 2016). Currently, the largest known viral genome belongs to the giant dsDNA virus *Pandoravirus salinus*, which contains 2,473,870 nucleotides (600,000 nucleotides larger than the *Haemophilus* bacterium) (Aherfi et al. 2018). These nucleotides encode 2541 proteins; however, the function of many of these gene products are unknown since homologous sequences have not been found in other genomes. DNA viruses have substantially larger genomes than RNA viruses. The median genome size of DNA viruses is 148.21 kilobases (kb) compared to only 13.32 kb for RNA viruses (Cui et al. 2014). While *Pandoravirus salinus* has a genome size of almost 2.5 megabases (Mb), the largest RNA virus genomes belong to members of the nidovirus family and range between only 33.5 and 41.1 kb in length (Saberi et al. 2018), for example, planarian secretory cell nidovirus, which infects secretory cells of flatworms, contains only 4.1 kb of ssRNA. The smaller size of RNA viral genomes is likely due to the more fragile nature of RNA molecules.

### 2.7.2 *What's Encoded in a Viral Genome?*

A wide range of proteins are encoded within a viral genome. While the genome of some viruses only encodes a single protein, others encode thousands of proteins. Viral genomes aren't strictly protein-coding; they also contain regulatory sequences for protein synthesis (Rampersad and Tennant 2018). The genomes of giant viruses even contain gene products such as transfer RNAs (tRNAs), tRNA synthetases, and ribosomal subunits that are involved in protein synthesis (Moreira et al. 2025). All RNA viruses, and some DNA viruses, contain RNA or DNA polymerases for replicating their genomes (Choi 2012). Genes are present for translating enzymes and structural proteins for assembling and packaging. Some viruses also contain genes that encode proteins who are responsible for temporally regulating the replication cycle (Bagga and Bouchard 2014). For protection, every known virus contains at least one gene that produces a product that antagonizes the host's immune response (Rouse and Sehrawat 2010). Finally, some virus gene products are responsible for stimulating the spread of the virus to other host cells and organisms.

What is not found in the genome of a virus are genes that encode the complete protein synthesis machinery—which is why viruses are obligate parasites. Viruses do not encode proteins that are capable of synthesizing membrane components like those lipids found in eukaryotic and prokaryotic cells. While these facts are true for the time at which this book is being printed, as more and more genomes of viruses are released, it is likely that additional proteins involved in protein and membrane biosynthesis, metabolism, etc. will be identified. In many of the newly discovered giant viruses, 90% of the nucleotide sequences are referred to as "dark matter" since scientists currently don't know their function (Santiago-Rodriguez and Hollister 2022).

## 2.8 Viral Infection

Viruses have something in common with vampires. For a vampire to enter an owned house, they must first be invited in by the owner. The owner is now their host, and the vampire is now able to control the home and enjoy its amenities. This situation is like a virus interacting with host cells. Viruses do not simply enter any cell they encounter; they only enter specific host cells that possess a receptor on their membrane surface that attaches to a protein within the capsid or envelope of the virus. While viral entry via attachment to a main receptor may seem straightforward, it is a complex process. Along with a main receptor, entry often involves co-receptors, accessory receptors, and attachment factors. Two examples of viruses that interact with multiple receptors are SARS-related and Zika viruses (Jeffers et al. 2004; Lee et al. 2018). Only if the specificity of the interaction with the main receptor is sufficiently high and all the necessary auxiliary receptors interact will the virus be able to enter the host cell. Once inside, the virus avails itself of the many amenities provided by the host cell.

When a virus is recognized by a receptor on the host cell, it either injects its genetic material into the cell or is brought into the cell via endocytosis. Bacteriophages (viruses that infect bacteria) commonly inject their genome into the cell and leave the capsid outside (Roos et al. 2007). Plant and animal viruses more commonly enter via endocytosis. Once inside the cell, the capsid disintegrates, and the viral genome is released into the host cell.

The viral genome provides the plans that describe the biomolecules (protein and nucleic acid) needed to reassemble multiple copies of the virus. What it doesn't provide are the tools for producing them. The process of replicating the necessary biomolecules depends on the type of genome the virus contains. DNA viruses most often use host cell proteins and enzymes to replicate the viral DNA and transcribe it into mRNA. This mRNA is directly used as the template to translate the viral proteins. If the infectious agent is an RNA virus, its RNA is used to synthesize both viral genomic RNA and mRNA. The mRNA is used to synthesize viral proteins, while the copied genomic RNA is repackaged into viral progeny. While viruses generally depend on the host cell machinery to produce its proteins, there are exceptions (Herschhorn and Hizi 2010). The RNA genome-containing viruses known as retroviruses initially transcribe their RNA into DNA. To convert RNA into DNA, they contain a gene that produces the enzyme reverse transcriptase (RT). While RT is transcribed by the host cell ribosomes, its gene is an original part of the virus genome. This DNA is then incorporated into the host cell's genome using integrase, another protein translated directly from the viral genome.

### 2.8.1 Viral Egress

Once all the parts of the virus have been reassembled, the last stage of the cycle is release of its progeny from the host cell. Viruses leave the host cell via one of two general mechanisms; they bud through the membrane, or the host cell dies, and the virions are directly released into the extracellular space (Welsch et al. 2007). While they are routinely released from host cells by budding at the plasma membrane, many enveloped viruses exit the cell through an internal compartment acquired from the membrane of an internal organelle such as the endoplasmic reticulum (Villanueva et al. 2005; Welsch et al. 2007). In this latter case, the virus, along with its envelope acquired from the organelle, exits the cell via a conventional secretory pathway. Non-enveloped viruses do not have an external membrane. The simplest way for them to pass through the cell membrane is to disrupt the integrity of that bilayer. While many people simply believe that the cell lyses because it becomes bloated with viral particles, there are several different mechanisms by which viruses cause the cell to lyse (Young 2014; Jayawardena et al. 2020). Some viruses incorporate their own proteins into host cell membranes or release enzymes that degrade the membranes resulting in its rupture (de Armas-Rillo et al. 2016). Some viruses cause excess amounts of water to accumulate internally by disrupting the cell's ability to regulate its water intake. Some viruses stimulate cell lysis by disrupting its overall structure through interfering with its cytoskeleton (Taylor et al. 2011). Virus can also directly or indirectly trigger apoptosis by either directing interacting with cell proteins within apoptotic pathways or activating sensor that trigger cell apoptosis (Neumann et al. 2015). Regardless of the method, the net result is the release of newly formed viral progeny that can infect nearby host cells.

## 2.9 Viral Replication

While a general framework for viral entry, replication, and exit from the host cell was provided above, the specific details of viral replication can differ significantly between viruses. The focus of this book is not on the specifics of viral replication, but I thought it would be helpful to describe in detail two high-profile viruses—HIV and influenza—to help better understand the intricacies of viral replication.

### 2.9.1 HIV Replication

HIV is an enveloped retrovirus that directly targets and infects T-helper ($T_h$) cells, also known as CD4+ cells (Gallo 2006; Scott and Worku 2024). HIV's envelope is comprised of a lipid membrane, which is derived from the host cell. Within the lipid membrane are glycoproteins, primarily glycoprotein 120 (gp120) and glycoprotein

41 (gp41) that are synthesized by the host cell based on information within HIV's genome. Within the envelope is a capsid comprised of a single protein known as the capsid protein or more commonly p24. The capsid surrounds two RNA molecules, and the enzymes reverse transcriptase, integrase, and a protease. Between the capsid and envelope is an inner coat comprised of matrix protein, which forms a structural shell on the inner surface of the envelope. The matrix plays a role in the early stages of HIV replication.

When it encounters a $T_h$ cell, gp120 molecules on the surface of HIV bind to CD4 receptors on the surface of the $T_h$ cell (see Fig. 2.6) (Wilen et al. 2012). This interaction results in a conformational change that allows a second receptor on the $T_h$ cell, the C-C chemokine receptor type 5 (CCR5), to bind to gp120. The stalk of gp120 punctures the host cell membrane bringing the virus closer to the host cell. When HIV and the host cell get close enough, their membranes fuse together, and the viral genetic material is injected into the host cell. Although the genetic material enters the cell enclosed within the capsid and matrix proteins, gp120 remains on the host cell's surface.

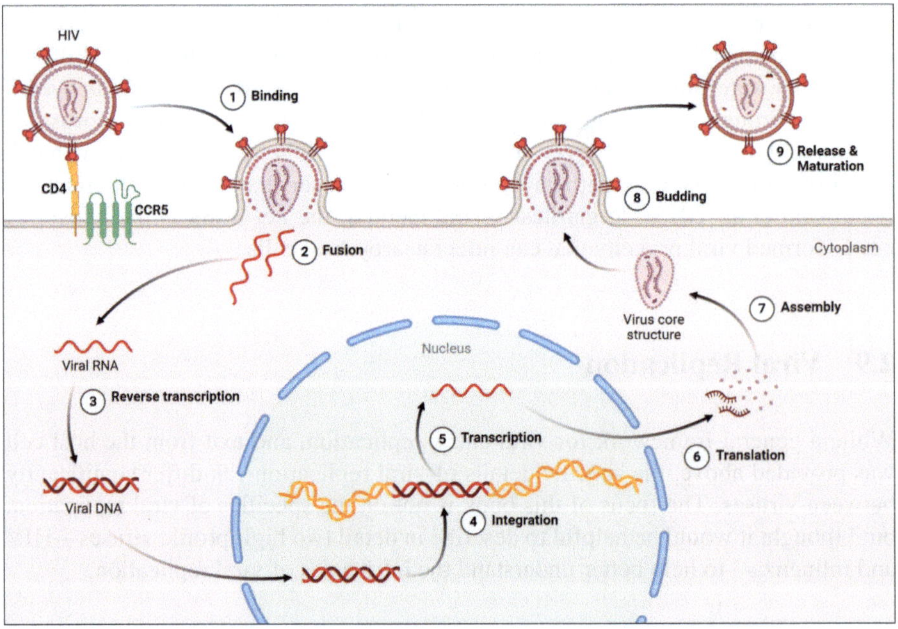


**Fig. 2.6** The life cycle of HIV-1. HIV-1 interacts with CD4 receptors and CCR5 co-receptors on the surface of T-helper cells resulting in the fusion of the virus with the cell membrane and the release of its core genome and proteins into the cytoplasm (2). Reverse transcriptase then converts the viral RNA into double-stranded DNA, which is transported into the nucleus and integrated into the cell's genome. The viral genes are then transcribed and translated by the host cell and cleaved into functional proteins by HIV's protease enzyme. The virus components are reassembled into progeny viruses that migrate to the cell membrane. As they bud from the cell, the progeny reclaim their envelope and are released into the extracellular matrix. (Reproduced by permission from Scott and Worku (2024), https://creativecommons.org/licenses/by-sa/4.0/)

HIV's genome is not fully released into the host cell until the protease enzyme proteolytically digests the matrix and capsid proteins. The HIV reverse transcriptase, which also entered the host cell within the capsid, now uses host cell nucleotides to convert the HIV RNA into a single strand of DNA. The single DNA strand is again acted upon by reverse transcriptase and turned into a double-stranded DNA molecule. HIV integrase, which was also released from the capsid into the host cell, takes this DNA and transports it into the nucleus. Once there, the integrase locates a chromosome and makes a nick in it. The HIV DNA is then inserted (i.e., integrated) into this region of the host chromosome. The challenge with having HIV genetic material insert into the host cell genome is that the result is lifelong infection, which can only be rectified by removing this foreign segment of DNA from each host cell.

As a side note, the insertion of HIV DNA into the host cell genome by HIV integrase is not entirely random (Tolomeo et al. 2025). While theoretically integrase can integrate the DNA at any location, it preferentially targets regions of the host genome that either are near the nuclear periphery or contain active genes. Along with this trend, integrase also favors open euchromatin regions for integrating HIV's genome since these regions are more accessible to the integration machinery. Another interesting feature of HIV integration is the presence of the host cell protein LEDGF (lens epithelium-derived growth factor), which is also known as p75 or PC4 (positive cofactor 4) and SRSF1 (splicing factor arginine/serine-rich 1) (De Luca et al. 2011). LEDGF is a chromatin-binding protein that is involved in several cellular processes, including repair of DNA double-strand breaks (Daugaard et al. 2012). Its overexpression has been implicated in the progression of leukemia and breast and prostate cancers, among others (Canella et al. 2022). LEDGF is a cofactor for HIV integrase and facilitates the integration of HIV genetic material in the host genome by binding the viral pre-integration complex onto the host cell's chromatin (Van Maele et al. 2006).

In the next phase, host cell RNA polymerase transcribes the viral DNA into mRNA that leaves the nucleus. The mRNA encodes the different proteins that are required to reproduce the virus. Interaction between the mRNA and ribosomes on the host cell's endoplasmic reticulum ends up producing these viral proteins. One of the proteins being produced is HIV envelope protein, which migrates to the surface of the host cell and embeds itself in the membrane. Other mRNAs are produced that are translated into other viral proteins. The viral proteins are made in a single long proprotein chain that moves to the cell surface along with a strand of viral RNA. The proprotein and viral RNA occupy a space near where the HIV envelope proteins are embedded. This collection of gp120, envelope proteins, proprotein, and viral RNA then buds off the cell surface and leaves the host cell.

Even though the viral particle has budded off the surface of the host cell with all the necessary components to reconstruct HIV, it is not yet a mature virion. Prior to becoming a mature HIV virion, the proprotein needs to be cut into its component parts. The HIV protease enzyme is responsible for cutting the proprotein into smaller functional proteins. These smaller proteins include the matrix protein (that lines the viral envelope), the capsid protein (that forms the outer shell of the HIV core region that houses the genomic material), the nucleocapsid protein (that coats the genomic RNA), and the p6 protein (that facilitates the release of the viral particles from host cells). This mature virion is now capable of infecting other cells and propagating the spread of HIV.

### 2.9.2 *Influenza A Replication*

In the first stage of replication, the influenza A virus attaches to the host cell via hemagglutinin (HA) on the surface of the virion (see Fig. 2.7) (Dou et al. 2018; Carter and Iqbal 2024; Yang et al. 2016). Hemagglutinin is a trimeric glycoprotein molecule composed of three identical subunits (Skehel and Wiley 2000). Each of the subunits is divided into two disulfide-linked polypeptides, HA1 and HA2. HA2 forms a stalk that anchors the hemagglutinin to the membrane of the virus, while

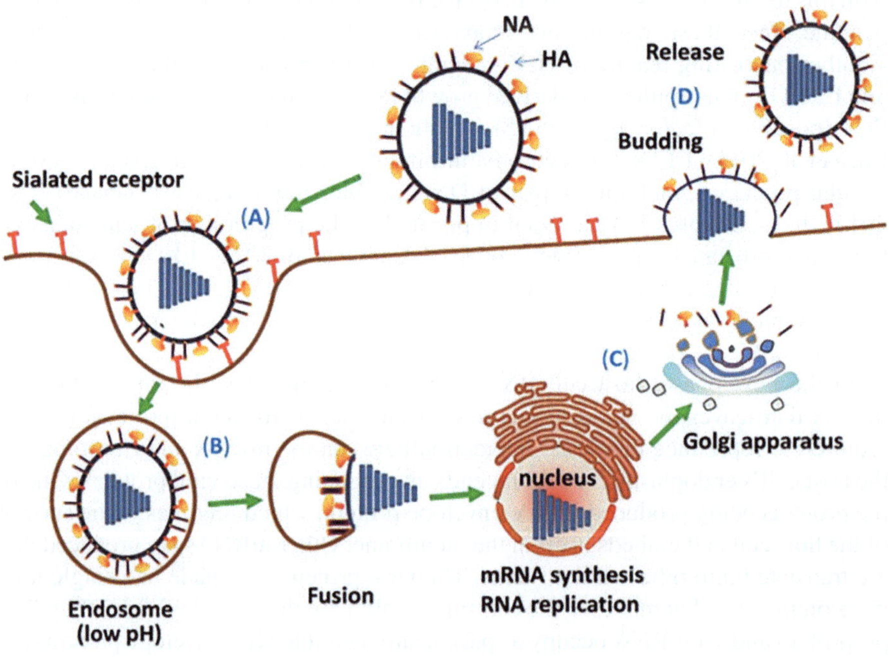


**Fig. 2.7** The influenza A virus life cycle (IAV). In the first step, hemagglutinin of influenza A binds to cell surface receptors containing α-sialic acid residues. The virus envelope fuses with the cell membrane and enters the cell within an endosome. The low pH of the endosome causes the influenza A envelope to fuse with the endosome membrane resulting in the release of the virus's genetic materials and associated proteins into the cytosol. The viral RNA and mRNA enter the nucleus and are replicated and translated into proteins, respectively. After exiting the nucleus, the viral components are reassembled at the cell membrane, where the envelope is reformed and the viral progeny bud, and are released to freely infect neighboring cells. (Reproduced by permission from Yang et al. (2016), https://creativecommons.org/licenses/by-sa/4.0/)

HA1 comprises the receptor-binding site that binds to polysaccharides on the surface of host cells that have terminal α-sialic acid residues. When influenza binds to its surface, the host cell takes it up via endocytosis. The vessel created via endocytosis, referred to as an endocytic vesicle, is made up of an influenza virion surrounded by a membrane. This vesicle then fuses with a lysosome, which contains proteolytic enzymes and an acidic interior. It is important to note that at this point the contents of the virus required for its replication (i.e., genome) are still encapsulated and cannot freely enter the host cell's cytoplasm. This vesicle fusion is the beginning of the process of releasing the virus's genetic material into the cytoplasm of the host cell.

The fused vesicle creates an acidic environment containing several proteolytic enzymes around the influenza virus (Luo 2012; Stauffer et al. 2014). Ion channels allow protons to enter the virion disrupting various protein-protein interactions, including those between matrix protein that surround the virus's genome, which is comprised of eight single-stranded, negative-sense RNA molecules. Disintegration of interactions that hold the matrix proteins together disrupts its interaction with the RNA genome. This process then allows the RNA to freely move within the viral particle. The low pH of the fusion vesicle also causes a structural change in hemagglutinin that causes it to insert into the endocytic membrane. The result is fusion between the envelope of the virus and the endocytic vesicle, which creates an opening allowing the viral RNA and other proteins to freely enter the host cell's cytoplasm.

The eight viral RNA molecules of the virus are still covered with nucleocapsid protein (NP) and attached to other proteins that form the RNA polymerase complex. In the next step toward replication, the RNA/protein complexes enter the host cell's nucleus via nuclear pores (Moreira et al. 2021). These RNAs are noncoding, negative-sense strands and need to be copied into positive-sense strands by the viral RNA polymerase prior to their translation. To prepare this positive-sense mRNA that can be used by the host cell's ribosomes, the viral RNA adds a 5′-cap to the beginning of the mRNA transcript (Gu et al. 2015). To do this, the virus's RNA polymerase removes a 5′cap from an mRNA molecule already in the host cell and attaches its own RNA. This process is termed "cap-snatching" (De Vlugt et al. 2018). The attachment of the 5′cap initiates transcription of the viral RNA, complete with the addition of a poly-A tail to increase the stability of the transcript.

Unlike HIV, the genomic material of influenza A is not integrated into the host cell genome. The influenza A mRNA molecules exit the nucleus and are translated into viral proteins by the host cell ribosomes that reside on the rough endoplasmic reticulum. The eight RNA segments code for between 10 and 14 proteins depending on the influenza strain (Chou et al. 2012; Bouvier and Palese 2008). These proteins include RNA polymerase subunits PA, PB1, and PB2, along with hemagglutinin, neuraminidase, the matrix proteins M1 and M2, the NP, and two nonstructural proteins (NS1 and NEP) (Bouvier and Palese 2008). While these proteins are being translated, newly formed NP reenters the nucleus and stabilizes the viral positive-sense RNA molecules that are being used to translate viral proteins (Te Velthuis and Fodor 2016). Viral RNA polymerase enzymes also enter the nucleus and generate negative-sense RNA strands that will serve as the genome of the newly formed influenza virions.

The newly formed RNA genome strands that are combined with RNA polymerase, NP, matrix proteins, and packing proteins are exported from the nucleus so that new viral particles can be assembled (Carter and Iqbal 2024; Hutchinson and Fodor 2013). Envelope proteins made on the rough endoplasmic reticulum are moved to the Golgi apparatus by transport vesicles. From there, they move to the plasma membrane and meet with the RNA-protein complexes where they are packaged into progeny viruses. They then bud from the cell membrane. It is during budding that the influenza protein neuraminidase plays a key role. Remember at the beginning of the infection of the host cell by the influenza virus that hemagglutinin bound to α-sialic acid residues on the cell surface. The assembled progeny viruses also encounter α-sialic acid-containing receptors as they bud off the host cell. Hemagglutinin proteins that bind to these receptors are unable to detach from the host cell and are therefore unable to infect other cells. Neuraminidase is a glycohydrolase (more specifically a sialidase) that cleaves the α-sialic acid residues, thereby allowing the newly budded influenza viruses to leave the host cell (Shtyrya et al. 2009). With the cleavage of the α-sialic acid residue, the newly formed influenza progeny are free to leave the host cell and infect other cells.

## 2.10 Conclusions

Compared to living organisms, most viruses are structurally simple. They have small genomes and contain only a few proteins that are encased within a simple protein shell or lipid-based membrane. They have five basic goals: bind to a host cell, enter its cytoplasm, replicate its genome and proteins, reassemble progeny, and leave the host cell. The virus does not have the resources to sustain itself, no metabolic pathways, no means of independent movement, etc. All these characteristics make it difficult to rationalize how viruses have such major impacts on the living world. Yet here we are just recovering from one of the greatest upheavals of the past century.

These first two chapters describing the basics of proteomic technology and viruses are in some way contradictory. The first chapter described technologies that can characterize thousands of proteins in a single experiment. The second chapter describes entities that contain a tiny genome and an even smaller protein complement. So, what is the need for a book that describes proteomic applications in virology? The answer lies in what viruses can do once they enter a host cell. Their ability to control the molecular pathways necessary for their propagation is quite astonishing. As will be described in various chapters of this book, not only are they able to use the host cell to replicate, but viruses can also control processes that mask their invasion from other cells and keep its host cell alive so that it can maximize its invasive effect. A second reason for using proteomics to study viruses is the sheer number of species. As of 2024, the ICTV has defined 16,125 different viral species (https://ictv.global/taxonomy). While only about 270 of these are known to cause disease in humans, they operate in a variety of different manners (Forni et al. 2022). They have different methods of entry, expressing their genomic information, and exiting the cell. While this chapter provided an overview of these processes, the mechanisms behind each are quite complicated and involve a significant portion of a cell's proteome beyond a single surface receptor and ribosomes. Now that the reader has a basic understanding of virus structure and pathogenesis, the hope is that they will come to appreciate both the complexity of viral infections and the value that proteomics has brought to our understanding of these complicated processes.

## References

Aherfi S, Andreani J, Baptiste E, Oumessoum A, Dornas FP, Andrade ACDSP et al (2018) A large open pangenome and a small core genome for giant pandoraviruses. Front Microbiol 9:1486–1498

Bagga S, Bouchard MJ (2014) Cell cycle regulation during viral infection. Methods Mol Biol 1170:165–227

Baltimore D (1971) Expression of animal virus genomes. Bacteriol Rev 35:235–241

Barthelemy R-M, Goto T, Faure E (2019) Serendipitous discovery in a marine invertebrate (Phylum Chaetognatha) of the longest giant viruses reported till date. Virology: Curr Res 3(1) https://creativecommons.org/licenses/by-sa/4.0/

Bosmon T, Abergel C, Claverie JM (2025) 20 years of research on giant viruses. NPJ Viruses 3:9–19

Bouvier NM, Palese P (2008) The biology of influenza viruses. Vaccine 26 Suppl 4(Suppl 4):D49–D53

Brachelente S, Galli A, Cervelli T (2023) Yeast and virus-like Particles: a perfect or imperfect couple? Appl Microbiol 3:805–825

Canella A, Van Belle S, Brouns T, Nigita G, Carlon MS, Christ F et al (2022) LEDGF/p75-mediated chemoresistance of mixed-lineage leukemia involves cell survival pathways and super enhancer activators. Cancer Gene Ther 29:133–140

Carter T, Iqbal M (2024) The influenza A virus replication cycle: a comprehensive review. Viruses 16:316–368

Choi KH (2012) Viral polymerases. Adv Exp Med Biol 726:267–304

Chou YY, Vafabakhsh R, Doğanay S, Gao Q, Ha T, Palese P (2012) One influenza virus particle packages eight unique viral RNAs as shown by FISH analysis. Proc Natl Acad Sci USA 109:9101–9106

Cui J, Schlub TE, Holmes EC (2014) An allometric relationship between the genome length and virion volume of viruses. J Virol 88:6403–6410

Daugaard M, Baude A, Fugger K, Povlsen LK, Beck H, Sørensen CS et al (2012) LEDGF (p75) promotes DNA-end resection and homologous recombination. Nat Struct Mol Biol 19:803–810

de Armas-Rillo L, Valera MS, Marrero-Hernández S, Valenzuela-Fernández A (2016) Membrane dynamics associated with viral infection. Rev Med Virol 26:146–160

De Luca L, Ferro S, Morreale F, De Grazia S, Chimirri A (2011) Inhibitors of the interactions between HIV-1 IN and the cofactor LEDGF/p75. ChemMedChem 6:1184–1191

De Vlugt C, Sikora D, Pelchat M (2018) Insight into influenza: a virus cap-snatching. Viruses 10:641–647

Dou D, Revol R, Östbye H, Wang H, Daniels R (2018) Influenza A virus cell entry, replication, virion assembly and movement. Front Immunol 9:1581–1597

Dupuy LC, Spiropoulou CF, Towner JS, Spengler JR, Sullivan NJ, Montgomery JM (2023) Filoviruses: scientific gaps and prototype pathogen recommendation. J Infect Dis 228:S446–S459

Fauquet CM, Fargette D (2005) International Committee on Taxonomy of Viruses and the 3,142 unassigned species. Virol J 2:64–73
Forni D, Cagliani R, Clerici M, Sironi M (2022) Disease-causing human viruses: novelty and legacy. Trends Microbiol 30:1232–1242
Forterre P (2016) To be or not to be alive: how recent discoveries challenge the traditional definitions of viruses and life. Stud Hist Phil Biol Biomed Sci 59:100–108
Gallo RC (2006) A reflection on HIV/AIDS research after 25 years. Retrovirology 3:72–78
Gelderblom HR (1996) Structure and classification of viruses. In: Baron S (ed) Medical Microbiology, 4th edn. University of Texas Medical Branch at Galveston, Galveston. Chapter 41
Gu W, Gallagher GR, Dai W, Liu P, Li R, Trombly MI et al (2015) Influenza A virus preferentially snatches noncoding RNA caps. RNA 21:2067–2075
Herschhorn A, Hizi A (2010) Retroviral reverse transcriptases. Cell Mol Life Sci 67:2717–2747
https://ictv.global/about/taxonomy. Accessed May 2025
https://ictv.global/taxonomy. Accessed May 2025
Hutchinson EC, Fodor E (2013) Transport of the influenza virus genome from nucleus to nucleus. Viruses 5:2424–2446
Jayawardena N, Poirier JT, Burga LN, Bostina M (2020) Virus-receptor interactions and virus neutralization: insights for oncolytic virus development. Oncolytic Virother 9:1–15
Jeffers SA, Tusell SM, Gillim-Ross L, Hemmila EM, Achenbach JE, Babcock GJ et al (2004) CD209L (L-SIGN) is a receptor for severe acute respiratory syndrome coronavirus. Proc Natl Acad Sci USA 101:15748–15753
Kikuchi Y, King J (1975) Assembly of the tail of bacteriophage T4. J Supramol Struct 3:24–38
Klasse PJ, Bron R, Marsh M (1998) Mechanisms of enveloped virus entry into animal cells. Adv Drug Deliv Rev 34:65–91
Krupovic M, Koonin EV (2017) Multiple origins of viral capsid proteins from cellular ancestors. Proc Natl Acad Sci USA 114:E2401–E2410
Lee I, Bos S, Li G, Wang S, Gadea G, Desprès P et al (2018) Probing molecular insights into Zika virus-host interactions. Viruses 10:233–258
Louten J (2016) Virus structure and classification. Essential Human Virology 6:19–29
Lu C, Wu Y, Zhang Z, Mao L, Ge X, Wu A et al (2025) Estimated size of the total genome and protein space of viruses. mSphere 10:e0068324–e0068333
Luo M (2012) Influenza virus entry. Adv Exp Med Biol 726:201–221
Mahmoudabadi G, Phillips R (2018) A comprehensive and quantitative exploration of thousands of viral genomes. elife 7:e31955–e31980
Más V, Melero JA (2013) Entry of enveloped viruses into host cells: membrane fusion. Subcell Biochem 68:467–487
Mihara T, Nishimura Y, Shimizu Y, Nishiyama H, Yoshikawa G, Uehara H et al (2016) Linking virus genomes with host taxonomy. Viruses 8:66–71
Moreira EA, Yamauchi Y, Matthias P (2021) How influenza virus uses host cell pathways during uncoating. Cells 10:1722–1744
Moreira TIR, Carvalho JVRP, Filho CAC, Souza JW, de Azevedo BL, Abrahão JS et al (2025) Investigations into the diversity and distribution of tRNA and phylogenetics of translation factors in amoebozoa-infecting *Nucleocytoviricota*. Viruses 17:328–350
Neumann S, El Maadidi S, Faletti L, Haun F, Labib S, Schejtman A et al (2015) How do viruses control mitochondria-mediated apoptosis? Virus Res 209:45–55
Ono A (2010) Viruses and lipids. Viruses 2:1236–1238
Parvez MK (2020) Geometric architecture of viruses. World J Virol 9:5–18
Rampersad S, Tennant P (2018) Replication and expression strategies of viruses. Viruses 30:55–82
Roos WH, Ivanovska IL, Evilevitch A, Wuite GJ (2007) Viral capsids: mechanical characteristics, genome packaging and delivery mechanisms. Cell Mol Life Sci 64:1484–1497
Rouse B, Sehrawat S (2010) Immunity and immunopathology to viruses: what decides the outcome? Nat Rev Immunol 10:514–526
Rymaszewska a, Piotrowski M (2024) *Rickettsia* species: genetic variability, vectors, and rickettsiosis-a review. Pathogens 13:661–673

Saberi A, Gulyaeva AA, Brubacher JL, Newmark PA, Gorbalenya AE (2018) A planarian nidovirus expands the limits of RNA genome size. PLoS Pathog 14:e1007314–e1007357

Sampath S, Khedr A, Qamar S, Tekin A, Singh R, Green R et al (2021) Pandemics throughout the history. Cureus 13:e18136–e18144

San Martín C (2013) Structure and assembly of complex viruses. Subcell Biochem 68:329–360

Sanger F, Air GM, Barrell BG, Brown NL, Coulson AR, Fiddes CA et al (1977) Nucleotide sequence of bacteriophage phi X174 DNA. Nature 265:687–695

Sanjuán R, Domingo-Calap P (2021) Genetic diversity and evolution of viral populations. Encyclopedia Virol 1:53–61

Santiago-Rodriguez TM, Hollister EB (2022) Unraveling the viral dark matter through viral metagenomics. Front Immunol 13:1005107–1005119

Schwemmlein N, Pippel J, Gazdag EM, Blankenfeldt W (2018) Crystal structures of R-type bacteriocin sheath and tube proteins CD1363 and CD1364 from *Clostridium difficile* in the preassembled state. Front Microbiol 9:1750–1771

Scott GY, Worku D (2024) HIV vaccination: navigating the path to a transformative breakthrough-a review of current evidence. Health Sci Rep 7:e70089–e70102

Sherman MB, Smith HQ, Smith TJ (2020) The dynamic life of virus capsids. Viruses 12:618–638

Shtyrya YA, Mochalova LV, Bovin NV (2009) Influenza virus neuraminidase: structure and function. Acta Nat 1:26–32

Skehel JJ, Wiley DC (2000) Receptor binding and membrane fusion in virus entry: the influenza hemagglutinin. Annu Rev Biochem 69:531–569

Stauffer S, Feng Y, Nebioglu F, Heilig R, Picotti P, Helenius A (2014) Stepwise priming by acidic pH and a high K+ concentration is required for efficient uncoating of influenza A virus cores after penetration. J Virol 88:13029–13046

Tarannum T, Ahmed S (2023) Recent development in antiviral surfaces: impact of topography and environmental conditions. Heliyon 9:e16698–e16711

Taylor MP, Koyuncu OO, Enquist LW (2011) Subversion of the actin cytoskeleton during viral infection. Nat Rev Microbiol 9:427–439

Te Velthuis AJ, Fodor E (2016) Influenza virus RNA polymerase: insights into the mechanisms of viral RNA synthesis. Nat Rev Microbiol 14:479–493

Tolomeo M, Tolomeo F, Cascio A (2025) The complex interactions between HIV-1 and human host cell genome: from molecular mechanisms to clinical practice. Int J Mol Sci 26:3184–3233

Van Maele B, Busschots K, Vandekerckhove L, Christ F, Debyser Z (2006) Cellular co-factors of HIV-1 integration. Trends Biochem Sci 31:98–105

Villanueva RA, Rouillé Y, Dubuisson J (2005) Interactions between virus proteins and host cell membranes during the viral life cycle. Int Rev Cytol 245:171–244. https://creativecommons.org/licenses/by-sa/4.0/

Villarreal LP (2004) Are viruses alive? Sci Am 291:100–105

Welsch S, Müller B, Kräusslich HG (2007) More than one door - budding of enveloped viruses through cellular membranes. FEBS Lett 581:2089–2097

Wilen CB, Tilton JC, Doms RW (2012) HIV: cell binding and entry. Cold Spring Harb Perspect Med 2:a006866–a006878

Wisskirchen K, Lucifora J, Michler T, Protzer U (2014) New pharmacological strategies to fight enveloped viruses. Trends Pharmacol Sci 35:470–478

Xian Y, Xiao C (2020) Current capsid assembly models of icosahedral nucleocytoviricota viruses. Adv Virus Res 108:275–313

Yang J, Liu S, Du L, Jiang S (2016) A new role of neuraminidase (NA) in the influenza virus life cycle: implication for developing NA inhibitors with novel mechanism of action. Rev Med Virol 26:242–250

Young R (2014) Phage lysis: three steps, three choices, one outcome. J Microbiol 52:243–258

Zeng L, Li J, Lv M, Li Z, Yao L, Gao J et al (2023) Environmental stability and transmissibility of enveloped viruses at varied animate and inanimate interfaces. Environ Health (Wash) 1:15–31

# Chapter 3
# Proteomics in the Diagnosis of Viral and Pathogenic Infections

**Abstract** While this book's theme is proteomics and virology, this chapter on the role of proteomics in clinical virology may seem unsatisfying to the reader. The reason is simple: proteomics has not made a major impact on the diagnosing of viral infections. However, it is still important to evaluate the present role of proteomics in clinical virology as it provides a vision of what is required if it is to ever have a major impact in diagnosing viral infections. This chapter will initially provide a background on current techniques for diagnosing viral infections. Although the names may not be familiar to the casual reader, the techniques will be as most individuals have been tested using one of these techniques during their lifetime. The chapter will later switch to describe proteomic methods for the direct detection of viruses in clinical samples. While outside the scope of this book, a proteomic method for bacterial identification is included because (i) it describes a proteomic technology that has been approved by the FDA for clinical use and (ii) since viruses and bacteria are both infectious agents it demonstrates the potential of proteomics for overall pathogen detection. To keep the topic in perspective, however, the advantages and disadvantages of using proteomics in clinical virology are compared to current existing methods particularly those that target the detection of genomic features within these pathogens.

**Keywords** Diagnosis • Clinical virology • Cytopathic • Antivirus antibodies • Enzyme-linked immunosorbent assay • Immunoassays • Nucleic acid amplification test • Clinical bacteriology • Microarrays • Multiple reaction monitoring mass spectrometry

## 3.1 Introduction

Correctly diagnosing a virus based solely on clinical information is unreliable since many viral infections result in overlapping symptoms. One exception to this rule would include epidemics or outbreaks in which many individuals are infected with a virus that produces a distinctive feature, for example, prior to its eradication,



T. D. Veenstra, *Proteomic Applications in Virology*, Advances in Experimental Medicine and Biology 1511, https://doi.org/10.1007/978-3-032-22340-1_3

smallpox could be recognized by the distinctive rash it produced. This recognition enabled medical workers to quickly activate containment measures to limit the virus spread. While not as reliable, measles may be diagnosed in people who have been in contact with an infected individual through the presence of facial red flat spots and oral Koplik's spots. Although different viruses can result in similar clinical symptoms in their host, it is important to accurately diagnose the specific virus that is causing the infection as early as possible (Metsky et al. 2022). Just like other conditions such as cancer, heart disease, and dementia, rapid diagnosis enables the proper treatment to begin promptly, which guarantees the best possible outcome. Another feature that makes rapid diagnosis critical is its ability of the virus to spread. By recognizing the virus quickly, isolation measures can be implemented to limit its further spread. Accurate and early diagnosis is always important, especially for pregnant women, immunocompromised individuals, transplant recipients, and those being treated for concurrent diseases (Billioux et al. 2022; Moreno-Sanchez and Gomez-Gomez 2022). Early, accurate detection is not simply important for treating and containing infected individuals. It is also critical for screening of tissues and blood products that are used for curative or prophylactic treatment (https://www.fda.gov/vaccines-blood-biologics/blood-donor-screening). As we all witnessed during the recent COVID-19 pandemic, accurate viral diagnosis plays a major role in the scientific community's ability to properly trace pandemic hot spots and implement necessary control measures (Pandit et al. 2022; Eyre 2022; Rosenfeld and Tibshirani 2021).

## 3.2 Clinical Virology

The field of clinical virology is responsible for developing, validating, and executing the diagnostic testing required to identify a virus (Trudy et al. 1994). Beyond these roles, some clinical virologists are involved in the development of antiviral therapies, containment strategies, and epidemiology studies designed to prevent, treat, and study the spread of viral infections. Meeting these goals requires a variety of scientists and healthcare specialists from different disciplines including molecular biologists, geneticists, microbiologists, analytical chemists, clinicians, pharmacologists, epidemiologists, and public health administrators. Beyond simple identification, diagnostic testing is conducted to confirm the cause of viral infections and whether they are acute or chronic. While tests are predominantly performed to identify the virus so that the best treatment can be administered, they also aid in the surveillance of these pathogens. Surveillance is critical as it assists in public health decisions to implement the best practices for infection avoidance as well as measuring treatment (e.g., vaccine) effectiveness, viral pathogenicity, and identifying the emergence of novel strains.

As mentioned, the primary focus of clinical virology laboratories is the diagnosis of viral infections. Owing to the COVID-19 pandemic, there has been a huge increase in the number of viral diagnostic tests performed. Viral detection methods

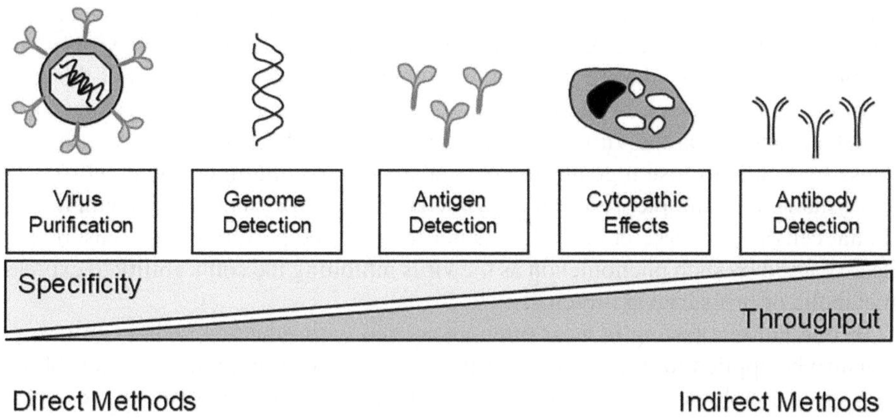


**Fig. 3.1** Comparison of specificity and throughput between various methods used to detect viral infection. While direct methods have greater specificity than indirect methods, they require more stringent sample preparation methods since they analyze the virus itself

can be grouped into two primary categories, either direct or indirect (see Fig. 3.1) (Teymoorian et al. 2021; Fathi Karkan et al. 2022; Pearson et al. 2021). Direct methods detect the virus itself, while indirect methods detect either the effects of the virus or its presence. Since they detect the virus itself, direct methods provide more specificity than indirect methods; however, indirect methods are higher throughput and enable the presence of the virus to be rapidly screened in thousands of samples. Indirect methods can require propagating the viral particles in a host cell line or testing for the presence of virus-specific antibodies in an infected individual. Direct methods assay for the presence of the virus itself without the need for in-cell propagation.

### *3.2.1 Indirect Detection of Viral Cytopathic Effects*

As mentioned above, indirect diagnostic methods measure the effects of the virus, which can include cell death, change in cell morphology, or the production of antibodies. Indirect testing requires the virus to be isolated so that effects from a single strain are observed. The virus is then infected into cultured cells so that its effects can be observed. Infecting cultured cells is also a useful method for amplifying the virus if a larger amount of sample is needed for additional diagnostic tests (Diallo et al. 1995; Cassedy et al. 2021). While culturing viruses is a relatively slow method, centrifuge-enhanced shell-vial methods have greatly shortened the required time (Jayakeerthi et al. 2006). By subjecting the virus sample to low-speed centrifugation in the presence of a suitable cell line, the virus is rapidly absorbed into the host cells. Multiple cell lines (i.e., 3–6) can be co-cultured to also decrease the time required for indirect diagnosis by increasing the probability that a suitable host is selected.

By plating the cells as a monolayer on glass slides, the infected cells can be readily analyzed using a light microscope. Light microscopy is used to observe virus cytopathic effects such as change in cell morphology, cell lysis, detachment of the cells from the culture vessel, or the presence of balloon-like vacuoles that can swell or shrink depending on the virus (Ji et al. 1991; Wang et al. 2020). Some viruses can cause host cells to fuse together to create a single giant multinucleated cell (i.e., a syncytium), for example, respiratory syncytial virus (RSV) induces syncytia in epithelial cells (Guo-Parke et al. 2013). Other tests can be performed to measure cell death caused by such phenomenon as the virus inhibiting the cell's ability to express metabolic or pro-survival proteins.

While indirect testing is most often associated with infecting cells in culture, it can also be applied to infected cells or tissues retrieved from patients via swabbing or biopsy. Cytological examination is performed on blood, lung washing, cerebral spinal fluid, or cell scrapings that are either preserved in liquid or smeared across a slide, for example, Papanicolaou (pap) smears conduct cytological examination of cells scraped from the cervix that are smeared on glass slides (Marshall 1983). Tissue samples for histological examination are often collected via biopsy. Since most cells and tissues collected from a patient are absent of color, stains are added to enable cytopathic effects to be visualized. Inclusion bodies, which are sites associated with viral replication in the cell, are another cytopathic effect that can be observed for some viruses via cytology or histology.

### 3.2.2 *Indirect Detection of Viral Antibodies*

When a host is infected with a virus, it generates an immune response to rid itself of the pathogenic invader. Enzyme-linked immunosorbent assays (ELISAs) are a common technology used in cellular and molecular biology that are often used to directly detect a specific molecule. Although they are based on the use of antibodies, ELISA methods can also be used to detect *antibodies* generated by the host against a virus. This detection of antibodies produced by the host falls within the parameters of an indirect method for determining if a patient is infected with a specific virus (Pan et al. 2023). An example of this type of ELISA is shown in Fig. 3.2 for the detection of SARS-CoV-2 antibodies within an infection patient. In this format, a viral antigen is coated onto the bottom of an ELISA plate followed by the addition of a patient sample. If the patient sample contains antibodies that recognize the antigen, they will bind to the plate and can be detected using a secondary antibody that recognizes the patient's antigen-specific antibody. The key that makes this an indirect detection method is that the virus itself is not being detected; it is the antibodies that are produced by the patient in response to the viral infection.

Lateral flow immunoassays (LFIAs) are also an indirect detection method that detect viral antibodies that are produced by the individual (Sajid et al. 2015). These assays are like ELISAs but operate on a much different, but very familiar, format (see Fig. 3.3). These immunoassays are familiar to most people as pregnancy tests

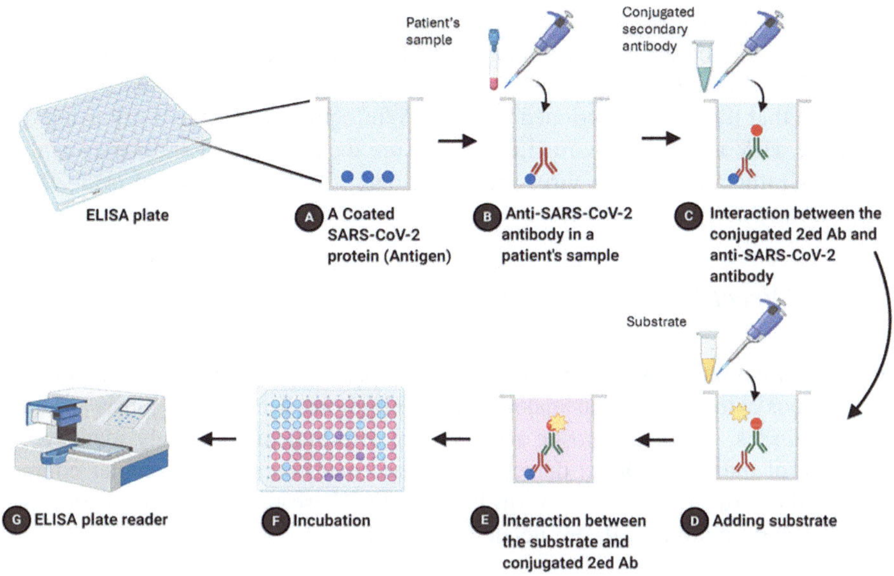


**Fig. 3.2** Diagram of enzyme-linked immunosorbent for the detection of SARS-CoV-2 antibodies in infected patient. (By Davidfay20 - (Metsky et al. 2022), © 2020 Dhamad and Abdal Rhida, CC BY-SA 4.0, https://commons.wikimedia.org/w/index.php?curid=95809228)

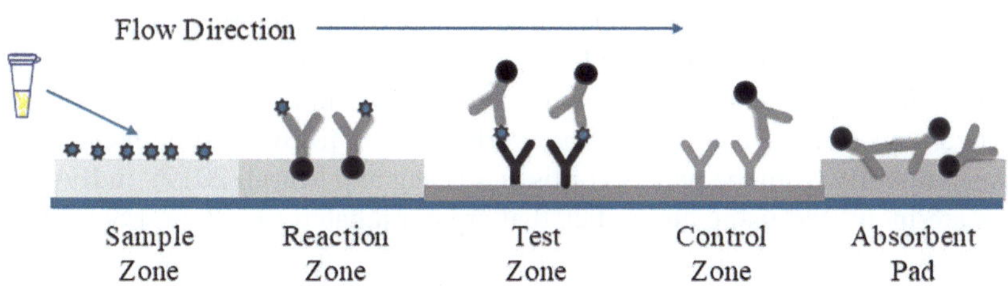


**Fig. 3.3** Diagram showing the mechanism of action of a lateral flow immunoassay

but are also available to test for HIV, dengue virus, influenza, and RSV. These tests are generally in the form of a dipstick or a small plastic test in which a drop of sample (usually urine or blood) is placed onto a sample zone. The liquid migrates via capillary action along the length of a test strip, which is usually composed of nitrocellulose. The test strip has three major zones: a reaction, test, and control zone. If the sample applied to the end of the test strip contains the target viral antigens, they react with free antibodies in the reaction zone, which are labeled with gold or latex beads that contain dye. These antibodies migrate to the test and control zones where antibodies that have bound to the viral antigen bind to antibodies in the test zone that are also specific for the same antigen. Antibodies that did not bind to the viral antigen proceed to the control zone that contains antibodies that bind to the free antibodies that have migrated from the reaction zone. In either case, antibodies labeled with gold or dye are visualized in the test and control zone. The visualization of the antibodies in the test zone indicates the presence of the targeted viral antigen in the sample, while visualization of the antibodies in the control zone demonstrates that the LFIA functioned properly. While pregnancy tests work using this principle, HIV and dengue virus LFIAs measure the presence of antibodies against the virus, rather than specific virus antigens (Louten 2016; Martiskainen et al. 2021).

### 3.2.3 *Direct Detection Methods*

While indirect methods have proven effective, some viruses provide overlapping or nonobvious cytopathic effects making an unambiguous, accurate diagnosis difficult. In these cases, direct diagnostic methods are required to accurately identify the presence of the virus. Direct detection methods measure some molecular component of the virus directly to confirm its identity. These tests roughly fall into two categories: immunological- or nucleic acid-based.

### 3.2.4 *Immunological-Based Direct Detection*

Immunological-based methods rely on antibodies that recognize a proteinaceous part of the virus, while nucleic acid-based methods identify viruses based on some part of their genome sequence. A common form of immunoassays is immunofluorescent assays (IFAs). These assays are performed on fixed cells (both cultured and obtained directly from the patient) or tissues (Ling and Murali 2019). In IFAs, specimens are incubated using antibodies that are conjugated to a fluorophore. Various viruses such as herpesviruses, influenza, adenovirus, mumps, and measles are often diagnosed using IFA (Louten 2016). Immunohistochemistry (IHC) is similar to IFA except it is used primarily for tissue sections (either fresh or formalin-fixed paraffin-embedded) and a different antibody-antigen binding visualization method (Obi et al. 2023). Both IFA and IHC can be completed in a few hours and can be scaled up to operate in multiplexed, high-throughput modes enabling thousands of samples to be analyzed simultaneously.

Although ELISAs were described as indirect detection methods, they can also be used in a direct detection format, for example, ELISAs can use an antibody to directly detect viruses within liquid samples (e.g., serum, urine, saliva, cell lysates, etc.) (McIntosh 1980). In this direct format, an ELISA plate is coated with an antibody specific to a targeted viral antigen first, and then the sample containing the virus of interest is added (Hayrapetyan et al. 2023). The first antibody captures the virus via a specific antigen. At this point, a second antibody is added, which binds to a different antigen within the virus. This method has greater specificity than a standard ELISA since two antibodies must recognize the virus to obtain a positive signal. While immunological-based detection methods are widely accepted for their accuracy, sensitivity, and throughput, they are dependent on the specificity of the antibodies for their intended viral target. Antibodies with low specificity can provide false-positive or false-negative results. Therefore, care must be taken to validate any antibody used for viral identification.

### 3.2.5 Nucleic Acid-Based Direct Detection

Nucleic acid amplification tests (NAATs) provide several advantages over immunological tests including sensitivity and accuracy (Cassedy et al. 2021; Yoo et al. 2021). Nucleic acid amplification tests can also detect viruses for which no other test currently exists. These tests are based on the use of polymerase chain reaction (PCR), which amplifies specific regions within DNA strands (Mullis et al. 1986). In its standard form, PCR is only able to amplify DNA, however, the addition of a reverse transcriptase, enables RNA to be detected as well. This addition enables the detection of both DNA and RNA viruses (i.e., retroviruses). While in its purest form PCR is a qualitative method, it can be made quantitative in the form of real time quantitative PCR (RT-qPCR) (VanGuilder et al. 2008). RT-qPCR uses DNA intercalating dyes that fluoresce when bound to double-stranded (ds) DNA as it is produced or TaqMan probes that fluoresce when the probe is cleaved during amplification. The fluorescent signal is measured over time during a PCR reaction and its quantity is directly proportional to the concentration of the virus in the sample.

There are several additional NAATs that use PCR for direct identification of viruses. These include loop-mediated isothermal amplification (LAMP), recombinase polymerase amplification (RPA), helicase-dependent amplification (HAD), and rolling circle amplification (RCA) (Cassedy et al. 2021). Even CRISPR-based systems have been developed that use guide RNAs to recognize specific viral genome sequences (Yang et al. 2023). Of all these techniques, RT-qPCR's sensitivity and accuracy made it the globally accepted standard for SARS-CoV-2 diagnosis (Dong et al. 2023). Viral RNA collected from a variety of samples can be used to perform this test, however, oropharyngeal (OPS) and nasopharyngeal swabs (NPS) are most widely used because these represent the locations where the SARS-CoV-2 virus is most likely to be present. Saliva has also been used owing to its ease of collection and transportation, lower storage costs, and decreased risk of transmission during sample collection. After the RNA has been extracted it is reverse transcribed and amplified using SARS-CoV-2 specific primers, whose amplification is detected in real time, primarily using TaqMan probes. Different sets of primer pairs are used depending on the source of the test. For example, primer pairs corresponding to open reading frame 1ab (ORF1ab), ORF8, RNA-dependent RNA polymerase (RdRp), hemagglutinin-esterase, spike genes, envelope genes, transmembrane genes, and the nucleocapsid genes N1 and N2 are currently used in different tests (Dutta et al. 2022). Human ribonuclease P primers and probe are added to some tests to assess sample extraction efficiency and specimen quality. Other controls include 2019-nCoV positive control and human specimen control (https://www.fda.

gov/media/134922/download), which consists of material obtained from non-infectious cultured human cells. The CDC recommends targeting N1 and N2, however, the WHO recommends targeting envelope genes, and using RdRp genes as confirmation of a positive test (Abbasi et al. 2022).

As mentioned, RT-qPCR is considered the gold standard molecular test for many viruses including SARS-CoV-2 because of its sensitivity, accuracy, throughput, and ability to conduct thousands of tests simultaneously. Related to sensitivity, a commonly used NAAT has a limit of detection (LOD) of 3.9 copies of virus per 5 μL (Corman et al. 2020). There are, however, certain limitations to NAATs. Beyond the need for highly trained personnel, the costs of instrumentation for conducting the tests are quite expensive. Owing to the potential for false-positive results, the laboratory environment must be clean and sterile. This requirement is partially due to the ubiquitous nature of viruses even in samples from recovered patients and the sensitivity of the RT-qPCR technique. To minimize the risk of errors leading to misdiagnosis, many companies have fully automated their methods. Automation not only serves to limit human contamination but also decreases the time to report diagnostic results, which is critical for containment of an outbreak, epidemic, or pandemic.

### 3.2.6 Direct Detection Using Microarrays

While many of the methods described above can be multiplexed by adding additional primer pairs, they are limited by the number of viruses that they can reasonably test for. In addition, these tests are largely hypothesis-driven since they require the use of customized reagents (e.g., primers, Taqman probes, etc.) that target a specific virus. DNA microarrays, however, allow for the detection of thousands of different viruses and strains within a single sample (Striebel et al. 2003). In a DNA microarray, thousands of short single-stranded pieces of DNA (i.e., oligos or oligo probes) are immobilized onto a solid surface usually composed of glass or silicon (Schena et al. 1995). Preparing a sample for microarray analysis requires amplifying all the DNA within the sample using sets of randomized primers. Fluorescent nucleotides are added to the mixture and become incorporated into the amplified pieces of DNA. The microarray is incubated with amplified fluorescent DNA, allowing complementary DNA segments to hybridize. After a series of washing steps to rinse away unhybridized DNA segments, the microarray is visualized using a fluorescent scanner. The presence of fluorescence at a particular spot indicates DNA hybridization, and since the sequence of the oligo attached to the slide at each position is known, the sequence of the amplified DNA segment can be determined. When using it to detect specific viruses, it is important to design the DNA microarray with oligo probes that are complementary to known virus sequences. While this requirement still makes DNA microarrays hypothesis-driven, the sheer number of DNA sequences that can be tested per experiment allows them to test for the presence of an extremely large number of diverse viruses giving it a hybrid discovery-driven character.

### 3.2.7 Future Nucleic Acid-Based Detection Methods

The diagnosis of viruses will continue to be dominated by NAATs and immunological-based methods for the foreseeable future. The reason is primarily due to the variety of reagents already available for these methods, familiarity with the techniques, and the reliable results they provide. Next-generation sequencing (NGS) platforms, however, will become increasingly important in clinical virology primarily owing to their ability to identify unknown and novel pathogens (Boonham et al. 2014; Cantalupo and Pipas 2019). Next-generation sequencing is a massively parallel sequencing technology that determines genomic sequences in a discovery-driven, extremely high-throughput manner. Since it is an unbiased approach, it does not require primers that target specific DNA sequences. This capability enables NGS to identify a viral DNA or RNA sequence in a background of host or other viral nucleic acids. This capacity makes it an extremely powerful diagnostic tool in cases where the causative pathogen is unknown or if the virus has not previously been discovered. Although established immunological and cell culture methods rely on hypothesis-driven or "process of elimination" strategies, they are currently more rapid in diagnosing viruses than NGS. While NGS methods may evolve to replace current diagnostic methods, they are presently critical for diagnosing previously unknown or unanticipated viruses within samples. Since NGS methods are not pathogen-specific or limited to a group of viruses, it can be used to passively monitor viruses in crowded areas such as schools, hospitals, transportation hubs, etc. Routine screening of animal viral vectors (e.g., mosquitos, bats, ticks, etc.) using NGS can aid in the surveillance of emerging zoonotic pathogens, enabling preemptive approaches to combat a potential epidemic or pandemic (Epstein and Anthony 2017).

As with any developing technology, NGS faces challenges to make it a routinely used tool for diagnosing viruses (Maliogka et al. 2018). On the scientific side, NGS has poor sensitivity for low viral load infections. In addition, the large dynamic size range between the host and virus genomes makes viral detection challenging. Their small size limits the number of virus-specific sequencing reads obtained compared to that of the host, making it difficult to characterize a novel virus. On the practical side, the equipment to perform NGS is expensive and requires highly trained staff that are skilled in either molecular biology or bioinformatics. But as we have seen repeatedly with scientific technologies, they continue to become progressively automated requiring less proficiency for their routine operation.

## 3.3 Proteomic Contributions to Clinical Virology

The detection and diagnosis of infectious agents require techniques that are rapid, sensitive, and specific. While each characteristic is important, speed is critical in a clinical setting so the best possible treatment (i.e., antiviral or antibiotic) can be administered prior to the pathogen overwhelming the patient. Rapid diagnosis is also critical to contain the spread of contagious pathogens. While proteomic technologies, specifically those that involve the use of mass spectrometry (MS), possess sensitivity and specificity, they presently lack the speed and throughput necessary to compete with existing immunoassays and PCR-based techniques. Although they cannot presently compete with these existing methods, it is still worth investigating the developments that have been made to bring MS methods to the clinical virology and bacteriology laboratory.

### *3.3.1 Proteomics in Clinical Bacteriology*

While the routine use of MS in clinical virology has not yet been realized, it is worth discussing its use in clinical bacteriology. One of the most successful clinical uses of MS is the identification of bacteria using matrix-assisted laser desorption ionization time-of-flight (MALDI-TOF) MS (Sandalakis et al. 2017; Alizadeh et al. 2021). While identification of bacteria using MS analysis of their lipid content dates to the early 1970s, it wasn't until the 1990s that the ability to use protein fingerprints to identify these pathogens was developed. Protein fingerprints are a list of masses generated by MS analysis of detectable bacterial proteins. By comparing this list of masses to those generated by an in silico analysis protein sequences of specific bacteria found within curated databases, the identity of the bacterial species can be determined. Most of the proteins detected using this method are derived from high-abundance proteins, including ribosomal proteins; however, there is generally enough variation in their sequences between bacterial species to make positive identifications (Patel 2016).

To prepare it for MS analysis, a single bacterial colony is grown on a solid medium that is smeared onto a metal MALDI plate (Abdolrasouli and Fraser 2022). A matrix solution, such as α-cyano-4-hydroxycinnamic acid (CHCA), is applied to the sample. Not only does the matrix assist in the ionization of the proteins, but it also lysis the cell wall. If the sample is a yeast or fungus, formic acid and acetonitrile are added to improve protein extraction and ionization during MALDI-TOF MS analysis. The MS spectra are acquired, and the list of recorded masses is searched against a suitable database to identify the bacteria.

To detect a blood-borne pathogen requires a short (i.e., ~2 min) incubation of approximately 1 mL of blood culture broth with an extraction strip that contains a lysis buffer (Ryzhov and Fenselau 2001). This treatment lyses red blood cells but keeps the bacterial cells intact. The bacterial cells are captured from the extraction solution via a membrane surface attached to a filter wand. After a series of washing steps, the bacteria are transferred to a MALDI plate. α-Cyano-4-hydroxycinnamic acid and formic acid are added, and the masses of the bacterial proteins are recorded using MALDI-TOF MS.

The use of MALDI-TOF MS in clinical bacteriology is rapidly expanding in the United States, Europe, and Asia. The two major commercial systems available for clinical use are the VITEK® MS Plus (bioMérieux) and Biotyper CA® (Bruker Daltonics) systems. The automation incorporated into current systems can identify bacteria in as little as 5 h. In 2021, Bruker announced that the FDA had cleared the use of MBT Sepsityper Kit US IVD for the rapid identification of more than 425 microorganisms from blood cultures when analyzed using their MALDI Biotyper CA System (https://ir.bruker.com/press-releases/press-release-details/2021/Bruker-Introduces-MBT-Sepsityper-Kit-US-IVD-for-Rapid-and-Affordable-Identification-of-over-400-Microorganisms-from-Positive-Blood-Cultures/default.aspx). In March of 2022, bioMérieux received FDA 510(k) clearance for the use of their VITEK MS Prime systems for the identification of more than 350 microbial species (https://www.biomerieux-usa.com/biomerieux-receives-fda-510k-clearance-its-vitek-ms-prime-new-maldi-tof-mass-spectrometry).

### *3.3.2 Mass Spectrometry in Clinical Virology*

Mass spectrometry has not made the types of advances necessary for routine clinical use in diagnosing viruses as it has for bacteria. Besides throughput issues, another major factor is the dynamic range of concentrations present in an unpurified viral sample. Ideally, MS could measure viruses in crude biological samples without the need for extensive purification; unfortunately, the number of different types and concentrations of host proteins dwarfs those of the virus. This large dynamic range of concentrations would make it extremely difficult to specifically measure the viral proteins in the background of the host proteome. Using a MALDI-MS approach that does not include offline fractionation would make distinguishing the viral and host proteins virtually impossible. Detecting viral proteins requires a method to pre-separate them so they can be identified within the background of surrounding signals. It is for this reason that clinical virology was developed using MS-based targeted proteomics (see Fig. 3.4). This technique uses the ability of the mass spectrometer itself to separate and isolate specific ions. Separating and isolating specific ions allow a small number of proteins to be measured within a complex mixture with high specificity. In brief, proteins are extracted from the sample and cut into peptides using a protease (Meng and Veenstra 2011). The peptides are separated using liquid chromatography (LC) prior to entering the mass spectrometer. The mass spectrometer is instructed to isolate and fragment specific peptides based on their mass-to-charge (*m/z*) ratios. The peptides are fragmented using tandem MS ($MS^2$), and they (and by extension the proteins) are identified by matching the $MS^2$ spectra to those found in a library of $MS^2$ spectra. Since this method is designed to

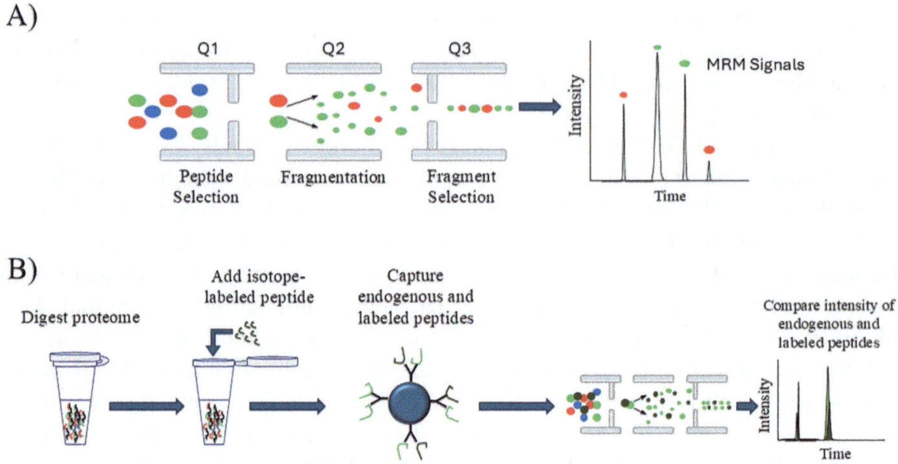


**Fig. 3.4** (**a**) Detection of SARS-CoV-2-specific peaks in nasal mucous secretions using matrix-assisted laser desorption/ionization mass spectrometry (MALDI-MS) and machine learning (ML). Samples taken from SARS-CoV-2-infected and healthy controls were processed and analyzed using MALDI-MS. Peaks distinctive to SARS-CoV-2 patients were identified using ML. (**b**) Diagnosis of SARS-CoV-2 using ML. Patient data were processed to generate intensity matrix containing a list of peaks and their intensities found in the MALDI-MS spectra. Feature selection was performed to identify SARS-CoV-2-specific peaks. Principal component analysis was used to identify patients as either true positives (TP), true negatives (TN), false positives (FP), or false negatives (FN) for a diagnosis of SARS-CoV-2 infection

measure a specific set of peptides within the sample, it is logically referred to as multiple reaction monitoring MS (or MRM-MS) or targeted proteomics. To increase the specificity of the experiment, antibodies can be added to the sample to extract peptides of interest. To quantitate the absolute concentration of the protein, known concentrations of stable isotope-labeled (SIL) internal standards with sequences that are identical to the targeted peptides are added to the sample. These peptides will co-elute and be isolated and fragmented by the mass spectrometer in an identical fashion as the wild-type peptides in the sample. Comparing the peak area of the isotope-labeled internal standard to the wild-type sample peptide will provide the sample peptide concentration. While this internal calibration provides more accurate concentration measurements, external calibration methods can also be used.

As mentioned previously, proteomics has yet to make a major impact in clinical virology that allows it to compete with technologies such as NAATs, ELISAs, or LFIAs. An example of some of the pioneering work applying proteomics to clinical virology was performed using human metapneumovirus (HMPV) isolated from nasopharyngeal airways (Foster et al. 2015). This virus is a leading cause of acute respiratory tract infections in young children and the elderly (Van Den Bergh et al. 2022). Since it replicates slowly, diagnosis through cytopathic effects caused by the virus in culture cells takes too long (Lee et al. 2019). This fact necessitates alternative methods for its diagnosis. One opportunity was the development of a targeted MRM-MS assay to measure the virus in clinical samples. Developing such an assay for any sample (including viruses) requires several preparation steps. The first step is to isolate the virus and identify peptides that are accurately identifiable using MS. To accomplish this step, LLC-MK2 cells were infected with HMPV reference strains. LLC-MK2 cells are an epithelial-like cell line isolated from 3- to 4-week-old male pigs. After allowing the virus to propagate, it was isolated from the cell supernatant. The virus sample was heat inactivated, buffer exchanged, and digested into tryptic peptides. The peptides were separated and directly analyzed using LC/$MS^2$. This analysis resulted in greater than 30% sequence coverage for five of the nine HMPV gene products. These five proteins and number of peptides identified for each were the nucleoprotein (20 peptides), phosphoprotein (9 peptides), matrix protein (12 peptides), fusion glycoprotein (22 peptides), and matrix M2-1 protein (5 peptides).

From these many peptide candidates, the final MRM assay was targeted to measure the most intense transitions from 11 HMPV peptides. Before selecting these peptides, a BLASTP analysis was conducted to show that no other human or human pathogen proteins contained these sequences. Although three peptides were identical to those found in avian MPV, this virus was incapable of infecting human cells and would therefore not be found in human samples. In addition, three peptides from human serum albumin were measured as part of the assay to act as normalization factors. These peptides are convenient choices since the assay is designed to analyze human samples that contain a high abundance of albumin. These normalization factors are important calibrants that can serve as quality controls to ensure the assay is performing properly. Three peptides from yeast alcohol dehydrogenase 1 (ADH1) were also monitored. These peptides were obtained by adding ADH1 protein to the clinical samples to serve as quality assurance standards to ensure the procedure was conducted properly. Finally, SIL peptides corresponding to the 11 viral peptides selected to measure HPMV were added for quantitation purposes. Not only do SIL peptides assist in quantitation, but they also provide the specificity that is critical for viral diagnostic assays.

The dynamic range and linearity of the HMPV MRM-MS assay were determined by spiking purified HPMV into nasopharyngeal aspirates (NPAs) at dilution ratios of 1:10, 1:100, and 1:1000 (w/w). Excellent signals were obtained for all 11 peptides at a 1:100 dilution ratio, and some peptides were detectable at 1:1000 dilution. For the final steps in the development of the assay, ten viral culture and eight NPA patient samples were analyzed. Seven of the peptides targeted for MRM-MS analysis showed excellent signal to noise. The average percent coefficient of variation (%CV) for quantitating these peptides across cultured virus samples was approximately 4%. Although the quantity of HPMV in the human NPA samples was less than that in the viral culture samples, their results were similar with a %CV of only 4% between replicate analyses.

Calibration curves were incorporated into the final version of the MRM-MS assays to determine the level of virus in samples based on plaque-forming units (PFUs). Cell culture supernatants containing known amounts of PFU/mL were serially diluted into NPA across four orders of magnitude. The samples were processed and analyzed using the MRM-MS assay. While most of the HPMV peptides demonstrated a good response curve over three orders of magnitude of concentration range, two peptides originating from the nucleocapsid and M21 proteins showed superior performance. Calibration curves for these two peptides were used to quantitate the PFUs in the viral culture and NPA samples. The amount of virus was above the limit of detection for all the viral culture samples and six of the eight NPA samples. The total time required for the MRM-MS analysis was 30 min, excluding the time required for sample preparation and data analysis.

MRM-MS assays had been previously used to verify the presence of influenza hemagglutinin and neuraminidase antigens during vaccine development (Santana et al. 2014; Williams et al. 2012); however, this study represented the first of its kind that could quantitate a virus in clinical samples. While this method cannot compete with the throughput of PCR-based methods, it is much faster than the standard culture time necessary for a positive indirect HPMV diagnosis (i.e., about 2 weeks). The MRM-MS method has advantages compared to immunology-based methods, for instance, the use of SIL peptides ensures that the correct viral peptide is being measured unlike immunoassays that can suffer from nonspecificity owing to antibody cross-reactivity. The specificity of MRM-MS assays also enables the genetic lineage of the virus within the clinical sample to be established. While MRM-MS assays are not rapid, they can be multiplexed. This feature would enable a single assay to test for the presence of many unique viruses, for example, a single assay could be designed to test for many blood-borne viruses. Since the technology is species-independent, an assay could be designed to identify the virus itself as well as host proteins that respond to the presence of the virus.

## 3.4 Clinical Diagnosis of SARS-CoV-2

The COVID-19 pandemic quickly spread around the world resulting in almost 800 million confirmed cases and 7 million deaths as of May 10, 2023 (https://covid19.who.int/). Owing to its transmission, a rapid test was needed to be quickly developed that could track and hopefully minimize SARS-CoV-2's spread (Mannino et al. 2023). Once the genome sequence of the virus was determined, a rapid RT-PCR test was developed, which became the gold standard for SARS-CoV-2 diagnosis (Mohit et al. 2021). As the rate of virus spread soared, the supply of reagents necessary to perform RT-PCR testing became limited (Wulff et al. 2022). To increase the testing supply and increase the number of individuals that could be screened, several point-of-care immunological tests for measuring SARS-CoV-2 antigens were developed (Trinh et al. 2023). While these tests are simple to perform, sensitive, rapid, and inexpensive, their specificity can vary owing to inherent challenges in antibody recognition (e.g., cross-reactivity, interference by other molecules, etc.).

### *3.4.1 Clinical Detection of SARS-CoV-2 Using Proteomic Patterns*

One of the first proteomic tests for detecting SARS-CoV-2 did not detect the virus directly but rather based its diagnostic on a collective pattern of peaks that would be observed in samples containing the virus. This approach used MALDI-MS to acquire the spectra and machine learning (ML) to analyze the data (see Fig. 3.5) (Nachtigall et al. 2020). This method analyzed nasal mucous secretions from 211 positive and 151 negative cases of SARS-CoV-2 (as determined by RT-PCR). The samples were acquired from three different laboratories in Peru, Argentina, and Chile during March and April of 2020. One microliter of each sample was spotted directly onto a MALDI plate and mixed with an equal volume of CHCA. After allowing the sample and matrix to air-dry, the spectra were acquired using a MALDI-TOF mass spectrometer.

The peaks most characteristic for distinguishing between SARS-CoV-2-positive and SARS-CoV-2-negative samples were identified using feature selection (FS) methods. These peaks were analyzed using principal component analysis (PCA) and six different ML algorithms to diagnose the patient samples. The ML algorithms were trained on the 362 mucous samples and tested on a set of 80 samples

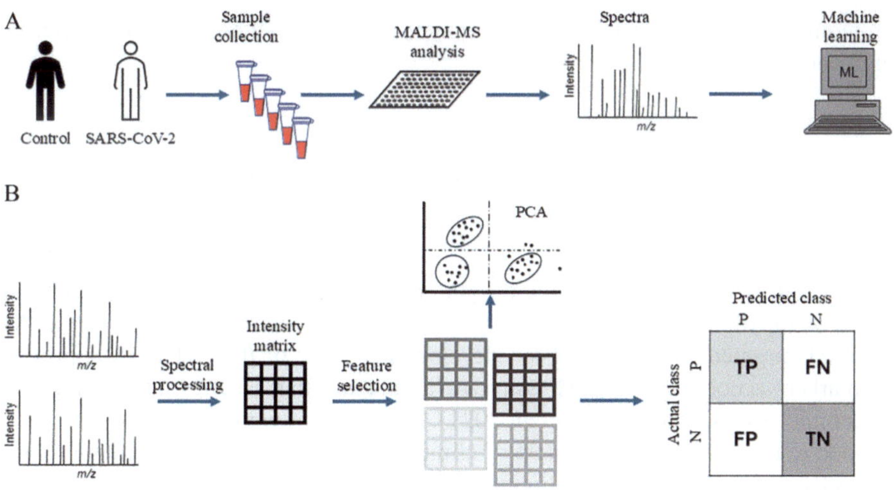


**Fig. 3.5** Diagram outlining the principle of multiple reaction monitoring mass spectrometry (MRM-MS) for quantitating specific peptides in clinical samples. (**a**) In MRM-MS, specific peptides are isolated in the mass spectrometer's first quadrupole (Q1). The peptides then pass into the second quadrupole (Q2) where they are fragmented. The mass spectrometer is programmed to only allow some of these specific fragments to pass through the third quadrupole (Q3) and onto the detector where their mass-to-charge ratio and intensity are measured. (**b**) To measure the abundance of a peptide, stable isotope-labeled peptides are added to the clinical sample corresponding to the targeted peptides being measured. To increase the sensitivity and specificity of the assay, an isolation step using antibodies that bind to the targeted peptides can be added to the assay

(i.e., 40 SARS-CoV-2-positive and 40 SARS-CoV-2-negative cases). The result was the ability to distinguish positive and negative SARS-CoV-2 patients with a sensitivity and specificity of 0.947 and 0.926, respectively.

The MALDI-MS method combined with ML has several positive attributes in diagnosing viral disease. The required equipment is readily available and easy to use. The sample preparation requires no more than simply spotting the sample onto a MALDI plate and mixing it with matrix (e.g., CHCA). The biggest drawback with this technique, however, is that the identity of the molecules that give rise to the distinguishing peaks is unknown. This type of diagnostic analysis is like "proteomic pattern" technology, which was developed in the early 2000s for diagnosing cancer using blood samples (Petricoin et al. 2002). The inability to identify the peaks within these proteomic patterns creates uncertainty as to what features the test is measuring to make its diagnosis. There was a great interest in this technology for many years, including a significant investment made by the US National Cancer Institute to develop and refine this method. Unfortunately, other reports have shown that samples could be distinguished based on subtle differences in sample collection or found the diagnostic peaks were different from study to study even when the exact same disease and sample type were being analyzed (Galvão et al. 2011). To date, this technology has not clearly proven whether proteomic patterns reveal inter-individual differences. These proteomic patterns, however, could be used as a confirmatory diagnosis to complement RT-PCR testing, but their future as a standalone diagnostic tool looks murky. This fact is unfortunate because of all the available proteomic tools available it possesses the type of throughput necessary for diagnosing pandemic events.

### *3.4.2 Clinical Diagnosis of SARS-CoV-2 Using Targeted Proteomics*

Targeted proteomics, unlike proteomic pattern technology, can diagnose a condition based on the direct measurement of a known protein(s) within a clinical sample. As of 2020, targeted assays have been developed for 222 of the 241 FDA-approved biomarkers according to MRMAssayDB (a database that collates lists of targeted proteomic assays) (Bhowmick et al. 2021). Targeted proteomic methods have been applied to the routine measurement of clinical biomarkers such as coagulation factors VIII and IX, prolactin, apolipoprotein E, C-reactive protein, troponin T, and a variety of complement factors. One often overlooked advantage of targeted proteomic methods compared to immunoassays is an MRM-MS assay is faster to develop than having to produce and qualify a new antibody.

The rapid development time and specificity of targeted proteomic assays perfectly aligned with the need generated by the SARS-CoV-2 pandemic. One of the early methods to develop a SARS-CoV-2 test analyzed a total of 985 respiratory tract samples (combined from nasopharyngeal and oropharyngeal swabs) that had been previously diagnosed using RT-PCR (540 positive and 445 negative for SARS-CoV-2) (Cardozo et al. 2020). The first step was to identify peptides that could be readily detected using MS (see Fig. 3.6). Ten samples that exhibited low cycle threshold (Ct), a measure of the number of RT-PCR cycles required to generate a fluorescent signal above background, values were selected based on their high viral load. Proteins were extracted from these samples, concentrated, and digested into tryptic peptides. The peptides were analyzed by LC-MS$^2$ using a hybrid quadrupole Orbitrap mass spectrometer operating in a data-dependent assay mode. This particular instrumental setup was chosen because the aim was to identify as many peptides as possible within the sample. A total of 119 unique peptides originating from 8 of the 14 predicted SARS-CoV-2 proteins listed in UniProt were identified. This list of peptides was analyzed using the software program Skyline to identify peptides that would be suitable in an MRM-MS assay. Skyline evaluates the peptides based on several user-defined factors including the presence of a no missed proteolytic cleavage sites in the peptide. The size and amino acid makeup of the chosen peptides are also often limited to those between 7 and 26 amino acids in length that do not contain certain prone-to-modification amino acids such as cysteine, methionine, and histidine. Last, but certainly not least, Skyline selects peptides that are unique to the target protein and not found in other proteins.

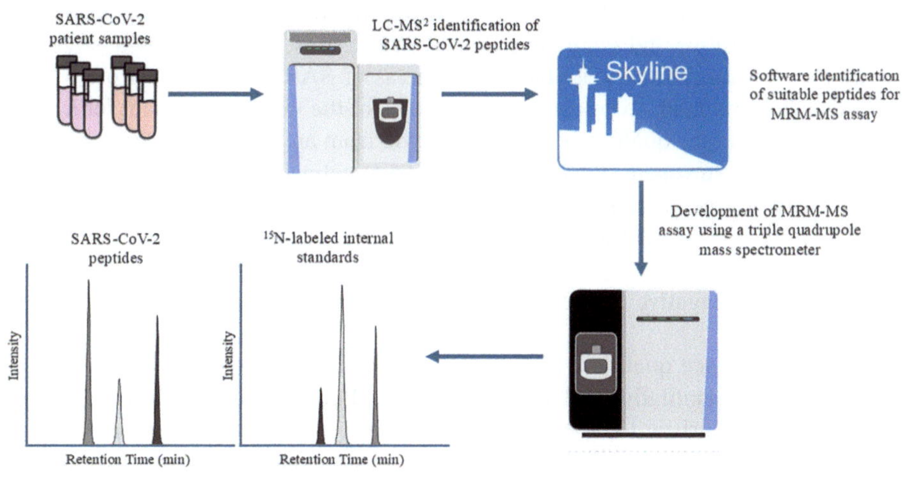


**Fig. 3.6** Steps in the development of a multiple reaction monitoring mass spectrometry (MRM-MS) assay for quantitating SARS-CoV-2. Viral and human proteins are extracted from confirmed SARS-CoV-2-infected patient samples. The proteins are digested into peptides and analysis via LC-MS2 using a hybrid quadrupole Orbitrap mass spectrometer to identify as many human and viral proteins as possible. Skyline is used to identify SARS-CoV-2 peptides that can potentially serve as surrogate peptides for the presence of the virus. Once peptides are chosen, a targeted quantitative MRM-MS assay is developed using stable isotope-labeled internal standards and a triple quadrupole mass spectrometer. (Reproduced with modifications from Cardozo et al. (2020). http://creativecommons.org.licencses/by/4.0/)

Seventeen peptides were selected from these proteins as targets in the first iteration of the MRM-MS assay: nine from the nucleoprotein, five from the spike glycoprotein, two from the membrane protein, and one from protein 3a. At this point, the specific type of mass spectrometer used was switched from a quadrupole Orbitrap to a triple-quadrupole mass spectrometer. Without getting into too much analytical detail, the reason is that triple quadrupoles provide superior sensitivity and specificity for quantifying trace analytes and are considered the gold standard for targeted analysis in forensics, food safety, and pharmacokinetics. In this initial assay, which had an overall MS analysis time of 60 min, most of the nucleoprotein peptides had an 80-fold higher relative intensity compared to other peptides. Therefore, the next iteration of the assay was substantially shortened (i.e., only 9 min) and targeted the eight nucleoprotein peptides, whose specificity was confirmed using BLASTP. Of these, the peptides IGMEVTPSGTWLTYTGAIK and DGIIWVATEGALNTPK were selected as the final surrogate peptides to measure SARS-CoV-2 in clinical samples. Although these peptides produced less intense signals than the other six peptides, they had better chromatographic features than the others. These features produced sharper peaks that led to better signal-to-noise ratios as measured using MS. This peptide selection process demonstrates the intricacies in developing a targeted proteomic assay: it is not simply about finding the peptides with the most abundant MS signals but rather those that provide reliable, consistent results.

This study added an interesting twist on target proteomic assays. While quantitation is generally understood to be part of the MRM-MS assay, this group developed separate quantitative and qualitative assays. While most biomarker assays require some form of absolution quantitation, diagnosis of SARS-CoV-2 most often does not consider viral load, so detecting the presence of the virus is sufficient for a positive diagnosis. In the qualitative assay, a peptide from an $^{15}$N-labeled chromogranin A was used as a surrogate standard (to assess tryptic digestion efficiency) because it has an elution time similar to SARS-CoV-2 peptides. For the quantitative assay, a $^{15}$N-labeled nucleoprotein was added at a concentration of 10 ng/mL to serve as an SIL internal standard. To generate calibration curves, recombinant nucleoprotein was added to negative pooled samples at concentrations ranging from 2 to 512 ng/mL.

The performance qualifiers final quantitative and qualitative assays were determined for both quantitative and qualitative assays. For the qualitative assay, receiver operating characteristic (ROC) curves were built to determine the performance of the qualifiers for classifying the samples. Since the qualitative assay was simply designed to determine the presence or absence of SARS-CoV-2, the limit of blank (LOB) was established for the two targeted peptides. The LOB is the highest obvious analyte concentration expected to be observed when replicates of a blank sample containing no analyte are tested. The LOB cutoffs for the two peptides were 1.65 (IGMEVTPSGTWLTYTGAIK) and 0.85 (DGIIWVATEGALNTPK). By using a combination of qualifiers, the assay was able to diagnose the presence of SARS-CoV-2 in human samples with a sensitivity of 83.6% and a specificity of 93.3%. For the quantitative assay, the LOD for the two peptides were 2.7 ng/mL

(IGMEVTPSGTWLTYTGAIK) and 3.2 ng/mL (DGIIWVATEGALNTPK). The limits of quantitation (LOQ) were 6.0 and 4.4 ng/mL for the IGMEVTPSGTWLTYTGAIK and DGIIWVATEGALNTPK peptides, respectively. These LOQs compare favorably with a targeted proteomic approach used to quantify serum troponin 1. By combining the results from both peptides, the sensitivity and the specificity of the quantitative assay were 78.2% and 96.8%, respectively.

For an assay to gain wide acceptance, it needs to be reproducible and robust. To evaluate these parameters, SARS-CoV-2-positive samples were analyzed over a 10-day period using both qualitative and quantitative assays. The CVs for these measurements were less than 20% for both assays. The LC retention times for the targeted peptides, as well as all SIL peptides and control peptides, were stable. The linearity of the quantitative assay ranged from 0.1 to 1000 ng/mL for viral samples containing the medium used to transport the sample. Results obtained for samples stored at temperatures ranging from +21 to −80 °C were compared to those stored in liquid nitrogen and showed no significant differences. Virus stored in sterile saline samples showed reproducible results even after being stored at room temperature for 30 days, while virus stored in transport medium required storage at 4 °C to produce reliable results. In an extreme test in which the samples were incubated at 90 °C, the targeted peptides were only stable for 5 min. The ability of the assays to detect SARS-CoV-2 in specimen that were infected with other types of viruses was also tested. A series of samples that were either mono-infected (coronavirus 229E, coronavirus NL63, parainfluenza 1, parainfluenza 4, influenza A/H1-2009, HMPV, and RSV) or co-infected (coronavirus HKU1/rhinovirus/enterovirus, rhinovirus/enterovirus, and rhinovirus/enterovirus/HMPV) with viruses were analyzed. SARS-CoV-2 was not detected in any of these samples using either qualitative or quantitative assays. This result suggests that the complexity added by additional viruses inhibited the ability to detect SARS-CoV-2-specific peptides through some undetermined mechanism.

While the MRM-MS component performed well, a key necessity of any clinical viral assay is throughput. This assay included an automated sample processing method that used paramagnetic beads to extract protein from the patient sample (see Fig. 3.7). The extracted proteins were extensively washed and tryptically digested. All steps were performed using a robotic liquid handler except for a single offline centrifugation step that accelerated the magnetic bead collection. Modern liquid handling stations, however, have centrifugation capabilities so this step could be easily included in the automation process. Performing the sample processing steps with a robotic liquid handler reduced processing time, analytical variation, and risk of infection for laboratory personnel. From an initial MRM-MS analysis time of 60 min, the assay was refined to be performed in about 2.5 min. These improvements along with automation enabled the complete analysis of over 500 samples in a 24-h period.

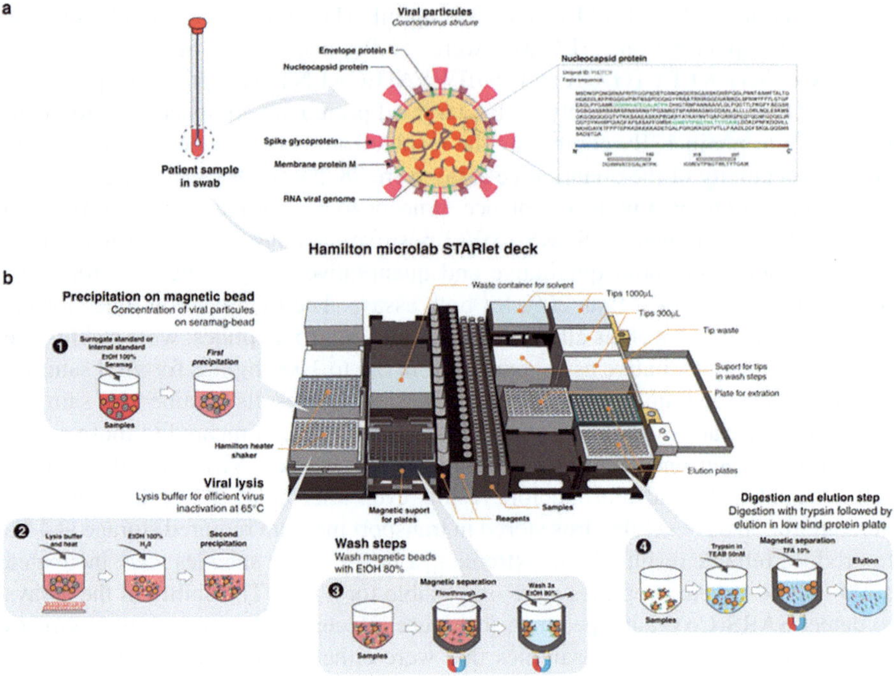


**Fig. 3.7** Development of automated high-throughput assay to measure SARS-CoV-2 in patient samples. After collecting samples with swabs, they are placed in storage medium prior to sample analysis. After addition of stable isotope-labeled internal standards, the virus is precipitated and lysed to release the nucleocapsid proteins. After a series of washing steps, the viral proteins are digested with trypsin and the resultant peptides extracted and eluted into wells of a sample plate. The targeted nucleoprotein peptides identified as SARS-CoV-2 surrogates, are then quantitated using the previously developed multiple reaction monitoring mass spectrometry (MRM-MS). (Reproduced with permission from Cardozo et al. (2020). http://creativecommons.org.licencses/by/4.0/)

## 3.5 Limitations of Targeted Proteomic Clinical Virology Assays

Just like RT-PCR tests, targeted proteomic assays can generate an increasingly higher number of false-negative diagnoses if specific mutations unknowingly occur within the coding regions of the targeted peptides. Mutations that modify the mass of the target peptide will make it undetectable using the current parameters of a targeted assay. Fortunately, a targeted proteomic assay can be redesigned to measure the mutated peptide by simply incorporating the new peptide mass and synthesizing new SIL peptide standards. Since thousands of SARS-CoV-2 genomes have been sequenced, areas of the genome with high mutation rates have been discovered (van Dorp et al. 2020). These regions include those responsible for encoding Orf1a, Nsp11, Nsp13, and the spike protein. Any mutations within the peptides measured in the above assay would be quickly recognized by the disappearance of one of the peptide's signals in a series of samples. The cause of the signal loss could be quickly confirmed as a mutation using RT-PCR.

Other limitations of targeted proteomic assays are also relevant to other types of viral tests. During pandemics or large epidemics, reagents and instruments (e.g., swabs, transport containers, etc.) may not be readily available because of demand. Also, it is impossible for a single site to analyze the number of samples required during an epidemic or pandemic, so any targeted proteomic diagnostic test would need to be standardized across many laboratories. This need would require reference materials and SIL standards to be centrally produced and shipped to various labs. In addition, a series of standard operating procedures would need to be produced, disseminated, and carefully followed by participating labs.

The obvious limitation, however, is throughput. No current proteomic method can come close to matching the throughput of NAATs, LFIAs, or even ELISAs. While huge throughput improvements have been realized, the need to chromatographically separate peptides prior to their entrance into the mass spectrometer is a major time-limiting step. Another major limitation that is obvious is that a mass spectrometer can only analyze a single sample at a time, while a 96-well plate analyzes that number of samples simultaneously. The problem is that the only way to get more samples analyzed is to get more mass spectrometers online and these are very expensive pieces of equipment. Although their operation is becoming increasingly easier and automated, they still require well-trained staff to ensure optimal performance.

## 3.6 Conclusions

Proteomics has existed from its inception primarily as a basic research tool. It has made incredible strides in this area as it has grown from a field only practiced by highly specialized labs to one that is essential for most medium-to-large-sized research institutes or universities. The first dedicated proteomic lab was founded in 1995, and now proteomic core labs are commonly found both within academia and industry. The current global proteomic market size is estimated to be just shy of 28 billion USD and is projected to grow at a compounded annual growth rate of 12.9% over the next 5 years. Several factors will contribute to this growth. These include continued technological advancements in the form of mass spectrometers, LC, bioinformatics, and consumables, which have historically driven the proteomic market. There is also a growing demand for personalized medicine that includes both genomic and proteomic profiles of patients and more specifically tumors. The World Health Organization predicts a 77% surge in the number of new cancer cases worldwide by 2050. Personalized treatment will become more urgent and will require molecular profiles generated via state-ot-the-art genomic and proteomic sequencing technology.

Another growth area is directly tied to the rise of chronic and infectious diseases. The push to develop MS tests that could be used for clinical diagnostics has been discussed for at least the past 25 years. The conversation intensified with the development of methods to identify biomarkers in samples such as serum and plasma. The COVID-19 pandemic has served to accelerate research into the development of proteomic diagnostic tools. While proteomics currently lacks the throughput of genomic-based detection methods, it does offer high specificity and the ability to easily modify existing methods to target different pathogens. Just like any important technology, it is important to project what its capabilities will be in the future. Considering the rapid growth of proteomics in the past 25 years and the current excitement that continues to surround the field, it is almost certain that proteomics will be a major clinical diagnostic tool in the future for a variety of conditions ranging from physiological to infectious diseases.

## References


Abbasi H, Tabaraei A, Hosseini SM, Khosravi A, Nikoo HR (2022) Real-time PCR Ct value in SARS-CoV-2 detection: RdRp or N gene? Infection 50:537–540

Abdolrasouli A, Fraser MA (2022) Candida auris identification and profiling by MALDI-ToF mass spectrometry. Methods Mol Biol 2517:21–32

Alizadeh M, Yousefi L, Pakdel F, Ghotaslou R, Rezaee MA, Khodadadi E et al (2021) MALDI-TOF mass spectroscopy applications in clinical microbiology. Adv Pharmacol Pharm Sci 2021:9928238

Bhowmick P, Roome S, Borchers CH, Goodlett DR, Mohammed Y (2021) An update on MRMAssayDB: a comprehensive resource for targeted proteomics assays in the community. J Proteome Res 20:2105–2115

Billioux BJ, Mbaya OT, Sejvar J, Nath A (2022) Neurologic complications of smallpox and monkeypox: a review. JAMA Neurol 79:1180–1186

Boonham N, Kreuze J, Winter S, van der Vlugt R, Bergervoet J, Tomlinson J et al (2014) Methods in virus diagnostics: from ELISA to next generation sequencing. Virus Res 186:20–31

Cantalupo PG, Pipas JM (2019) Detecting viral sequences in NGS data. Curr Opin Virol 39:41–48

Cardozo KHM, Lebkuchen A, Okai GG, Schuch RA, Viana LG, Olive AN et al (2020) Establishing a mass spectrometry-based system for rapid detection of SARS-CoV-2 in large clinical sample cohorts. Nat Commun 11:6201–6213

Cassedy A, Parle-McDermott A, O'Kennedy R (2021) Virus detection: a review of the current and emerging molecular immunological methods. Front Mol Biosci 8:637559

Corman VM, Landt O, Kaiser M, Molenkamp R, Meijer A, Chu DK et al (2020) Detection of 2019 novel coronavirus (2019-nCoV) by real-time RT-PCR. Euro Surveill 25:2000045

Diallo A, Libeau G, Couacy-Hymann E, Barbron M (1995) Amplicor enterovirus polymerase chain reaction in patients with aseptic meningitis: a sensitive test limited by amplification inhibitors. Vet Microbiol 44:307–317

Dong H, Zhang K, Zhang J, Xiao Y, Zhang F, Wang M et al (2023) A fast RT-qPCR system significantly shortens the time for SARS-CoV-2 nucleic acid test. Drug Discov Ther 17:37–44

Dutta D, Naiyer S, Mansuri S, Soni N, Singh V, Bhat KH et al (2022) COVID-19 diagnosis: a comprehensive review of the RT-qPCR method for detection of SARS-CoV-2. Diagnostics (Basel) 12:1503

Epstein JH, Anthony SJ (2017) Viral discovery as a tool for pandemic preparedness. Rev Sci Tech 36:499–512

Eyre DW (2022) Infection prevention and control insights from a decade of pathogen whole-genome sequencing. J Hosp Infect 122:180–186

Fathi Karkan S, Maleki Baladi R, Shahgolzari M, Gholizadeh M, Shayegh F, Arashkia A (2022) The evolving direct and indirect platforms for the detection of SARS-CoV-2. J Virol Methods 300:114381

Foster MW, Gerhardt G, Robitaille L, Plante PL, Boivin G, Corbeil J et al (2015) Targeted proteomics of human metapneumovirus in clinical samples and viral cultures. Anal Chem 87:10247–10254

Galvão ER, Martins LM, Ibiapina JO, Andrade HM, Monte SJ (2011) Breast cancer proteomics: a review for clinicians. J Cancer Res Clin Oncol 137:915–925

Guo-Parke H, Canning P, Douglas I, Villenave R, Heaney LG, Coyle PV et al (2013) Relative respiratory syncytial virus cytopathogenesis in upper and lower respiratory tract epithelium. Am J Respir Crit Care Med 188:842–851

Hayrapetyan H, Tran T, Tellez-Corrales E, Madiraju C (2023) Enzyme-linked immunosorbent assay: type and applications. Methods Mol Biol 2612:1–17

https://covid19.who.int/. Accessed June 2023

https://ir.bruker.com/press-releases/press-release-details/2021/Bruker-Introduces-MBT-Sepsityper-Kit-US-IVD-for-Rapid-and-Affordable-Identification-of-over-400-Microorganisms-from-Positive-Blood-Cultures/default.aspx. Accessed July 2025

https://www.biomerieux-usa.com/biomerieux-receives-fda-510k-clearance-its-vitek-ms-prime-new-maldi-tof-mass-spectrometry. Accessed July 2025

https://www.fda.gov/media/134922/download. Accessed July 2023

https://www.fda.gov/vaccines-blood-biologics/blood-donor-screening. Accessed June 2023

Jayakeerthi RS, Potula RV, Srinivasan S, Badrinath S (2006) Shell vial culture assay for the rapid diagnosis of Japanese encephalitis, West Nile and Dengue-2 viral encephalitis. Virol J 3:2

Ji CY, Du NX, Xu WY (1991) Adaption of the viral haemorrhagic disease virus of rabbits to the DJRK cell strain. Rev Sci Tech 10:337–345

Lee H, Woo HM, Kim K, Park S, Park MS, Kim SS et al (2019) Improving pneumovirus isolation using centrifugation and AZD1480 combined method. J Microbiol Biotechnol 29:2006–2013

Ling M, Murali M (2019) Antinuclear antibody tests. Clin Lab Med 39:513–524

Louten J (2016) Detection and diagnosis of viral infections. Essent Human Virol 2016:111–132

Maliogka VI, Minafra A, Saldarelli P, Ruiz-García AB, Glasa M, Katis N et al (2018) Recent advances on detection and characterization of fruit tree viruses using high-throughput sequencing technologies. Viruses 10:436

Mannino RG, Nehl EJ, Farmer S, Peagler AF, Parsell MC, Claveria V et al (2023) The critical role of engineering in the rapid development of COVID-19 diagnostics: lessons from the RADx tech test verification core. Sci Adv 9:eade4962

Marshall PN (1983) Papanicolaou staining – a review. Microsc Acta 87:233–243

Martiskainen I, Juntunen E, Salminen T, Vuorenpää K, Bayoumy S, Vuorinen T et al (2021) Double-antigen lateral flow immunoassay for the detection of anti-HIV-1 and -2 antibodies using upconverting nanoparticle receptors. Sensors (Basel) 21:330

McIntosh K (1980) Recent advances in viral diagnosis. Arch Pathol Lab Med 104:3–6

Meng Z, Veenstra TD (2011) Targeted mass spectrometry approaches for protein biomarker verification. J Proteome 74:2650–2659

Metsky HC, Welch NL, Pillai PP, Haradhvala NJ, Rumker L, Mantena S (2022) Designing sensitive viral diagnostics with machine learning. Nat Biotechnol 40:1123–1131

Mohit E, Rostami Z, Vahidi H (2021) A comparative review of immunoassays for COVID-19 detection. Expert Rev Clin Immunol 17:573–599

Moreno-Sanchez F, Gomez-Gomez B (2022) Antibiotic management of patients with hematologic malignancies: from prophylaxis to unusual infections. Curr Oncol Rep 24:835–842

Mullis K, Faloona F, Scharf S, Saiki R, Horn G, Erlich H (1986) Specific enzymatic amplication of DNA in vitro: the polymerase chain reaction. Cold Spring Harb Symp Quant Biol 51(Pt 1):263–273

Nachtigall FM, Pereira A, Trofymchuk OS, Santos LS (2020) Detection of SARS-CoV-2 in nasal swabs using MALDI-MS. Nat Biotechnol 38:1168–1173
Obi EN, Tellock DA, Thomas GJ, Veenstra TD (2023) Biomarker analysis of formalin-fixed paraffin-embedded clinical tissues using proteomics. Biomolecules 13:96
Pan D, Wang W, Cheng T (2023) Current methods for the detection of antibodies of Varicella-Zoster virus: a review. Microorganisms 11:519
Pandit JA, Radin JM, Quer G, Topol EJ (2022) Smartphone apps in the COVID-19 pandemic. Nat Biotechnol 40:1013–1022
Patel R (2016) New developments in clinical bacteriology laboratories. Mayo Clin Proc 91:1448–1459
Pearson JD, Trcka D, Lu S, Hyduk SJ, Jen M, Aynaud MM et al (2021) Comparison of SARS-CoV-2 indirect and direct RT-qPCR detection methods. Virol J 18:99
Petricoin EF, Ardekani AM, Hitt BA, Levine PJ, Fusaro VA, Steinberg SM et al (2002) Use of proteomic patterns in serum to identify ovarian cancer. Lancet 359:572–577
Rosenfeld R, Tibshirani RJ (2021) Epidemic tracking and forecasting: lessons learned from a tumultuous year. Proc Natl Acad Sci USA 118:e2111456118
Ryzhov V, Fenselau C (2001) Characterization of the protein subset desorbed by MALDI from whole bacterial cells. Anal Chem 73:746–750
Sajid M, Kawde AN, Daud M (2015) Designs, formats and applications of lateral flow assay: a literature review. J Saudi Chem Soc 19:689–715
Sandalakis V, Goniotakis I, Vranakis I, Chochlakis D, Psaroulaki A (2017) Use of MALDI-TOF mass spectrometry in the battle against bacterial infectious diseases: recent advances and future perspectives. Expert Rev Proteomics 14:253–267
Santana WI, Williams TL, Winne EK, Pirkle JL, Barr JR (2014) Quantification of viral proteins of the avian H7 subtype of influenza virus: an isotope dilution mass spectrometry method applicable for producing more rapid vaccines in the case of an influenza pandemic. Anal Chem 86:4088–4095
Schena M, Shalon D, Davis RW, Brown PO (1995) Quantitative monitoring of gene expression patterns with a complementary DNA microarray. Science 270:467–470
Striebel HM, Birch-Hirschfeld E, Egerer R, Földes-Papp Z (2003) Virus diagnostics on microarrays. Curr Pharm Biotechnol 4:401–154
Teymoorian T, Teymourian T, Kowsari E, Ramakrishna S (2021) Direct and indirect effects of SARS-CoV-2 on wastewater treatment. J Water Process Eng 42:102193
Trinh KTL, Do HDK, Lee NY (2023) Recent advancements in molecular and immunological diagnostic platform for virus detection: a review. Biosensors (Basel) 13:490
Trudy A, Larson MD, Laurie A, MLS P (1994) Reviews in medical virology. JAMA J Am Med Assoc 271:73
Van Den Bergh A, Bailly B, Guillon P, von Itzstein M, Dirr L (2022) Antiviral strategies against human metapneumovirus: targeting the fusion protein. Antivir Res 207:105405
van Dorp L, Acman M, Richard D, Shaw LP, Ford CE, Ormond L et al (2020) Emergence of genomic diversity and recurrent mutations in SARS-CoV-2. Infect Genet Evol 83:104351
VanGuilder HD, Vrana KE, Freeman WM (2008) Twenty-five years of quantitative PCR for gene expression analysis. BioTechniques 44:619–626
Wang TE, Chao TL, Tsai HT, Lin PH, Tsai YL, Chang SY (2020) Differentiation of cytopathic effects (CPE) induced by influenza virus infection using deep convolutional neural networks (CNN). PLoS Comput Biol 16:e1007883
Williams TL, Pirkle JL, Barr JR (2012) Simultaneous quantification of hemagglutinin and neuraminidase of influenza virus using isotope dilution mass spectrometry. Vaccine 30:2475–2482
Wulff RT, Qiu Y, Wu C, Calfee DP, Singh HK, Hatch I et al (2022) Laboratory interventions to eliminate unnecessary rapid COVID-19 testing during a reagent shortage. Am J Clin Pathol 158:401–408
Yang H, Zhang Y, Teng X, Hou H, Deng R, Li J (2023) CRISP-based nucleic acid diagnostics for pathogens. Trends Analyt Chem 160:116980
Yoo HM, Kim IH, Kim S (2021) Nucleic acid testing of SARS-CoV-2. Int J Mol Sci 22:6150

# Chapter 4
# Viral Prognosis Using Proteomics

**Abstract** So much was learned during the COVID-19 pandemic of 2020–2023. Some of the things that were learned were obvious. Many people in the public learned about social distancing, personal hygiene, and how to conduct a COVID-19 diagnostic test. Pharmaceutical companies and government organizations learned to efficiently work together to produce and distribute vaccines in record time. Although it may have generated controversy, the value in getting vaccinated was revalidated. There were other things that the world learned, although it was not as obvious. The effect of vaccination rate on preventing virus mutation became evident during the pandemic. The original SARS-CoV-2 virus that started the pandemic mutated several times over the course of the pandemic. One thing that became clear during the pandemic was the lack of knowledge on how SARS-CoV-2 infection would affect individuals during the course of the disease. This inability to accurately prognose the disease made it difficult to personalize treatments for specific individuals and the distribution of resources to protect the most vulnerable populations challenging. What is required to increase the accuracy of prognosis is more knowledge about how dynamic changes occur within the host cell during viral infection. Since many proteins become dysregulated during infection, proteomics is a prime technology for gaining this knowledge to increase prognostic capabilities and enable personalized treatments that alleviate the suffering viruses cause on humanity.

**Keywords** Prognosis • Disease severity • Viral factors • Host factors • Prognostic markers of COVID-19 • Patient stratification • Prognostic biomarkers • Biomarkers of long COVID • Biological pathways

## 4.1 Introduction

As of late 2022, it was estimated that there are approximately 270 known viruses that infect humans (Forni et al. 2022). This number has undoubtedly increased over the past few years and will continue to grow over time as novel viruses arise. Many viral infections are well-controlled by the immune system resulting in little damage



T. D. Veenstra, *Proteomic Applications in Virology*, Advances in Experimental Medicine and Biology 1511, https://doi.org/10.1007/978-3-032-22340-1_4

to host cells or tissues of most people. These viruses include *Human cytomegalovirus* (HCMV), adenoviruses, *Pegivirus*, herpes simplex virus, rhinoviruses, and many of the coronaviruses. Some viruses, however, cause noticeable damage to the host, which can be either acute or chronic. These viruses include human immunodeficiency virus (HIV), hantavirus, influenza, Ebola, Marburg, and severe acute respiratory syndrome coronavirus 2 (SARS-CoV-2). The severity of damage is primarily determined by the properties of the infecting virus. One of these properties is human familiarity (Sumbria et al. 2019). Human infection by an exotic virus derived from another species, known as a zoonotic virus, can often result in severe clinical symptoms that can sometimes lead to death. The most famous exotic infection was HIV in the early 1980s, which was derived from a nonhuman primate and led to the acquired immunodeficiency syndrome (AIDS) pandemic. More recent zoonotic viruses include bat-derived Ebola and nonhuman primate Zika viruses (Marie and Gordon 2023). Fortunately, in most instances, the exotic virus-host interaction causes genetic changes in both parties that make the viral infection less damaging (Sumbria et al. 2019). Humans can also more readily adapt if the exotic viral infection induces a highly protective immune response. Unfortunately, adapting to some persistent zoonotic infections such as HIV is difficult since they do not induce protective immunity but rather are immunosuppressive.

## 4.2 Viral Characteristics That Affect Severity of Disease

There are several factors that the severity of the disease is caused by a viral infection (see Fig. 4.1). While the type of infecting virus is paramount, its ability to evade components of the host's immune defense is also critical to determining disease severity. The targeting of CD4+ T cells by HIV is the most well-known example of this phenomenon (Cohen et al. 1997), but there are several other examples. Cowpox produces a homolog of a complement regulatory protein, termed inflammation modulatory protein, which inhibits production of the chemoattractant factors C3a and C5a by macrophages (Kotwal 2000). *Human cytomegalovirus* has several mechanisms for blunting immune defense systems and utilizes these to establish a period of latency where it exists within the host for an extended period without damaging host tissues but maintains its invisibility to the immune system (Forte et al. 2020). The virus can exit latency during periods when the host immune system is compromised resulting in greater tissue damage.

The dose and route of infection also play a role in the damage caused by a viral infection (Sumbria et al. 2019; Rouse and Sehrawat 2010). Control studies using animals have shown that mice given high doses of influenza virus all developed disease that could be lethal (Legge and Braciale 2005). At lower doses, virus infection was controlled by innate and adaptive immune responses. This observation suggests that limiting human interactions may not eliminate acquiring influenza but limiting close, affectionate interactions may reduce the severity of the disease. The route of exposure also affects the severity of damage to the host. Mice infected with herpes simplex virus either subcutaneously or intramuscularly usually suffer only

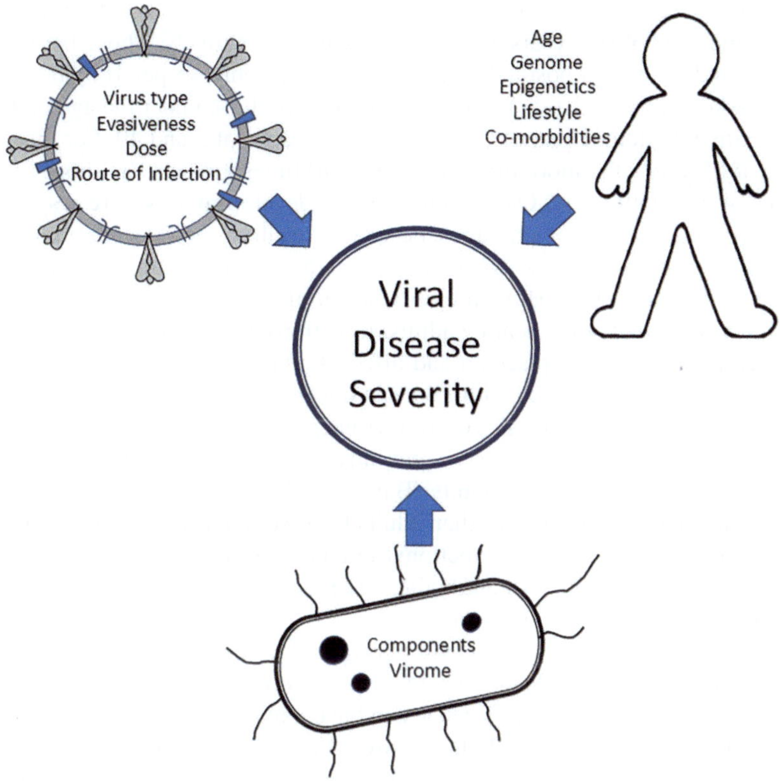


**Fig. 4.1** Factors that affect disease severity from virus infection. The severity of disease caused by a viral infection is multifactorial and includes characteristics of the virus, host, the host's microbiome, as well as the virome within the host's microbiome

subclinical effects, and no adaptive immune responses are inhibited. Intradermal infection on the flank, however, results in a skin rash or lesions that appear in a dermatomal pattern, like what occurs in human shingles (Simmons and Nash 1984). For most viruses, infections via the mucosa often result in minimal tissue damage because of the distinct properties of the mucosal immune system, for instance, the mucosal immune system can distinguish between harmless substances and dangerous pathogens, secretory immunoglobulins A (SIgA) in the mucosa are effective at neutralizing viruses, and regulatory T cells and anti-inflammatory cytokines are induced during mucosal infection (Gao et al. 2024; Traxinger et al. 2022). All these factors serve to minimize inflammation and reduce tissue damage.

## 4.3 Host Characteristics That Affect Severity of Disease

The age of the host also plays a major role in the severity of disease or tissue damage caused by a viral infection. In general, damage is most severe in very young children and elderly people, for example, the rate of hospitalization per 100,000 individuals infected with influenza is in the range of 125.7 and 332.4 for patients between the ages of 0–4 years and over 65 years, respectively. Those between the ages of 5 and 49 have a hospitalization rate of about only 45 per 100,000 (https://media.market.us/flu-statistics/). The mortality rate of those under the age of 4 years is approximately double that of those between the ages of 5 and 49. For those over the age of 65 years, the mortality rate is about 40 times greater than those between the ages of 5 and 49 years. Older patients often develop more severe responses to virus infections because of immunosenescence, resulting in a decreased ability of the body to detect infectious agents as well as a slower immunological response (Theodorakis et al. 2024). Human embryos can also suffer greater tissue damage from infectious agents than healthy adults. Infectious agents, such as Zika, HCMV, and rubella, can cross the placenta and affect the human embryo (Gordon-Lipkin et al. 2021). Embryos infected with Zika can develop hydrocephaly, a condition not observed in adults (Culjat et al. 2016). Cytomegalovirus and rubella significantly affect the nervous system of embryos, but only in rare circumstances do they cause harm to this system in healthy adults (Bale Jr 2009). Viral infection often causes greater tissue damage in embryos than adults for several reasons, primarily because the immune system is not fully functional in utero, for instance, while the embryo can produce some of their own antibodies, it relies heavily on antibodies produced by its mother that cross the placenta.

The host's genetic composition also plays a role in the severity of a viral infection (Anacleto et al. 2019). While the genomes of any two humans are estimated to be 99.9% identical, the 0.1% difference can have a major impact on the function of various functional pathways including those within the immune system, for example, a deletion of 32 base pairs in the *CCR-5 chemokine receptor* gene is associated with greater resistance to HIV infection (Allers and Schneider 2015). Genetic differences in *human leukocyte antigens* (*HLAs*) can affect individual susceptibility to autoimmune disease. There are over 20 genetic variants that can cause severe combined immunodeficiency (SCID), a condition that severely impacts the development of T and B lymphocytes (Kumrah et al. 2019). While not affecting their overall capabilities, genetic mutations in the integrin gene *CD18* can impair the ability of leukocytes to migrate to sites of infection (Das et al. 2019). While not as well understood, epigenetics also plays a role in viral infections, for example, the Epstein-Barr virus (EBV) utilizes DNA methylation to evade immune detection and prolong its persistence as an episome or as an integrated section within the host's genome (Tao and Robertson 2003).

The microbiome has recently been recognized as an influencer of individual health including immune response (Yousefi et al. 2024). This influence is not surprising considering the microbiome dwarfs its human host in terms of cell number and genetic complexity. While bacteria are its most well-known components, the microbiome also contains several families of fungi, protozoa, helminths, and viruses (Human Microbiome Project Consortium 2012). The microbiome's composition changes over an individual's lifetime and is heavily influenced by things such as diet, genetics, environment, lifestyle, age, and exposure to previous infections. While the mechanisms by which it affects the severity of a virus infection is not clear, some hypotheses include the expression of cell receptors on the surface of microbes that impact viral entry, products secreted by the microbiome that bind to viruses and impact their antigenicity, and the overall influence of the microbiome on the immune responsiveness of the mucosal lining to pathogens (Hou et al. 2022). For example, poliovirus infectivity is enhanced by binding to surface polysaccharides and peptidoglycans of bacteria (Kuss et al. 2011). This interaction also stabilizes the virus and increases its transmissibility. The benefit of the microbiome in inhibiting infection is illustrated by *Lactobacillus casei*, which releases factors that bind to surface glycans and inhibit rotaviruses from entering cells. A study done almost 20 years ago showed that consumption of a probiotic drink containing *Lactobacillus casei*, *Lactobacillus bulgaricus*, and *Streptococcus thermophilus* reduced the incidence of antibiotic- and *Clostridioides difficile*-associated diarrhea by almost one-third (Hickson et al. 2007).

The obvious mechanistic explanation of how the microbiome influences viral progression is its composition, which effects both innate and adaptive immune responses. In turn, the microbiome's composition is affected by several factors including diet, exercise, age, and even route of delivery at birth as babies delivered by Caesarean section lack maternal birth canal microorganisms (Inchingolo et al. 2024; Redondo-Useros et al. 2020; Ghosh et al. 2022). Diets rich in fiber produce short-chain fatty acids that favor systemic regulatory T-cell expansion especially within the mucosa, while diets low in fiber increase the growth of *Clostridia bacteria* that induce proinflammatory T-cell responses (Ney et al. 2023; Lee and Kim 2017). Evidence also suggests that the upper airway immune responses of obese individuals on high-fat diets are dysregulated, making them more susceptible to influenza (Almond et al. 2023).

There is increasing evidence that the virome component of the microbiome influences the damage caused by a viral infection (Cao et al. 2022). Most of the viruses within the microbiome are bacteriophages, while many of the rest are plant and animal viruses. Endogenous retroviruses account for up to 8% of the human genome, and there are several types of viruses (e.g., herpesviruses) that chronically reside within the body but do not cause disease (Mao et al. 2021). The most obvious mechanism by which tissue damage is impacted by the virome would be the cumulative effect of chronic and acute infections. Persistent nonpathogenic norovirus infections can reduce CD8 T-cell activation, which is a major defense against several types of viruses (Hassan and Baldridge 2019). Chronic HCMV infection can impact the maturation of the immune system making it unprepared to battle a novel acute infection (Vescovini et al. 2016).

A current viral infection can be significantly affected by a previous or concurrent infection. Infections occurring closely together stimulate innate and adaptive responses that affect the damage caused by each pathogen. The immune response to infections that are further apart in time will cause the adaptive response to the first infection to play a major role in the response to the following infection. Most closely spaced infections involve a viral and nonviral pathogen (i.e., bacteria). A common circumstance is influenza accompanied by a respiratory bacterial infection that often leads to severe tissue damage and possible lethality (Klomp et al. 2021). When co-infections solely involve viruses, the interference phenomenon affects the eventual prognosis. Interference involves interferons that influence the function of immune cells through their activation, modulation of signaling pathways, and increased expression of specific genes (Popovic and Minceva 2021). While it is logical to assume that co-infections can compound and cause worse symptoms than each on their own, sometimes viruses negatively impact each other, minimizing their overall effect on their shared host (Piret and Boivin 2022).

## 4.4 The Value of Prognosis

In its general sense, prognosis is the prediction of an outcome. In medicine, prognosis is the prediction of the course that a disease or injury will take and the eventual outcome. As discussed above, accurately prognosing a viral infection requires considering factors beyond simply identifying the virus. Host factors, such as age, lifestyle, weight, genetics, microbiota, etc., greatly affect an infected patient's outcome.

It is self-evident why accurate disease diagnosis is important, but why is accurate prognosis important? Diagnosis and prognosis are intimately connected in that one cannot make an accurate prognosis without making an accurate diagnosis. Prognosis takes diagnosis a step further. Prognosis is a key element in determining the appropriate treatment as well as describing to the patient the anticipated progress that will occur during treatment. In a sense, making an accurate prognosis is more challenging than making an accurate diagnosis as there are patients with different characteristics and stage of disease that fall within a gray area that make it difficult to determine if a specific treatment will be effective. Accurate prognosis not only helps determine the optimal course of treatment but also predicts the physical capabilities of the patient after primary treatment has ended. Early and accurate prognosis may enable the patient to plan for significant lifestyle changes including diet and exercise, long-term therapy, and/or change in occupation.

### *4.4.1 Prognosis of Virus Infections*

The first part of this chapter described some viral and host factors that affect the prognosis of an infection's severity. Unfortunately, these factors are either not specific enough or too difficult to quantify with the degree of specificity to provide the accuracy necessary to prognose the course and outcome of a viral infection. Honestly, selecting the course of treatment and predicting the outcome for most viral infections are done based on historical precedence. But for those individuals who may be more susceptible to severe damage (i.e., elderly, immunocompromised, co-infected, etc.), the biological factors needed to accurately prognose their disease are not readily or easily quantified.

Diagnosing the virus responsible for an infection is straightforward for many pathogens as genetic and antibody-based tests have remarkable sensitivity and specificity. Prognosis is a whole other story. While a complete blood count (CBC) can provide useful indicators of disease, it is not particularly useful for prognosis. While a viral infection typically shows a normal or increased white blood cell (WBC) count, it can sometimes show a decreased count. A lower neutrophil to lymphocyte ratio (NLR) is indicative of a viral infection, while a higher NLR is symptomatic of a bacterial infection (Ng et al. 2022). There are cases, however, of viral infections that also show higher NLRs (Simadibrata et al. 2021). While viral infections can lead to thrombocytopenia or thrombocytosis, a single strain of virus can cause either depending on the physiological mechanisms it impacts.

The current technological age with its ability to conduct high-throughput measurements of genes, transcripts, proteins, and metabolites has put medicine on the cusp of generating personalized prognosis for patients with all sorts of diseases. Just like it has led the omics revolution, genetics is leading the way in identifying molecular determinants that can prognose viral infections, for example, gene variants in the form of polymorphisms have been associated with resistance to infections (Kenney et al. 2017). These polymorphisms have been found primarily in genes involved in host receptors and co-receptors, for example, a genome-wide association study (GWAS) identified a variant of angiotensin-converting enzyme 2 (ACE2) that downregulates this receptor's expression and reduces the risk of SARS-CoV-2 infection by 40% (Horowitz et al. 2022). Polymorphisms within infected host cells have also been linked to prognosis, for example, a genetic variant that upregulates the *leucine zipper transcription factor like 1* (*LZTFL1*) gene is associated with greater epithelial-mesenchymal transition (EMT) in SARS-CoV-2-infected pulmonary epithelial cells (Downes et al. 2021). This EMT process weakens the cell adhesion properties of these pulmonary epithelial cells allowing them to migrate and invade distant tissues and form mesenchymal cells (Yang et al. 2020). Their migration enhances the inflammatory response induced by the virus resulting in poorer outcomes in the patient. Variations in the amount of cytokines produced by host cells and a dysregulation in their immune cell response have also been associated with polymorphisms that result in greater disease severity and poorer outcomes (Varzaneh et al. 2014).

Prognosis using genomic information is invaluable particularly in oncology. The utility of applied genomics will continue to increase and spread to other diseases when (and if) all individuals have their genomes sequenced at birth. Unfortunately, the fact that very few individuals have had their genomes sequenced combined with the rarity of specific polymorphisms makes it difficult to use this information for viral prognosis. Since the protein composition of humans is very similar, changes in the levels of specific proteins can make them more useful prognostic indicators. For COVID-19, the goal would be to measure specific protein biomarkers concurrently with a positive SARS-CoV-2 test that dictates the course of treatment required to minimize disease severity.

### *4.4.2 Prognosis Based on Modified World Health Organization Score*

A few years ago, the WHO developed the WHO Clinical Progression Scale (CPS) that ranked the severity of COVID-19 from a score of 0 to 10 based on a variety of factors (see Table 4.1) (World Health Organization (WHO) Working Group on the Clinical Characterisation and Management of COVID-19 infection 2020). These factors were mostly related to the presence of symptoms and treatment and were used to recommend specific treatments. Subsequently, a modified version known as the modified WHO (mWHO) score was developed (Garry et al. 2022). It restricted the severity levels only to hospitalized COVID-19 patients and collapsed these

**Table 4.1** Comparison of WHO Clinical Progression Scale (CPS) with modified WHO score

| **WHO Clinical Progression Scale** | | | **Modified WHO Score** |
|---|---|---|---|
| Patient state | Description | Score | Inpatient COVID-19 Severity Level |
| Uninfected | Uninfected; no viral RNA detected | 0 | Not applicable for inpatient |
| Ambulatory mild disease | Asymptomatic; viral RNA detected | 1 | |
| | Symptomatic; independent | 2 | |
| | Symptomatic; assistance needed | 3 | |
| Hospitalized: moderate disease | Hospitalized; no oxygen therapy | 4 | NEITHER |
| | Hospitalized; oxygen by mask or nasal prongs | 5 | $O_2$/NIV (supplemental oxygen or non-invasive ventilation) |
| Hospitalized: severe disease | Hospitalized; oxygen by NIV or high flow | 6 | |
| | Intubation and mechanical ventilation, $pO2/FiO_2 \geq 150$ or $SpO_2/FiO_2 \geq 200$ | 7 | IMV (invasive mechanical ventilation) |
| | Mechanical ventilation $pO_2/FIO2 < 150$ ($SpO2/FiO_2 < 200$) or vasopressors | 8 | |
| | Mechanical ventilation $pO_2/FiO_2 < 150$ and vasopressors, dialysis, or ECMO | 9 | |
| Dead | Dead | 10 | Not applicable for baseline severity |

*$FiO_2$* fraction of inspired oxygen, *NIV* noninvasive ventilation, *$pO_2$* partial pressure of oxygen, *$SpO_2$* oxygen saturation

The modified WHO score only applies to inpatient individuals and simplifies the WHO CPS from 11 scores to only 3 patients into three categories: those not requiring supplement oxygen or noninvasive ventilation or intubation ($O_2$/NIV nor IMV), those requiring $O_2$ or NIV without intubation ($O_2$ or NIV without IMV), and finally those requiring IMV with or without additional support. Patients in the $O_2$/NIV nor IMV category correspond to WHO CPS level 4, those in the $O_2$ or NIV without IMV category correspond to WHO CPS levels 5 and 6, and those that require IMV with or without additional support correspond to WHO CPS levels 7–9.

If organs begin to fail in severe cases, support to help the heart and lungs pump $O_2$ into the blood in the form of extracorporeal membrane oxygenation (ECMO) may be added to IMV (World Health Organization (WHO) Working Group on the Clinical Characterisation and Management of COVID-19 infection 2020). Other supporting therapies would include renal replacement therapy to help the kidneys with filtration and the administration of a vasopressor to improve blood and $O_2$ delivery to vital organs. While $O_2$, NIV, and IMV respiratory support requirements at admission may be useful in prognosis of COVID-19 infection outcome, they do not provide much specificity related to treatment. These categories only provide an indication of the severity of the course of the disease and the possibility that it could lead to death. While such measures may be useful for stratifying patients within clinical trials, their value in prognosing a COVID-19-infected patient in terms of disease progression and outcome is limited.

## 4.5 Plasma Analysis to Identify Prognostic Markers of COVID-19

As discussed throughout this book, proteomic technologies have grown exponentially over the past two to three decades. Where the capability to measure a few hundred proteins in a small number (i.e., <10) of blood samples was considered monumental 20 years ago, present studies are quantifying thousands of proteins in over a thousand samples today. What is also relevant to this chapter is the overlap between the growth of proteomic technology and the COVID-19 pandemic. This overlap resulted in most proteomic viral studies focusing on SARS-CoV-2 over the past few years. This disparity is particularly evident for studies examining disease prognosis. For this reason, all the studies described in this chapter focus on the use of proteomics for identifying prognostic markers related to SARS-CoV-2.

While most individuals infected with SARS-CoV-2 either are asymptomatic or exhibit mild symptoms such as cough, fever, and throat irritation, the virus can also cause severe symptoms such as acute respiratory distress syndrome (ARDS), thrombosis, and multiorgan dysfunction that can lead to death (Rabaan et al. 2023). The immune response invoked by SARS-CoV-2 is now well understood and can help provide information to predict disease outcome. Many proteomic studies have described notable differences in response to immune-related proteins between hospitalized cases and controls (Mohammed et al. 2022; COVID-19 Multi-omics Blood

ATlas (COMBAT) Consortium 2022; Arthur et al. 2021). While some proinflammatory cytokines that are associated with severe outcomes are elevated in COVID-19, the virus also suppresses type I interferon responses (Zhang et al. 2021). This action serves to depress host antiviral defenses and the adaptive immune response. SARS--CoV-2 infection can also lead to excessive activation of neutrophils including the release of neutrophil extracellular traps (NETs) (Prokop et al. 2021). In addition, dysregulation of the complement system leads to adverse events that can result in hospitalization (Yu et al. 2022). All these effects can lead to inflammation, immune thrombosis, organ damage, and possible death. While this information is useful, even combined with other factors related to the physical characteristics of the patient, it does not fully predict the disease's clinical trajectory, severity, or outcome. Accurate prognosis of COVID-19 (or any other virus) progression requires more detailed information about the molecular changes that occur during the infection.

It is not simply enough to be able to identify molecular changes that occur due to viral infection to make an accurate prognosis. These changes need to be measurable in an easily accessible patient sample (e.g., serum, plasma, urine, etc.). The reason for this requirement is because viral infections can rapidly cause deleterious effects in the patient, which makes the need for sample acquisition to be fast and simple. If these molecular changes are only measurable in infected cells (e.g., infected respiratory cells), then acquiring specimen for testing would be overly invasive and time-consuming. A particularly attractive sample is plasma owing to its ease of acquisition, protein complexity, and intimate relationship with the entire body. Changes in the plasma proteome of COVID-19 patients can guide physicians on the proper treatments and allow the accurate prognosis of the disease trajectory and outcome.

### 4.5.1 Stratification of Patients with COVID-19

To identify easily monitored prognostic markers, proteomes were measured in 3000 plasma samples acquired from 1117 hospitalized patients with COVID-19. An LC/MS proteomic platform was used to identify temporal changes in highly abundant proteins associated with disease severity (Viode et al. 2024). Up to six samples were acquired from each patient (for a total of 2934 samples) who were distributed across 15 different clinical centers. Prior to LC/MS analysis, the most abundant plasma proteins were depleted so that lower abundant proteins could be measured. The plasma samples were stratified into five clinical trajectory groups (TGs) from hospitalized COVID-19 patients based on the severity of disease (see Fig. 4.2a). The TGs ranged from those with a hospital stay of less than 3 days (TG1) to those who died from COVID-19 within 28 days (TG5). Besides having a short hospital stay, TG1 patients had only mild respiratory disease. Although they required more

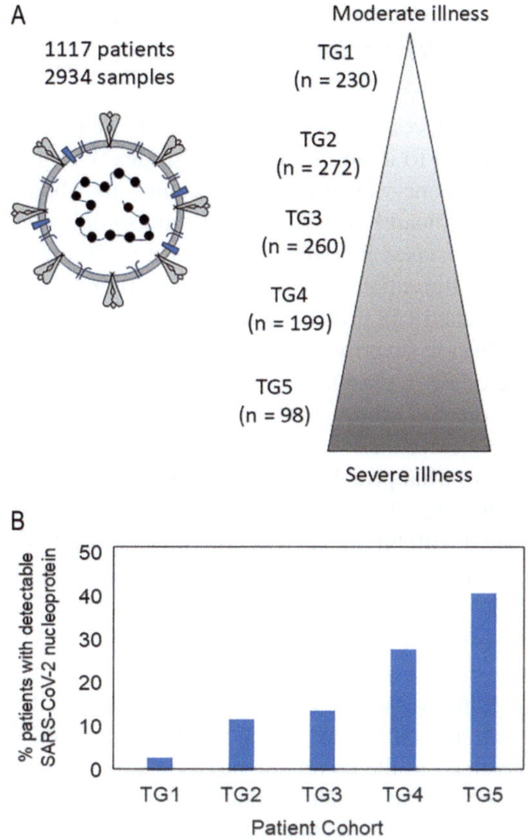


**Fig. 4.2** (**a**) Classification of patients into trajectory groups (TGs) based on COVID-19 disease severity. Patients ranged from moderate (TG1) to severe (TG5) illness. Patients in TG5 died within 28 days of being admitted into the hospital. (**b**) Percentage of patients in each TG who had detectable levels of SARS-CoV-2 nucleoprotein

respiratory support and had a longer hospital stay than TG1 patients, TG2 patients were also discharged without limitation. While TG3 and TG2 patients had similar respiratory support requirements and hospital stay lengths, TG3 patients had post-discharge limitations. TG4 patients required both prolonged hospital stays and more aggressive respiratory support. Some patients within the TG4 cohort eventually died but only after 28 days post-admission. Individuals within the TG5 cohort required high respiratory support during a prolong hospital stay and died within 28 days of admission. Another interesting indicator of disease severity in these cohorts was the detection of SARS-CoV-2 nucleoprotein (NP) in the plasma of the patients (see Fig. 4.2b). The percentage of patients with detectable levels of SARS-CoV-2 NP correlated with disease severity. The outcome observed in these patients may thus be related to the dose of virus received at the point of infection. Other studies have shown a correlation between increased plasma viral RNA levels and disease severity and risk of mortality (Fajnzylber et al. 2020; Bermejo-Martin et al. 2020). While NP and viral RNA are different biomolecules, they both originate from the virus so it may not be surprising that their levels directly correlate.

### 4.5.2 Protein Clusters Identified in Different COVID-19 Trajectory Groups

The data, comprised of an average of 576 proteins identified per plasma sample, was analyzed to identify temporal changes within each of the TGs (Viode et al. 2024). Using a one-way analysis of variance, the abundances of 305 proteins were found to be significantly different in plasma samples taken from the patients at admission. Unsupervised hierarchical clustering of these proteins revealed three main clusters that differentiated the TG groups at time of admission. The first cluster (cluster 1) contained 158 proteins whose abundance levels directly correlated with an increase in disease severity. Cluster 2 was comprised of 58 proteins whose abundance levels were low in plasma obtained from patients in the TG1 cohort compared to those in cohorts TG2 through TG5. Cluster 3 contained 89 proteins whose abundances negatively correlated with increasing disease severity.

So, what proteins typified each of these clusters? Cluster 1, which included proteins upregulated in patients with more severe disease, was linked to proteins associated with NETs, such as core histones (H2AC8) and cathepsin G (CTSG), as well as extracellular matrix (ECM) organization and degradation, such as metallopeptidase 14 (MMP14) and elastin (ELN). An additional group of proteins, including myosin 7 (MYH7) and tropomyosin 2 (TPM3), was associated with hypertrophic and dilated cardiomyopathy.

Cluster 2 contained proteins associated with acute infection. The abundances of these proteins were lower in TG1 compared to all other TGs. Protein pathways involved in the complement cascade, innate immune response, formation of fibrin clot, and platelet degranulation were part of this cluster. Cluster 3, which contained proteins whose abundances decreased as disease severity increased, was populated with proteins related to cholesterol metabolism (e.g., apolipoprotein A1), cell adhesion (e.g., E-selectin), platelet degranulation (e.g., platelet basic protein), and viral protein interactions with cytokine and cytokine receptor (e.g., platelet factor 4).

### 4.5.3 Biological Pathways Associated with Negative Outcomes

Biological pathways associated with poorer outcome were found to be linked to inflammation and innate immune responses, including complement and coagulation cascades, NETs, ECM degradation, and heart damage (see Fig. 4.3). The abundances of complement and coagulation cascade pathway proteins were elevated within TG2–TG5 compared to TG1 at the time of admission (see Fig. 4.3a). On day 10, however, their abundances started to diverge between patients with mild and severe COVID-19 symptoms. Levels of these proteins in TG1, TG2, and TG3 continuously decreased over time but converged at the end of the hospital phase of the study (i.e., 28 days). The levels of proteins in these pathways also decreased in TG4; however, their levels remained higher than within TG1–TG3 at each time point

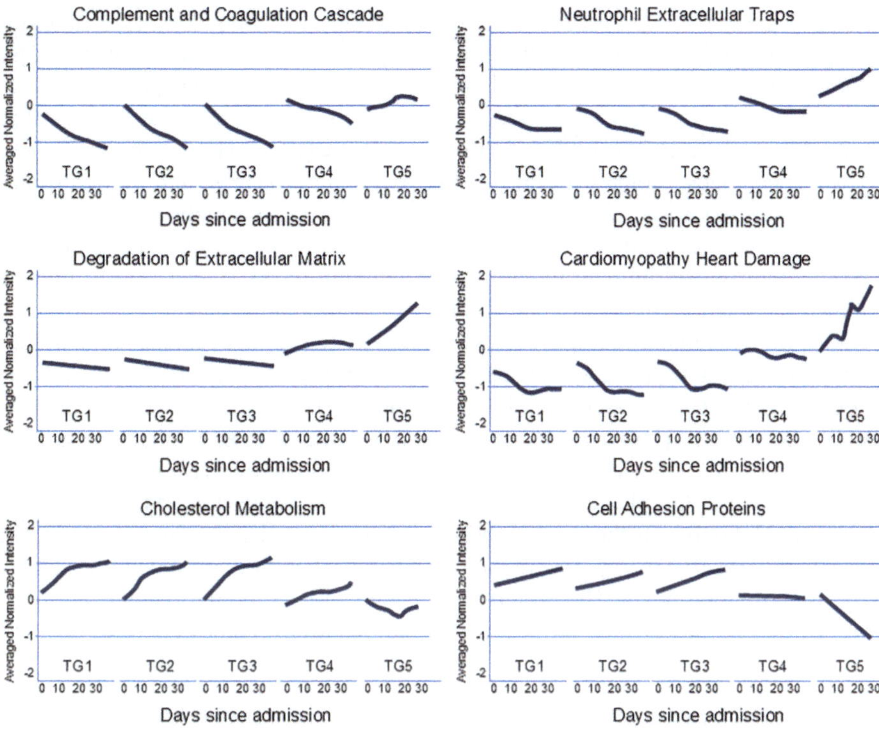


**Fig. 4.3** Time course analysis of selected biological pathways observed in the plasma proteome in trajectory groups ranging from moderate (TG1) to severe (TG5) COVID-19 illness. While complement and coagulation cascades, neutrophil extracellular traps, degradation extracellular matrix, and cardiomyopathy heart damage showed decreasing levels in TG1–TG4, they all increased in TG5. Conversely, cholesterol metabolism and cell adhesion protein pathways decreased in TG5 but increased in TG1–TG4

measured. Patients in TG5 showed lower levels of complement and coagulation cascade-associated proteins than even TG2 and TG3 patients at admission, but they quickly increased in abundance to a point they surpassed those within TG4 about 10 days post-admission. While activation of the complement system is protective in viral infection as it functions to detect and eliminate pathogens from the body, hyperactivation of this system propagates disease and is associated with increased COVID-19 severity primarily through its role in tissue damage (Silva et al. 2023). The data showing that sustained upregulation of complement activity occurred within 10 days of admission for patients that died suggests that therapies targeting this system should be administered within the first week of patients that do not show a decrease in complement activity within the first couple of days.

The trajectories for NET-associated proteins were like that observed for those associated with complement and coagulation cascade (see Fig. 4.3b). There was a consistent decrease in the abundance of NET proteins from admission to release for cohorts TG1–TG4, but those in TG5 showed an elevated level of these proteins that continued to increase up until the time of their death. A striking difference between TG5 patients and those in the other four groups was the sharp increase in NET--associated proteins that was detectable shortly after admission. One protein specifically, histone H1.5, showed a steady increase in TG5, while the groups that eventually recovered (TG1–TG4) showed a decrease in H1.5 over time (see Fig. 4.4). Histones, along with chromatin, oxidizing enzymes, and antimicrobial proteins, are associated with NETs (Mócsai 2013). The presentation of NETs outside the cell serves to trap, inactivate, and kill circulating pathogens. Unfortunately, owing to their indiscriminate targeting, upregulation of NETs can lead to uncontrolled inflammation, microvascular thrombosis, and organ damage.

Some of the damage caused by SARS-CoV-2 infection was evident in the trajectories observed for proteins associated with the degradation of the ECM and cardiomyopathy heart damage (see Fig. 4.3c, d). Cohorts TG1–TG3 showed continuous decrease in levels of proteins associated with ECM degradation, while those in TG4 and TG5 showed increases over time (see Fig. 4.3c). The increase in the levels of these proteins in TG4 plateaued at about 20 days post-admission and then slightly decreased. Not only was the level of these proteins elevated at the time of admission in the TG5 compared to any time in the other groups, but they continued to dramatically rise until their death. The proteins associated with cardiomyopathy heart damage showed similar trends as those for ECM degradation. While TG1–TG3 showed

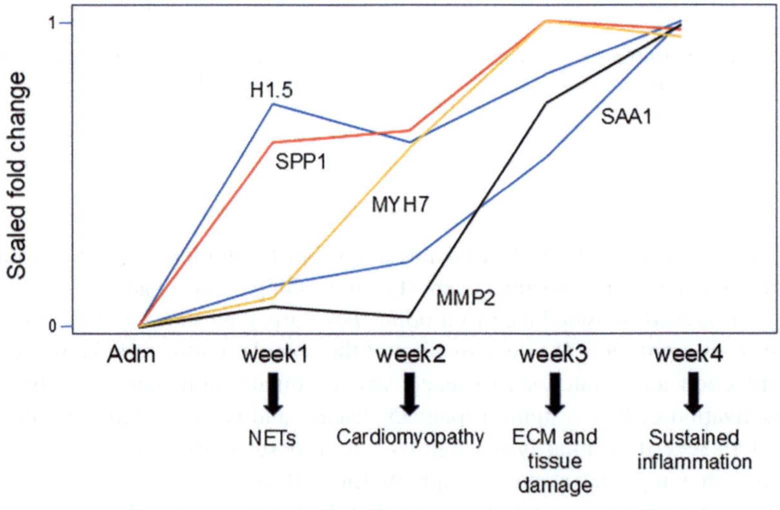


**Fig. 4.4** Longitudinal changes between patients within trajectory group 4 (TG4) who survived or died from COVID-19. While the levels of histone H1.5 (H1.5), osteopontin (SPP1), myosin 7 (MYH7), matrix metallopeptidase 2 (MMP2), and serum amyloid A1 (SAA1) followed individual trajectories, they all showed an increase in patients that eventually perished from COVID-19. Increases in H1.5 and SPP1, which correlated with death, were measured after 1 week of hospitalization. Besides showing the time of sample analysis, the *x* axis shows the biological pathway impacted at each time point

a steady drop in the levels of these proteins during the first 20 days post-admission, TG4 had a slight rise in their levels over the first 10 days followed by a subtle, but steady, decline during the rest of the time points. Of all the protein network trajectories identified in this study, the TG5 trend associated with cardiomyopathy heart damage is most striking. This group presents with high levels of these proteins at admission (especially compared to TG1–TG3), which continue to rise in a steep upward fashion throughout their hospital stay. These results corroborate those that show COVID-19 results in long-term heart damage (Xie et al. 2022). It is possible that cardiac damage associated with TG5 directly correlates with the increase in complement activation and NETs also seen in this group.

### *4.5.4 Biomarkers Associated with Death in Trajectory Group 4 Patients*

Cohort TG4 was unique in that some of the patients survived COVID-19, while others succumbed to the infection. Within the patients that perished, elevated levels of H1.5 were detected. Also found in higher abundance in TG4 patients that died were cardiac tissue damage markers such as MYH7 and ECM degradation markers (see Fig. 4.4). Correlating these findings to trends observed in TG5 allows sequences of events that lead to morbidity to be discerned. While complement proteins increased in abundance in TG5, their rate of increase during the days right after admission was much less than that observed for NETs in the same cohort and within TG4 patients that died. These results suggest that NET-associated proteins are early prognostic markers for survival versus death. While it does not exclude treating uncontrolled complement activation, preventing hyperactivation of NETs may be more critical in preventing circulating NETs from causing vessel damage, cardiomyopathy, ECM degradation, and lung injury.

### *4.5.5 Biological Pathways Associated with Positive Outcomes*

There were protein networks identified that were associated with positive prognosis and recovery from SARS-CoV-2 infection. These two networks included both an increase in cholesterol metabolism and cell adhesion proteins (see Fig. 4.3e, f). While cholesterol metabolism proteins showed similar levels for all TGs at admission, they steadily increased for groups 1–3 and more subtly for TG4, demonstrating a recovery in liver function. In contrast, the levels of cholesterol metabolism proteins showed a consistent decrease during the first 20 days of admission in TG5 and then moderately recovered prior to death. Similar profiles were seen in the TGs for proteins associated with cell adhesion. The levels of these proteins in TG1–TG3 were slightly higher than those in TG4 and TG5 at admission and continued to increase during their time within this study. The increase in cholesterol metabolism and cell adhesion pathways in TG1–TG3 suggests these markers can be monitored as prognosticators of a positive outcome and recovery.

### *4.5.6 Specific Proteins for COVID-19 Prognosis in Trajectory Group 4*

Understanding protein pathways that are prognostic of COVID-19 severity that can predict outcome and determine an appropriate treatment plan is an astonishing development. What is lacking, however, is an easy translation to a clinical assay that can rapidly assess the patient's status. It is impractical to conduct a complete proteomic analysis of patient plasma similar to what was done in the above study. Ultimately, the goal is to identify a small number of proteins that can provide correct diagnostic and prognostic information about the patient's condition. Turning a small number of proteins into a diagnostic or prognostic test would be doable and akin to what is already done for liver function tests or complete metabolic panels.

The large number of samples and the longitudinal sampling of patients with COVID-19 who had different outcomes (ranging from mild symptoms to death) make this an opportune study for identifying specific markers. Another unique aspect of this study was the fate of the participants in TG4. As described earlier, 16% of the individuals in TG4 passed away during the study period after the initial 28 days, so this cohort provide a unique situation in which a single group resulted in two different outcomes (i.e., survival versus death). To search for specific markers, t tests were conducted comparing the proteome results between TG4 patients that survived versus those that died at admission and through weeks 1–4. While no significant differences were observed between protein abundances at admission, individuals in the TG4 group that died had significantly higher levels of H1.5 and SPP1 (osteopontin) by the end of their first week after admission (see Fig. 4.4). Two weeks after admission, this group had higher levels of circulating MYH7, suggesting the presence of cardiomyopathy. Over the course of the next week, increased levels of MMP2 were observed in TG4 patients that perished pointing toward an increase in ECM and tissue damage. By the end of the fourth week, these individuals displayed chronically high levels of the acute-phase protein serum amyloid 1A1 (SAA1), indicating an increased inflammatory response to the SARS-CoV-2 infection (Zinellu et al. 2021). This comparison filtered the number of proteins necessary to measure to predict the outcomes for this group down to only five, which makes the development of a clinical assay manageable. As individual immunoassays for these proteins presently exist, it would not be difficult to develop a multiplexed ELISA kit to measure each in the plasma of hospitalized COVID-19 patients.

### 4.5.7 *Prognostic Severity Biomarker Outcome Across All COVID-19 Cohorts*

While the five proteins identified above showed promise in being able to predict the outcome of a specific group (i.e., TG4), this outcome still fails to address most of the patients that are hospitalized with SARS-CoV-2 infection. Fortunately, the large number of patients and plethora of proteomic data allowed the prognostication of the clinical disease course and outcomes for all the COVID-19 hospitalized patients. A logistic regression at time of admission was applied to half of the plasma proteomic data for training the algorithm, and the second half of the data was used to validate the prognostic model (Viode et al. 2024). The algorithm identified four proteins, interleukin 1 receptor like 1 (ILRL1), platelet factor 4 (PF4), elastin (ELN), and α-1-antichymotrypsin (SERPINA3), that were able to predict death at the time of hospitalization with an area under the receiver operating characteristic curve (AUROC) value of 0.85 in both the training set and validation set.

The measurement of two proteins at admission were found to be prognostic for indicating an escalation of care in the form of ECMO, and possibly invasive mechanical ventilation would be required during hospitalization. These proteins were desmoyokin (AHNAK) and H2A clustered histone 20 (H2AC20). Desmoyokin is associated with cell adhesion and organization of the cytoskeleton, while H2AC20 is an H2A variant associated with DNA packaging and chromatin organization. H2A is a well-known component of NETs, which have been associated with COVID-19 infection in this and other studies. The increased presence of both proteins in plasma of patients that require aggressive care could be related to cell damage caused by either SARS-CoV-2 infection or the resultant cytokine storm unleased in response to the virus. Using an identical training and validation strategy resulted in an AUROC of 0.79 and 0.70 for the testing and validation sets, respectively, for predicting a necessary escalation of care.

It is worth summarizing the results from these two analyses not only because of their potential prognostic value but also because they demonstrate how far proteomics has evolved. One protein panel had high prognostic value for predicting death at the time of hospitalization, and the other could predict the need for aggressive treatment in the form of ECMO or mechanical ventilation. The ability to predict the need for mechanical ventilation harkens back to the early days of the COVID-19 pandemic when this equipment was in short supply. When supply is low, it is critical to recognize patients that are most in need of ventilators. While being able to predict death at the time of hospitalization may sound like a morbid result, what it really dictates is the level and urgency of treatment that will be required to attempt to save the patient's life. Having this information will allow a more scientifically based triage of patients so that the most vulnerable can be treated quickly and appropriately.

While this study does not provide the ultimate prognostic tests for COVID-19, it demonstrates how far and fast proteomic analysis has developed. This study does not simply provide a list of key proteins or pathways that are divergent in healthy and infected patients. What it shows is the potential to distinguish closely related groups of patients early in the course of the disease. The results show that proteomics will soon fulfil its goal of enabling the accurate diagnosis and prognosis of diseases, providing the opportunity for providing effective therapies for patients soon after they are hospitalized.

## 4.6 Search for Biomarkers That Indicate the Need for Intensive Care

When many people hear the word "proteomics," the first thing that comes to mind is studies attempting to measure thousands of proteins and then assembling this information into comprehensible patterns of protein expression that explain something about the state of a cell or organism. While originally it was defined as the large-scale characterization of proteins, the term proteomics has evolved to refer to any type of research focused on measuring proteins. In graduate school, my focus was determining the three-dimensional structures of proteins using nuclear magnetic resonance spectroscopy. We would refer to it as three-dimensional protein structure determination, but now it is referred to as structural proteomics. Identifying phosphorylation sites on proteins, regardless of the number being analyzed, is now called phosphoproteomics. The point is that it is not necessary to characterize as many proteins as possible to conduct proteomics; simply focusing on a few, potentially interesting targets is also considered proteomics.

This strategy of building upon established information to generate hypotheses to focus on proteins believed to be impacted by a condition has been used to identify biomarkers that can assist in the prognosis of COVID-19. As discussed previously, most individuals with COVID-19 are either asymptomatic or develop mild symptoms. To older individuals, infants, and patients with hypertension or chronic inflammatory conditions, COVID-19 is life-threatening. What are not well known are the underlying host biomarkers that can predict the severity of the infection and inform the physician what course of drug regiment is needed or if ICU admission is necessary.

Most COVID-19 patients admitted to the ICU display rapid lung function deterioration and an overactive immune response. One uncertainty related to the overactive immune response, however, is the association between disease severity and increases in specific cytokine levels. Using this information, a group studied plasma samples obtained from hospitalized patients who did ($n = 15$) or did not ($n = 25$) require ICU admission (Haljasmägi et al. 2020). Mechanical ventilation was required for most of the patients that required ICU treatment within the first 10 days after the onset of symptoms.

Blood inflammation markers, antibodies, and 101 plasma proteins of hospitalized COVID-19 patients who did or did not require ICU admission were measured. The study used a variety of different platforms to measure these proteins including affinity-based techniques such as ELISA and flow cytometry along with a proximity extension assay (PEA) that measures 92 inflammation-related protein biomarkers. While this strategy only surveys a very small percentage of the plasma proteome (i.e., <1%), each protein targeted for measurement is known to be associated with viral infection. ICU patients exhibited a rapid upregulation of selective inflammatory markers including IL-6, CXCL10, CXCL11, IFNγ, IL-10, and monocyte-attracting chemokines CCL2, CCL7, and CCL8. A sustained upregulation of apoptosis-associated proteins CASP8, TNFSF14, HGF, and TGFB1 was also observed. Of these apoptotic markers, variations in HGF levels were able to discriminate plasma samples acquired from ICU and non-ICU patients. Overall, these results suggest that specific inflammatory and apoptotic markers could serve as prognostic markers for COVID-19 patients that require a more aggressive treatment regimen.

## 4.7 Long COVID

Early during the SARS-CoV-2 pandemic, many people reported suffering from symptoms several weeks or months following the onset of COVID-19. In the spring of 2020, this phenomenon was termed "long COVID" also known as post-acute sequelae COVID-19 (PASC) or post-COVID-19 (Ely et al. 2024). While the exact numbers are uncertain, about one-third of nonhospitalized patients do not fully recover 2–6 weeks from onset of COVID-19 symptoms or a positive test for the virus. Between 11% and 24% report persistent symptoms 3 months after onset of illness (Alwan and Johnson 2021). While long COVID is associated with all ages, it is most prevalent in patients between 36 and 50 years of age (Mallick et al. 2023). Most cases are associated with nonhospitalized patients (due to their sheer number); however, an astonishing 50–70% of hospitalized patients report having long COVID. Vaccination status does not prevent long COVID as 8–12% of those vaccinated against SARS-CoV-2 suffer with the condition (Al-Aly et al. 2022).

While they vary, most patients with long COVID experience multiple symptoms across various organ systems. There are several adverse outcomes associated with long COVID including cardiovascular, cerebrovascular, thrombotic, dysautonomia, myalgic encephalomyelitis/chronic fatigue syndrome (ME/CFS), and type 2 diabetes. One of the most reported symptoms is postural orthostatic tachycardia syndrome (POTS) (Davis et al. 2023). These symptoms can last for years, with dysautonomia and new onset ME/CFS lasting for the rest of the patient's life. The financial burden of long COVID has been estimated at $3.7 trillion, which includes loss of earnings, productivity, and quality of life along with increased medical expenses. This cost works out to be between $5084 and $11,646 annually per patient (https://cutler.scholars.harvard.edu//news/long-covid). There are multiple risk factors that may increase the likelihood of developing PASC. These include age, sex, preexisting conditions (e.g., type II diabetes), autoimmunity, environmental factors (e.g., gut microbiome dysregulation), EBV and SARS-CoV-2 viremia, vaccination

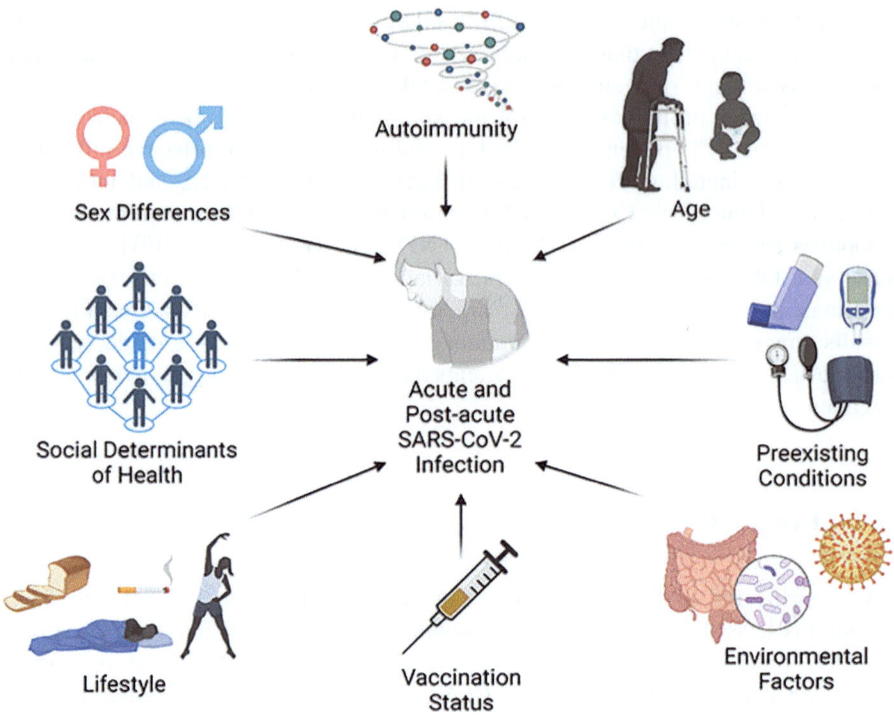


**Fig. 4.5** Panel showing possible factors that have been associated with developing post-acute sequelae COVID-19 (PASC). (This work is made available under the terms of reproduction from Wang et al. (2023) under Creative Commons 4.0)

status, and lifestyle habits such as smoking, exercise, diet, and alcohol, among others. Additionally, social determinants of health serve as a modifying risk factor (see Fig. 4.5) (Wang et al. 2023; Sherif et al. 2023).

## *4.7.1 Identification of Biomarkers Prognostic of Long COVID*

There are currently no effective treatments for long COVID that have been validated. Finding a treatment requires understanding the molecular mechanism of the disease. Biomarkers that can diagnose long COVID and serve as prognosticators of the disease progression need to be sensitive and specific, easily measured, and found within a clinical sample that can be acquired relatively noninvasively. This final characteristic would generally exclude tissue samples or any sample that would require an endoscopy to acquire. Another important factor would be the ability to identify these biomarkers soon after SARS-CoV-2 infection so that the appropriate treatments could be instituted quickly.

To explore long COVID disease pathogenesis at a broader proteomic level, a study compared plasma acquired from equal numbers ($n = 51$) of healthy convalescent and individuals with long COVID. Protein levels were quantitated using the Olink Explore 3072 version of a PEA, which measures the abundance of 2943 individual proteins (Gao et al. 2025). The proteins measured using this version of the PEA included those associated with inflammation, neurology, oncology, and cardiology. The donor samples were stratified based on breathlessness as severe breathlessness is associated with the activation of several pathways that characterize the proteomic signature of long COVID.

While most of the measured plasma proteins were skewed toward a positive correlation with severity of breathlessness, isopentenyl-diphosphate δ-isomerase 2 and small proline-rich protein 3 were among the most positively correlated. On the other hand, ectonucleotide pyrophosphatase/phosphodiesterase family member 5 and olfactory marker protein were among the most negatively correlated. All proteins were ranked according to their degree of correlation with breathlessness and analyzed using gene set enrichment analysis (GSEA) to identify protein pathways associated with long COVID. The GSEA revealed dysregulation of pathways involved in atelectasis (lung collapse), tachypnea (rapid breathing), cell cycle progression, platelet activation (including the proteins thromboxane 2 and platelet-derived growth factor receptor β), and inflammation. Additional analysis using the protein network tool Cytoscape identified a protein network centered around the protein CD40. As seen in subsequent validation studies, this result turned out to be of particular interest.

CD40 is a transmembrane receptor that is found on the surface of antigen-presenting cells (APCs) and is required for their activation (Elgueta et al. 2009). Besides expression on APCs, it is also found on endothelial and epithelial cells. When activated, CD40 triggers signaling pathways that promote inflammation. Proteins within the CD40 network that were upregulated in patients with severe breathlessness and long COVID included CCL3, CCL4, IKBKG, and IL-18 (Gao et al. 2025). Both CCL3 and CCL4 (also known as MIP-1α and MIP-1β, respectively) are chemokines that play significant roles in the immune response. They are pro-inflammatory and act as chemoattractants that recruit immune cells to the site of infection. IKBKG (inhibitor of nuclear factor κB kinase regulatory subunit γ) is a regulatory subunit that activates NF-κB resulting in the expression of genes involved in various immune responses including inflammation (Israël 2010). IL-18 is a pro-inflammatory cytokine that contributes to IFN-γ production and can also contribute to certain autoimmunity diseases.

To validate the CD40-associated network results, a secondary cohort of plasma samples was analyzed to identify biomarkers associated with long COVID. This analysis focused on long COVID patients that were classified according to the Borg CR10 scale as having low (0–2, $n = 60$) or moderate (3–7, $n = 35$) breathing difficulty. The Borg CR10 scale is a subjective assessment tool that relies on the patient's rating of their own breathing difficulty and ranges from 0 (no breathlessness) to 10 (worst imaginable breathlessness) (Buckley and Borg 2011). It is used primarily on patients with conditions like asthma and chronic obstructive pulmonary disease to assess the severity of dyspnea during a specified activity. No single upregulated protein was able to achieve significance in recognizing patients with long COVID; however, GSEA analysis identified enriched signaling pathways that were also observed in patients with moderate BORG CR10 scores (i.e., moderate breathing difficulty) in the primary cohort of samples including MET and PI3K/AKT/MTOR (see Fig. 4.6a). Combined analysis of both cohorts provided the most encouraging results. The most confident network connections for predicting if an individual will

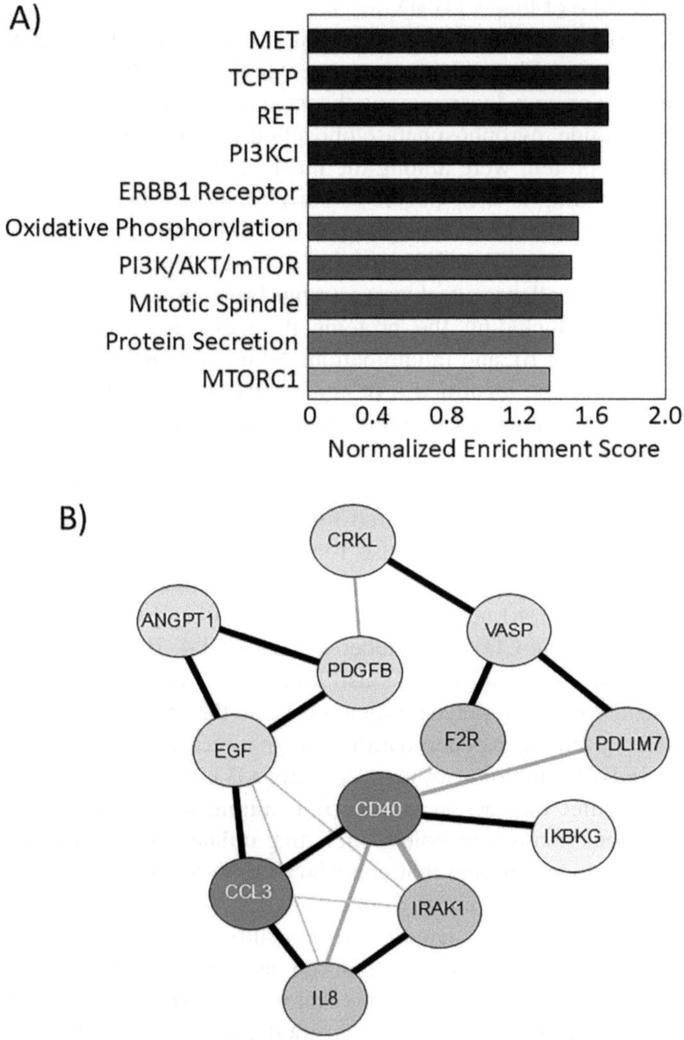


**Fig. 4.6** (**a**) Gene set enrichment analysis (GSEA) shows the highest ranked differentially expressed protein pathways that are positively associated with Borg CR10 scores, which rank patients based on their own perceived breathing difficulties. Proteins associated with the MET pathway were most highly associated with severe breathing difficulties. (**b**) Many of the proteins showing differential abundance in plasma were associated with CD40, suggesting that long COVID and breathlessness are linked to inflammation

develop long COVID were centered around CD40, CCL3, IKBKG, IL-18, and IRAK1 (see Fig. 4.6b). Upregulations of CD40, CCL3, IKBKG, and IL-8 were also seen in the primary cohort. Taken together, they suggest that this network is a prime candidate for the discovery of prognostic biomarkers of long COVID.

While the results of this study did not find a single specific protein that could be monitored to predict long COVID, they were consistent with other studies that show chronic inflammation as a key prognostic indicator (Peluso et al. 2021; Yin et al. 2024; Talla et al. 2023; Phetsouphanh et al. 2022). Many of these studies identified protein networks centered around CD40, which collectively suggest that long COVID and breathlessness are linked to SARS-CoV-2-induced inflammation. While not fully understood, an interesting finding is the dysregulation of platelet activation factors, such as PDGFRB and TXA2, in patients with long COVID. The combination of increased inflammatory responses and platelet activation may result in hypercoagulation and thromboinflammation that could cause lung damage. Lung damage would impair oxygen exchange and lead to the breathlessness observed in patients with long COVID.

## 4.8 Conclusion

When it comes to viral infections, prognosis is underappreciated by the general public. People are most familiar with prognosis when it concerns a disease such as cancer. Indeed, one of the most common questions that cancer patients have after diagnosis is their prognosis. People who go to see a physician with a viral infection usually don't ask what their prognosis is as the expectation is that they will be treated and recover in a few days. All of that changed in 2020. Although it is an unscientific poll, there are over 15,000 articles listed on PubMed during the years 2020 and 2021 that contained the keywords "virus" and "prognosis." Of these articles, almost 57% also contained the keyword "COVID." Compare these statistics to the year 2011, which had the highest number of articles containing the keywords "virus," "prognosis," and "influenza." The keyword influenza was found in less than 10% of articles that contained the keywords "virus" and "prognosis." Even in its most active year (i.e., 2014), the keyword "HIV" was found in less than 23% of articles associated with "virus" and "prognosis."

Since it was caused by a novel human virus, the COVID-19 pandemic brought great uncertainty. People of all ages, not just the elderly, were frightened to contract the virus because of the unpredictability of their prognosis. Testing facilities and hospitals became overrun with patients. It seemed that during the early stages of the pandemic, every patient's prognosis was dire. This scenario is illustrated in the predicted demand for ventilators (Branson and Rodriquez Jr 2023). US officials frantically scrambled to acquire ventilators from the Strategic National Stockpile while invoking the Defense Production Act to compel car manufacturers to collaborate with ventilator manufacturers to increase the speed of production. Unfortunately, this resulted in nonexperts making ventilators for the first time. Another key factor that was missing in this frenzy was the trained staff with expertise to properly operate the ventilators.

Over 200,000 ventilators were purchased by the US government, states, cities, health systems, and individuals (Branson et al. 2021). The use of "splitters" that permitted shared ventilation was granted emergency use authorization (EUA) by the FDA (https://www.fda.gov/media/136423/download). Beyond the impracticality of bringing individuals close enough together to utilize a single ventilation during the time of a highly contagious virus, shared ventilation resulted in several inaccurate prognoses as physiological problems were often mistaken as plumbing problems (Branson and Rodriquez Jr 2023). Many of the splitter devices approved under EUA where later rescinded. The need for rationing ventilators was a major topic of conversation. This demand was based on the prognosis that there would be a huge surge in hospitalized COVID-19 patients that would require ventilators. Unfortunately, a lot of the ventilators went unused as many were not capable of helping patients with severe conditions such as ARDS (https://www.reuters.com/article/business/healthcare-pharmaceuticals/the-us-has-spent-billions-stockpiling-ventilators-but-many-wont-save-critica-idUSKBN28C1N5/). As more became known about the progression of COVID-19 and improved treatment protocols were developed, fewer ventilators were required than initially anticipated. Unfortunately, the country was left with a plethora of ventilators that require ongoing storage and maintenance.

The frenzy in ventilator manufacturing was due to the medical community's inability to accurately prognose COVID-19 outcomes. While it may be excusable considering that SARS-CoV-2 is a novel human virus, it is reflective on how poor prognosis can lead to an inefficient use of treatment resources. More efficient use of treatments is going to require biomarkers that can predict disease progression. This strategy is widely used in the treatment of cancer but is lacking in viral infection. While they are common in oncology, genetic markers for predicting infectious disease severity are rare and only impact a very small percentage of the population. Proteomics is going to be key to identifying biomarkers that can predict the severity of an infectious disease and monitor its progression. These biomarkers are present in all individuals, and their measurement would enable personalized medicine to be applied to patient treatment. While COVID-19 has brought unimaginable worldwide suffering, it has been the catalyst for proteomic scientist to develop research strategies for finding prognostic biomarkers that can be used to guide treatment for this virus and others in the future.

## References

Al-Aly Z, Bowe B, Xie Y (2022) Long COVID after breakthrough SARS-CoV-2 infection. Nat Med 28:1461–1467

Allers K, Schneider T (2015) CCR5Δ32 mutation and HIV infection: basis for curative HIV therapy. Curr Opin Virol 14:24–29

Almond M, Farne HA, Jackson MM, Jha A, Katsoulis O, Pitts O et al (2023) Obesity dysregulates the pulmonary antiviral immune response. Nat Commun 14:6607–6620

Alwan NA, Johnson L (2021) Defining long COVID: going back to the start. Med 2:501–504
Anacleto O, Cabaleiro S, Villanueva B, Saura M, Houston RD, Woolliams JA et al (2019) Genetic differences in host infectivity affect disease spread and survival in epidemics. Sci Rep 9:4924–4935
Arthur L, Esaulova E, Mogilenko DA, Tsurinov P, Burdess S, Laha A et al (2021) Cellular and plasma proteomic determinants of COVID-19 and non-COVID-19 pulmonary diseases relative to healthy aging. Nat Aging 1:535–549
Bale JF Jr (2009) Fetal infections and brain development. Clin Perinatol 36:639–653
Bermejo-Martin JF, Gonzalez-Rivera M, Almansa R, Micheloud D, Tedim AP, Dominguez-Gil M et al (2020) Viral RNA load in plasma is associated with critical illness and a dysregulated host response in COVID-19. Crit Care 24:691
Branson RD, Rodriquez D Jr (2023) COVID-19 lessons learned: response to the anticipated ventilator shortage. Respir Care 68:129–150
Branson R, Dichter JR, Feldman H, Devereaux A, Dries D, Benditt J et al (2021) The US strategic national stockpile ventilators in coronavirus disease 2019: a comparison of functionality and analysis regarding the emergency purchase of 200,000 devices. Chest 159:634–652
Buckley JP, Borg GA (2011) Borg's scales in strength training; from theory to practice in young and older adults. Appl Physiol Nutr Metab 36:682–692
Cao Z, Sugimura N, Burgermeister E, Ebert MP, Zuo T, Lan P (2022) The gut virome: a new microbiome component in health and disease. EBioMedicine 81:104113–104129
Cohen OJ, Kinter A, Fauci AS (1997) Host factors in the pathogenesis of HIV disease. Immunol Rev 159:31–48
COVID-19 Multi-omics Blood ATlas (COMBAT) Consortium (2022) A blood atlas of COVID-19 defines hallmarks of disease severity and specificity. Cell 185:916–938
Culjat M, Darling SE, Nerurkar VR, Ching N, Kumar M, Min SK et al (2016) Clinical and imaging findings in an infant with Zika embryopathy. Clin Infect Dis 63:805–811
Das J, Sharma A, Jindal A, Aggarwal V, Rawat A (2019) Leukocyte adhesion defect: where do we stand circa 2019? Genes Dis 7:107–114
Davis HE, McCorkell L, Vogel JM, Topol EJ (2023) Long COVID: major findings, mechanisms and recommendations. Nat Rev Microbiol 21:133–146
Downes DJ, Cross AR, Hua P, Roberts N, Schwessinger R, Cutler AJ et al (2021) Identification of LZTFL1 as a candidate effector gene at a COVID-19 risk locus. Nat Genet 53:1606–1615
Elgueta R, Benson MJ, de Vries VC, Wasiuk A, Guo Y, Noelle RJ (2009) Molecular mechanism and function of CD40/CD40L engagement in the immune system. Immunol Rev 229:152–172
Ely EW, Brown LM, Fineberg HV, National Academies of Sciences, Engineering, and Medicine Committee on Examining the Working Definition for Long COVID (2024) Long COVID defined. N Engl J Med 391:1746–1753
Fajnzylber J, Regan J, Coxen K, Corry H, Wong C, Rosenthal A et al (2020) SARS-CoV-2 viral load is associated with increased disease severity and mortality. Nat Commun 11:5493
Forni D, Cagliani R, Clerici M, Sironi M (2022) Disease-causing human viruses: novelty and legacy. Trends Microbiol 30:1232–1242
Forte E, Zhang Z, Thorp EB, Hummel M (2020) Cytomegalovirus latency and reactivation: an intricate interplay with the host immune response. Front Cell Infect Microbiol 10:130–147
Gao P, Morita N, Shinkura R (2024) Role of mucosal IgA antibodies as novel therapies to enhance mucosal barriers. Semin Immunopathol 47:1–10
Gao Y, Cai C, Adamo S, Biteus E, Kamal H, Dager L et al (2025) Identification of soluble biomarkers that associate with distinct manifestations of long COVID. Nat Immunol 26:692–705
Garry EM, Weckstein AR, Quinto K, Bradley MC, Lasky T, Chakravarty A et al (2022) Categorization of COVID-19 severity to determine mortality risk. Pharmacoepidemiol Drug Saf 31:721–728
Ghosh TS, Shanahan F, O'Toole PW (2022) The gut microbiome as a modulator of healthy ageing. Nat Rev Gastroenterol Hepatol 19:565–584

Gordon-Lipkin E, Hoon A, Pardo CA (2021) Prenatal cytomegalovirus, rubella, and Zika virus infections associated with developmental disabilities: past, present, and future. Dev Med Child Neurol 63:135–143

Haljasmägi L, Salumets A, Rumm AP, Jürgenson M, Krassohhina E, Remm A et al (2020) Longitudinal proteomic profiling reveals increased early inflammation and sustained apoptosis proteins in severe COVID-19. Sci Rep 10:20533

Hassan E, Baldridge MT (2019) Norovirus encounters in the gut: multifaceted interactions and disease outcomes. Mucosal Immunol 12:1259–1267

Hickson M, D'Souza AL, Muthu N, Rogers TR, Want S, Rajkumar C et al (2007) Use of probiotic Lactobacillus preparation to prevent diarrhoea associated with antibiotics: randomised double blind placebo controlled trial. BMJ 335:80–84

Horowitz JE, Kosmicki JA, Damask A, Sharma D, Roberts GHL, Justice AE et al (2022) Genome-wide analysis provides genetic evidence that ACE2 influences COVID-19 risk and yields risk scores associated with severe disease. Nat Genet 54:382–392

Hou K, Wu ZX, Chen XY, Wang JQ, Zhang D, Xiao C et al (2022) Microbiota in health and diseases. Signal Transduct Target Ther 7:135–158

https://cutler.scholars.harvard.edu//news/long-covid. Accessed July 2025

https://media.market.us/flu-statistics/ Accessed July 2025

https://www.fda.gov/media/136423/download. Accessed August 2025

https://www.reuters.com/article/business/healthcare-pharmaceuticals/the-us-has-spent--billions-stockpiling-ventilators-but-many-wont-save-critica-idUSKBN28C1N5/. Accessed August 2025

Human Microbiome Project Consortium (2012) Structure, function and diversity of the healthy human microbiome. Nature 486:207–214

Inchingolo F, Inchingolo AD, Palumbo I, Trilli I, Guglielmo M, Mancini A et al (2024) The impact of cesarean section delivery on intestinal microbiota: mechanisms, consequences, and perspectives-a systematic review. Int J Mol Sci 25:1055–1074

Israël A (2010) The IKK complex, a central regulator of NF-kappaB activation. Cold Spring Harb Perspect Biol 2010(2):a000158

Kenney AD, Dowdle JA, Bozzacco L, McMichael TM, St Gelais C, Panfil AR et al (2017) Human genetic determinants of viral diseases. Annu Rev Genet 51:241–263

Klein J, Wood J, Jaycox JR, Dhodapkar RM, Lu P, Gehlhausen JR et al (2023) Distinguishing features of long COVID identified through immune profiling. Nature 623:139–148

Klomp M, Ghosh S, Mohammed S, Nadeem Khan M (2021) From virus to inflammation, how influenza promotes lung damage. J Leukoc Biol 110:115–122

Kotwal GJ (2000) Poxviral mimicry of complement and chemokine system components: what's the end game? Immunol Today 21:242–248

Kumrah R, Vignesh P, Patra P, Singh A, Anjani G, Saini P et al (2019) Genetics of severe combined immunodeficiency. Genes Dis 7:52–61

Kuss SK, Best GT, Etheredge CA, Pruijssers AJ, Frierson JM et al (2011) Intestinal microbiota promote enteric virus replication and systemic pathogenesis. Science 334:249–252

Lee N, Kim WU (2017) Microbiota in T-cell homeostasis and inflammatory diseases. Exp Mol Med 49:e340–e349

Legge KL, Braciale TJ (2005) Lymph node dendritic cells control CD8+ T cell responses through regulated FasL expression. Immunity 23:649–659

Mallick D, Goyal L, Chourasia P, Zapata MR, Yashi K, Surani S (2023) COVID-19 induced postural orthostatic tachycardia syndrome (POTS): a review. Cureus 15:e36955

Mao J, Zhang Q, Cong YS (2021) Human endogenous retroviruses in development and disease. Comput Struct Biotechnol J 19:5978–5986

Marie V, Gordon ML (2023) The (re-)emergence and spread of viral zoonotic disease: a perfect storm of human ingenuity and stupidity. Viruses 15:1638–1653

Mócsai A (2013) Diverse novel functions of neutrophils in immunity, inflammation, and beyond. J Exp Med 210:1283–1299

Mohammed Y, Goodlett DR, Cheng MP, Vinh DC, Lee TC, McGeer A et al (2022) Longitudinal plasma proteomics analysis reveals novel candidate biomarkers in acute COVID-19. J Proteome Res 21:975–992
Ney LM, Wipplinger M, Grossmann M, Engert N, Wegner VD, Mosig AS (2023) Short chain fatty acids: key regulators of the local and systemic immune response in inflammatory diseases and infections. Open Biol 13:230014–230027
Ng WW, Lam SM, Yan WW, Shum HP (2022) NLR, MLR, PLR and RDW to predict outcome and differentiate between viral and bacterial pneumonia in the intensive care unit. Sci Rep 12:15974–15985
Peluso MJ, Lu S, Tang AF, Durstenfeld MS, Ho HE, Goldberg SA et al (2021) Markers of immune activation and inflammation in individuals with postacute sequelae of severe acute respiratory syndrome coronavirus 2 infection. J Infect Dis 224:1839–1848
Phetsouphanh C, Darley DR, Wilson DB, Howe A, Munier CML, Patel SK et al (2022) Immunological dysfunction persists for 8 months following initial mild-to-moderate SARS-CoV-2 infection. Nat Immunol 23:210–216
Piret J, Boivin G (2022) Viral interference between respiratory viruses. Emerg Infect Dis 28:273–281
Popovic M, Minceva M (2021) Coinfection and interference phenomena are the results of multiple thermodynamic competitive interactions. Microorganisms 9:2060–2077
Prokop JW, Hartog NL, Chesla D, Faber W, Love CP, Karam R et al (2021) High-density blood transcriptomics reveals precision immune signatures of SARS-CoV-2 infection in hospitalized individuals. Front Immunol 12:694243
Rabaan AA, Smajlović S, Tombuloglu H, Ćordić S, Hajdarević A, Kudić N et al (2023) SARS-CoV-2 infection and multi-organ system damage: a review. Biomol Biomed 23:37–52
Redondo-Useros N, Nova E, González-Zancada N, Díaz LE, Gómez-Martínez S, Marcos A (2020) Microbiota and lifestyle: a special focus on diet. Nutrients 12:1776–1822
Rouse BT, Sehrawat S (2010) Immunity and immunopathology to viruses: what decides the outcome? Nat Rev Immunol 10:514–526
Sherif ZA, Gomez CR, Connors TJ, Henrich TJ, Reeves WB, Mechanistic Pathway Task Force RECOVER (2023) Pathogenic mechanisms of post-acute sequelae of SARS-CoV-2 infection (PASC). elife 12:e86002
Silva MJA, Ribeiro LR, Gouveia MIM, Marcelino BDR, Santos CSD, Lima KVB et al (2023) Hyperinflammatory response in COVID-19: a systematic review. Viruses 15:553
Simadibrata DM, Calvin J, Wijaya AD, Ibrahim NAA (2021) Neutrophil-to-lymphocyte ratio on admission to predict the severity and mortality of COVID-19 patients: a meta-analysis. Am J Emerg Med 42:60–69
Simmons A, Nash AA (1984) Zosteriform spread of herpes simplex virus as a model of recrudescence and its use to investigate the role of immune cells in prevention of recurrent disease. J Virol 52:816–821
Sumbria D, Berber E, Rouse BT (2019) Factors affecting the tissue damaging consequences of viral infections. Front Microbiol 10:2314–2326
Talla A, Vasaikar SV, Szeto GL, Lemos MP, Czartoski JL, MacMillan H et al (2023) Persistent serum protein signatures define an inflammatory subcategory of long COVID. Nat Commun 14:3417
Tao Q, Robertson KD (2003) Stealth technology: how Epstein-Barr virus utilizes DNA methylation to cloak itself from immune detection. Clin Immunol 109:53–63
Theodorakis N, Feretzakis G, Hitas C, Kreouzi M, Kalantzi S, Spyridaki A et al (2024) Immunosenescence: how aging increases susceptibility to bacterial infections and virulence factors. Microorganisms 12:2052–2069
Traxinger BR, Richert-Spuhler LE, Lund JM (2022) Mucosal tissue regulatory T cells are integral in balancing immunity and tolerance at portals of antigen entry. Mucosal Immunol 15:398–407
Varzaneh FN, Keller B, Unger S, Aghamohammadi A, Warnatz K, Rezaei N (2014) Cytokines in common variable immunodeficiency as signs of immune dysregulation and potential therapeutic targets - a review of the current knowledge. J Clin Immunol 34:524–543

Vescovini R, Telera AR, Pedrazzoni M, Abbate B, Rossetti P, Verzicco I et al (2016) Impact of persistent cytomegalovirus infection on dynamic changes in human immune system profile. PLoS One 11:e0151965–e0151978

Viode A, Smolen KK, van Zalm P, Stevenson D, Jha M, Parker K et al (2024) Longitudinal plasma proteomic analysis of 1117 hospitalized patients with COVID-19 identifies features associated with severity and outcomes. Sci Adv 10:eadl5762

Wang C, Ramasamy A, Verduzco-Gutierrez M, Brode WM, Melamed E (2023) acute and post-acute sequelae of SARS-CoV-2 infection: a review of risk factors and social determinants. Virol J 20:124

World Health Organization (WHO) Working Group on the Clinical Characterisation and Management of COVID-19 infection (2020) A minimal common outcome measure set for COVID-19 clinical research. Lancet Infect Dis 20:e192–e197

Xie Y, Xu E, Bowe B, Al-Aly Z (2022) Long-term cardiovascular outcomes of COVID-19. Nat Med 28:583–590

Yang J, Antin P, Berx G, Blanpain C, Brabletz T, Bronner M et al (2020) Guidelines and definitions for research on epithelial-mesenchymal transition. Nat Rev Mol Cell Biol 21:341–352

Yin K, Peluso MJ, Luo X, Thomas R, Shin MG, Neidleman J et al (2024) Long COVID manifests with T cell dysregulation, inflammation and an uncoordinated adaptive immune response to SARS-CoV-2. Nat Immunol 25:218–225

Yousefi Y, Baines KJ, Maleki Vareki S (2024) Microbiome bacterial influencers of host immunity and response to immunotherapy. Cell Rep Med 5:101487–101501

Yu J, Gerber GF, Chen H, Yuan X, Chaturvedi S, Braunstein EM et al (2022) Complement dysregulation is associated with severe COVID-19 illness. Haematologica 107:1095–1105

Zhang J, Zhao C, Zhao W (2021) Virus caused imbalance of type I IFN responses and inflammation in COVID-19. Front Immunol 12:633769

Zinellu A, Paliogiannis P, Carru C, Mangoni AA (2021) Serum amyloid A concentrations, COVID-19 severity and mortality: an updated systematic review and meta-analysis. Int J Infect Dis 105:668–674

# Chapter 5
# Identification of Virus-Host Interactions

**Abstract** No protein acts on its own to carry out its designated function. Proteins interact with other biomolecules to form complexes that perform a coordinated function. While proteins most often form complexes involving other proteins, they also interact with nucleic acids, lipids, carbohydrates, ions, and other small molecules and metabolites (Wang et al., Proteomics 22(15–16):e2100190, 2022; Veenstra, Proteomics 21:e2000235, 2021). While the early days of proteomics were focused on identifying and quantifying as many proteins as possible, scientists quickly recognized the power of proteomics to rapidly identify interactions between specific groups of proteins. All the necessary techniques have already been developed: methods to extract protein complexes (e.g., immunoaffinity purification under denaturing conditions), fractionate the complexes (e.g., chromatography and/or gel electrophoresis), identify the proteins that made up the complex (e.g., mass spectrometry (MS)), and validate the results (e.g., Western blotting, etc.). Indeed, one of the great accomplishments of the current proteomic era has been the speed at which protein complexes have been characterized (Labeau et al., Cell Rep 39:11074, 2022; May et al., Viruses 14:611, 2022; Havugimana et al., Nat Commun 13:4043, 2022), for example, as of 2022, the CORUM database (CORUM 3.1) contains data on 5134 mammalian protein complexes (Giurgiu et al., Nucleic Acids Res 47:D559–D563, 2019). The latest version of the Human Protein Complex Map (hu.MAP 2.0) contains information on over 7000 protein complexes (Drew et al., Mol Syst Biol 17:e10016, 2021). The IntAct Molecular Interaction Database contains over 700,000 binary molecular interactions curated from approximately 74,000 published experiments (Orchard et al., Nucleic Acids Res 42(Database issue):D358–D363, 2014). These databases show the amount of effort that continues to be expended in identifying biomolecular interactions, which demonstrates their importance in understanding how they affect cell function.

**Keywords** Virus-host interactions • Identification of protein interactions • Host cell receptors • Sample preparation • Yeast two-hybrid assay • Pull-down assay • Immunoprecipitation • Protein-protein interaction databases



T. D. Veenstra, *Proteomic Applications in Virology*, Advances in Experimental Medicine and Biology 1511, https://doi.org/10.1007/978-3-032-22340-1_5

## 5.1 Introduction

As described in Chap. 2, viruses have extremely small genomes and therefore contain many fewer proteins compared to living cells. In a simplistic view, the genome of a virus is sufficient for producing proteins that protect its genome and provide a mechanism for the genome to enter a host cell and the means for transforming its genetic material so that it can be replicated and translated into the proteins necessary for making multiple copies of the original virus. What the viral genome lacks, however, are the components necessary to produce the machinery (e.g., ribosomes, etc.) for making all the proteins required for complete viral replication. Therefore, viruses rely heavily on host cells to produce proteins necessary to complete its reformation cycle and continue propagating. Probably the most well-known virus-host cell interactions are those that occur between virus and host cell surface molecules that allow for viral entry and those between the host cell ribosomal proteins and viral genomic material to produce the necessary proteins to reform the virus (Huang et al. 2022; Maginnis 2018; Jaafar and Kieft 2019; Foxman and Iwasaki 2011). Understanding virus-host cell protein-protein interactions (PPIs) is important for understanding the viral life cycle and designing antiviral medications that can be used to impair viral entry, propagation, or regress (Chang et al. 2021; Gordon et al. 2020).

## 5.2 Virus-Host Cell Interactions

Interactions between virions and host cells are highly dependent on protein complexes. While interactions that occur between viruses and host cell surface proteins are probably the most well studied, viruses have found many ways to interact with host cell proteins to give themselves a survival advantage. For example, many retroviruses have developed tools to manipulate the eukaryotic initiation factor 4F (eIF4F) complex to ensure their propagation (Walsh and Mohr 2011). In some cases, the retroviral proteases can denature the eIF4F, thereby suppressing host cells from expression genes related to host immune responses. Translation of the retroviral mRNA is not inhibited, however, as they contain internal ribosome entry sites that are not dependent on the same type of eIF4F recognition. The human immunodeficiency virus-1 (HIV-1) proteins Vpr (viral protein R) and Vif (viral infectivity factor) have been shown to interact with Cdc2/Cdk1 in T cells, causing them to remain within the G2 phase of the cell cycle for an extended period (Wang et al. 2011). Remaining in the G2 phase provides the virus with the machinery needed for replication for a greater length of time. While the functions of many virus-host cell interactions are not exactly known, identifying the members of a virus-host cell complex can help infer their overall function. Therefore, using proteomic techniques to identify virus-host interactions provides a high-throughput tool for increasing the systematic understanding of viral infection and more targets for intelligent antiviral drug design.

### 5.2.1 Viral Entry

The viral life cycle can be roughly broken down into three stages: entry, genome and protein replication, and exit (Jones et al. 2021). The entry stage can be further broken down into three phases: attachment, penetration, and uncoating (Dimitrov 2004; Sobhy 2017; Kumar et al. 2018; Yamauchi and Greber 2016). Attachment involves the virion attaching to the host cell via a receptor on its surface. Penetration involves components of the virion reaching the cytoplasm. Uncoating, which occurs with both enveloped and non-enveloped virions, requires the shedding of the viral capsid giving the viral components access to host cell proteins. The attachment between the virus with a receptor on the host cell often determines the viral tropism (i.e., the virus's ability to infect specific cells or tissues) (Nomaguchi et al. 2012).

It was approximately 100 years ago that the specific attachment of a virus to a cell was first observed by the French-Canadian microbiologist Félix d'Hérelle, the co-discoverer of bacteriophages (d'Hérelle 1926). He cultured *Shigella* on solid agar medium in the presence of an infected bacteria. Within the background of *Shigella*, clear spots—corresponding to lysed bacteria—were observed. Although d'Hérelle did not know the nature of the agent that was lysing the bacteria, he named it bacteriophage ("bacteria eater"). Using co-sedimentation experiments, he demonstrated that attachment of the virus to the host cell is required to cause *Shigella* lysis. He also showed that lysis only occurred in certain bacteria, demonstrating that the range of hosts that a virus could infect was determined by the initial attachment step.

The first virus receptor found on the surface of a host cell was not a protein but rather a glycan. In 1981, it was discovered that fowl plague virus (an avian influenza A virus) binds to sialic acid-containing molecules on Madin-Darby canine kidney (MDCK) cells (Matlin et al. 1981). This result should not be surprising considering the variety of glycans found on the surface of cells attached to proteins (i.e., glycoproteins) or lipids (i.e., glycolipids). More than half of viruses use glycans as attachment proteins to initiate entry into mammalian cells. While adding proteases had no effect on this virus/MDCK cell interaction, the sialic acid-cleaving enzyme neuraminidase prevented the virus from binding to the cell surface.

Owing to the technology available at the time, most early studies identified a single host cell receptor as the site of interaction with the virus, for example, cluster of differentiation 4 (CD4, referred to as T4 in early manuscripts) was identified as the primary receptor of the HIV on CD4+ T-helper cells almost 40 years ago (Maddon et al. 1986). This interaction was identified by immunoprecipitating the viral envelope protein gp120 and the CD4 receptor from infected cells. Both proteins were found to immunoprecipitate only from HIV-infected CD4+ T-helper cells expressing CD4. CD4 receptors were detected using autoradiographic imaging since the transfected cells were surface labeled using $^{125}I$. What became evident as further studies were conducted is that viral entry is not a simple interaction between a single viral protein and a single host cell receptor. For example, while CD4 is the primary host cell receptor that recognizes gp120 on HIV, there are several co-receptors that facilitate viral entry. Both CXCR4 and CCR5 also bind to gp120 to facilitate gp41 exposure and subsequent fusion between the virus and $CD4^+$ T-helper cell (Dimitrov 1997). Other CD4+ T-helper cell receptors, such as CD26 and the integrin $\alpha 4\beta 7$, are known to catalyze interactions that include gp120 and gp41 with CD4, CXCR5, and CCR5 (Arthos et al. 2008). Continuing studies have shown that viral entry is a more complex process than previously thought and involves many different virus and host cell proteins. In this chapter, I discuss how technological advances in proteomics have enabled the discovery of virus-host cell interactions and increased our knowledge of these complex processes.

## 5.3 Proteomic Methods for Identifying Virus-Host Interactions

Historically, interactions between viruses and host cells were identified using techniques like those used to identify any type of PPI. These techniques were based on virus overlay protein blot assays (VOPBA), which share many similarities to Western blotting. In VOPBA, a complex protein mixture (e.g., cell lysate) is fractionated using sodium dodecyl sulfate polyacrylamide gel electrophoresis (SDS--PAGE). The separated proteins are transferred to a nitrocellulose membrane which is then incubated with virus particles (Karger and Mettenleiter 1996). In theory, the virus particles bind to specific proteins that are attached to the membrane. The position of the virus particles on the membrane are detected either using a virus-specific antibody or by radiolabeling of the virus. While these methods reveal the *position* of the virus and the binding protein, the *identity* of the binding protein is still unknown. Identification requires a hypothesis-driven approach in which an antibody that is believed to bind to the interacting host protein is used to probe the nitrocellulose membrane. The obvious challenge is selecting the correct antibody. Incorrect antibody selection can end up costing a signification amount of time and money. A study used VOPBA to identify chicken gizzard membrane proteins that bind to fowl adenovirus (FAV) serotype 1 strain JM1/1. The membrane proteins were separated using SDS-PAGE and transferred onto a PVDF membrane. Virus-binding proteins were detected using a FAV JM1/1. The binding of the virus was detected using an HRP-labeled secondary antibody. The study was able to identify a 200-kDa protein that is specifically bound to the virus (Taharaguchi et al. 2007). Beyond being a protein with a molecular weight of approximately 200 kDa, its identity could not be determined.

An obvious advantage of VOPBA is multiple cell lines can be tested simultaneously against a single virus, for example, four different cell lines were tested for their ability to bind dengue virus serotype 3 (DENV-3) (Wilm et al. 1996). Membrane proteins were extracted from the SH-SY5Y, U-87 MG, CHME-3, and C6 neural cell lines and separated using SDS-PAGE (see Fig. 5.1, panel I). The C6 cell line served as a negative control since DENV-3 is not known to infect this cell type. The resolved proteins were transferred to a nitrocellulose membrane, which was then

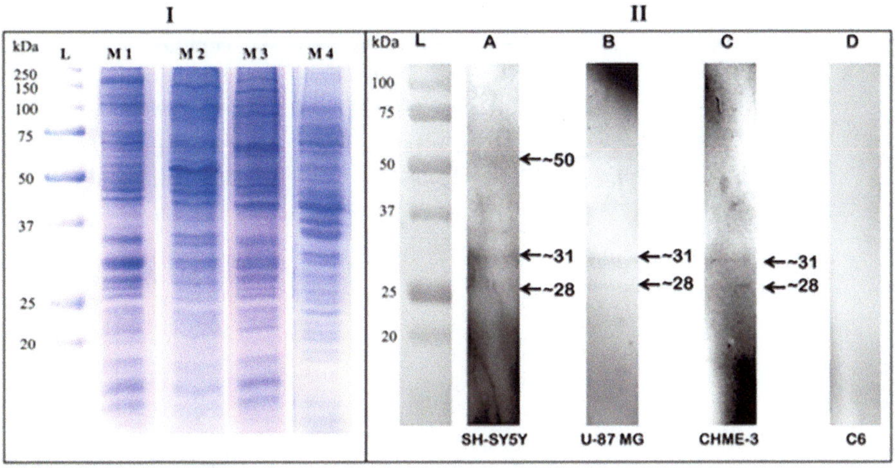


**Fig. 5.1** Virus overlay protein-binding assay (VOPBA) to identify proteins that mediate entry of Dengue-3 virus (DENV-3) into host cells. Membrane proteins from SH-SY5Y, U-87 MG, CHME-3, and C6 neural cells were resolved by SDS-PAGE (panel **I**) and transferred to nitrocellulose membrane. After incubating the membrane with DENV-3, interacting proteins were identified using a DENV-3-specific monoclonal antibody (panel **II**)

incubated with DENV-3. After washing the membrane, it was incubated with an anti-flavivirus (the family to which DENV-3 belongs) monoclonal antibody. The positions of the antibody were then visualized using chemiluminescence. Three bands at ~50, ~31, and ~28 kDa that interacted with DENV-3 were observed within the SH-SY5Y cell membrane lysate (see Fig. 5.1, panel II). Both bands at ~31 kDa and ~28 kDa were also detected in both U-87 MG and CHME-3 membrane lysates. No interaction between DENV-3 and the C6 membrane lysate was observed.

### *5.3.1 Viral/Host Cell Receptor Interaction Identification Using Mass Spectrometry*

One disadvantage of VOPBA is that it provides confirmation of an interaction between a virus and a host cell protein but only provides the molecular weight of that protein. The lack of resolution afforded using SDS-PAGE combined with the molecular weight degeneracy of the human proteome makes it impossible to identify the host protein based on the VOPBA data. Fortunately, in the early 1990s, methods were developed to enable proteins to be extracted from SDS-PAGE gels (Sharma et al. 2020) and polyvinylidene difluoride membranes (PVDF) (Mørtz et al. 1994) in forms that made them identifiable using MS (see Fig. 5.2). This capability, particularly from SDS-PAGE gels, has become so commonplace it is easy to overlook how monumental its development was. In the above study, the protein bands were excised from the gel and washed extensively to remove SDS and

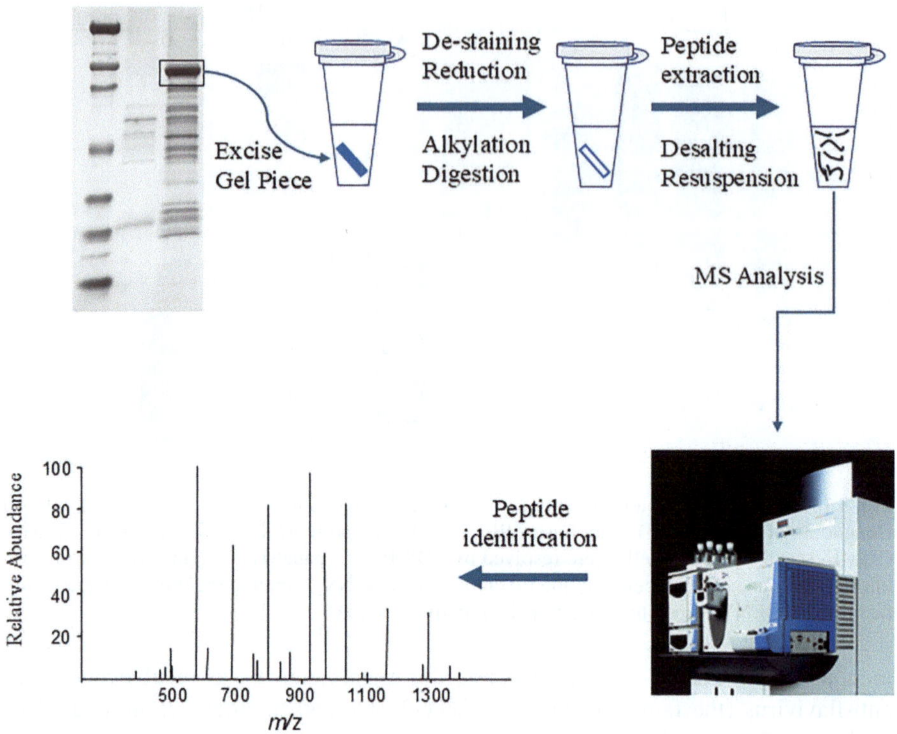


**Fig. 5.2** Illustration of method for identifying SDS-PAGE fractionated proteins using SDS-PAGE using LC/MS

Coomassie stain. The gel pieces are dehydrated using acetonitrile, and a reducing solution (e.g., dithiothreitol) is added to reduce protein disulfide bonds, which are subsequently alkylated by the addition of an iodoacetamide solution. Following another dehydration reaction using acetonitrile, a solution containing a proteolytic enzyme (e.g., trypsin, Lys-C) is added. Incubation of the gel piece in this enzymatic solution cuts the proteins into peptide, which are extracted from the gel using an acetonitrile solution and desalted. After vacuum-drying this solution, the peptides are resuspended in an appropriate buffer for MS analysis.

This method was used to identify the 31- and 28-kDa bands observed on the SDS-PAGE gel that corresponded to the positions seen on the VOPBA gel using extracts from all three cell lines. The low resolution of SDS-PAGE fractionation results in many proteins being extracted from a single gel piece and identified using LC/MS. The key to narrowing down the protein that is interacting with the virus is acquiring enough circumstantial evidence that will lead to the correct protein. In this study of three cell lines, prohibitin 2 (PHB2) isoform 1 and prohibitin 1 (PHB1) isoform 1 were identified in the bands at 31 and 28 kDa, respectively. While several other proteins were identified in these bands, none of them were observed with the consistency and coverage observed from PHB2 and PHB1. Further studies to decipher the role of these proteins in DENV-3 infection showed that while they were not necessary for viral attachment PHB2 and PHB1 were required for viral entry.

A study with a similar goal was done not long after the DENV-3 study to identify proteins that assisted the entrance of chikungunya virus (CHIKV) into host cells. This study extracted two identical membrane fractions from HEK-293 and Vero-E6 cells and fractionated them using SDS-PAGE (Paingankar and Arankalle 2014). One of the SDS-PAGE gels was transferred to a PVDF membrane, which was probed with CHIKV. Regions where CHIKV bound to the membrane were identified by incubating the blot using a peroxidase-labeled rabbit anti-mouse IgG. These bands were visualized using $H_2O_2$ and diaminobenzidine tetrachloride. The location of these bands was lined up with the SDS-PAGE gel that had not been transferred to the PVDF membrane. The corresponding regions were cut from the SDS-PAGE gel and the proteins extracted and identified using MS. The analysis identified HSP70 and actin as virus-binding proteins in both cell lines, with STAT-2 also identified as binding to Vero-E6 cells (Paingankar and Arankalle 2014). Functional validation studies showed that incubating the cells with anti-HSP70 antibody reduced the CHIKV titer up to 12 hours postinfection.

While these studies demonstrate the power of MS for identifying interacting proteins, these results only began to scratch the surface of the potential of this technique. As discussed in Chap. 1, a key attribute of MS is its ability to rapidly identify proteins using a discovery-driven approach. While not much different from the previously discussed study, one in which *Bombyx mori* (i.e., silk moth) nucleopolyhedrovirus (BmNPV)-binding proteins were identified illustrates the throughput of using MS for protein identification. In this study, lipid rafts were extracted from *Bombyx mori* and divided into two equal aliquots. Both aliquots were fractionated using SDS-PAGE. The proteins from one SDS-PAGE gel were transferred to a PVDF membrane that was incubated with BmNPV. The locations that the virus bound were visualized using a mouse polyclonal anti-BmNPV antibody followed by a peroxidase-labeled anti-mouse IgG (Hu et al. 2017). Areas corresponding to BmNPV-binding sites were excised from the replicate SDS-PAGE gel and the proteins in these slices extracted and identified using a high-performance hybrid quadrupole-Orbitrap mass spectrometer. A total of 65 proteins were identified. This stage is where the proteomic data requires careful assessment and validation. Since SDS-PAGE is a low-resolution separation technique, there are many proteins within a single region of the gel. With MS being a non-biased method, it will simply identify as many proteins as it can detect without regard for their function. The presumed functions of the 65 identified proteins were deciphered using gene ontology (GO) (Thomas et al. 2022) and Kyoto Encyclopedia of Genes and Genomes (KEGG) (Kanehisa et al. 2023) pathway analysis and by comparing their sequences to other proteins with known functions (i.e., homology comparison). Based on these results, seven of the proteins were considered putative BmNPV-binding proteins that mediate virus entry into the host cells. These proteins included actin, kinesin light chain--like isoform X2, annexin B13, heat-shock protein 90, barrier-to-autointegration factor (BAF) B-like, and serine/arginine-rich splicing factor 1 (SRSF1) A-like.

While not directly related to proteomics, this study treated the cells with MβCD (methyl-β-cyclodextrin), which depletes the cholesterol content of cell's membranes while leaving lipid rafts intact (Abe and Kobayashi 2021). Treatment with MβCD decreased the infectivity of BmNPV into *Bombyx mori* cells. Adding cholesterol back to the membranes restored *BmNPV* infectivity. In summary, this study identified seven proteins implicated in BmNPV entry into *Bombyx mori* cells and showed the important role that cholesterol plays in the process.

### 5.3.2 Sample Preparation Techniques

The race to identify as many proteins as possible in complex samples began in the 1990s. While there were several technological improvements that spurred this race, the main one was the development of mass spectrometers with greater sensitivity and resolution and shorter duty cycles. All three of these attributes not only increased the number of proteins that could be identified but also shortened the experimental time required. As the number of proteins identified started to enter the hundreds, the focus started to shift toward sample preparation techniques. Not to say MS technology development ceased; in fact, it continues to accelerate to a point where thousands of proteins can be identified in hours. Many proteomic scientists realized that instead of trying to identify everything a better strategy was to identify only what was relevant to their research. This focus required developing sample preparation techniques that were often customized to isolate specific groups of proteins.

While VOPBA meets this criterion by using virions to pinpoint specific host cell interacting proteins, it still leads to only a single protein being identified per MS analysis. To better integrate MS and sample preparation techniques required a movement away from gel-based to solution-based fractionation methods. Fortunately, the fundamental methods for making this advance were already in place. Among these methods were ways to simplify the sample so that it focused on PPIs involving viruses and host proteins. While many methods are available to focus on PPIs, the key to all of them was using proper bait.

Another key decision that needs to be made prior to selecting a method to identify virus-host interactions is whether to conduct the study ex situ or in situ. Ex situ (or ex vivo) studies measure interactions that are occurring either in a nonhost cell or in solution, while in situ (or in vivo) studies measure interactions that are occurring inside the host cell. While in situ represents the ideal scenario, it has limited throughput as only a few bait proteins can be studied per experiment. While ex vivo allows many bait proteins to be screened simultaneously, it is more prone to false positives (i.e., artifacts) since proteins can be forced together or misfolded under the experimental conditions. Owing to the potential for artifacts, ex vivo results require a greater degree of validation studies.

### 5.3.3 Yeast Two-Hybrid Identification of Protein-Protein Interactions

The yeast two-hybrid (Y2H) assay is the most common type of ex vivo experiment for identifying protein interactions (Fields and Song 1989). The basics of the Y2H assay are described in Chap. 1. While not generally considered high-throughput, Y2H assays can screen libraries containing thousands of proteins against a single bait protein. An added feature of the Y2H assay is its use in drug discovery. Once a PPI is discovered, the effect of various drug treatments to disrupt this interaction can be tested (Hamdi and Colas 2012). While cloning proteins in the proper transcription and activation domain plasmids is straightforward, adding the required extra sequences to the bait and target proteins that enable Y2H to visualize an interaction can affect how they naturally interact. Also, obtaining a positive result in a Y2H screen requires the two proteins to interact within the nucleus of the yeast cell (Galletta and Rusan 2015). This region may not represent the native environment of the proteins being tested, leading to a false-positive or false-negative result. In addition, Y2H assays are more suited for analyzing interactions between soluble proteins, limiting their effectiveness in identifying interactions involving membrane proteins, which often initiate the infection of a host cell by a virus. To overcome this limitation, the split ubiquitin membrane Y2H (MYTH) assay was developed (Dutta and Smith 2023). The MYTH assay tests proteins bound to either the amino- (Nub) or carboxy-terminal (Cub) region of ubiquitin (see Fig. 5.3). If the two test proteins bind, the Nub and Cub moieties join to form a functional ubiquitin protein that activates a protease that cleaves a reporter protein. The MYTH assay can test membrane proteins as the protein-specific regions can localize within their native membrane regions, while the Nub and Cub regions are able to interact within the cytoplasm of the cell.

While greater details will be presented in other chapters using specific examples, the Y2H assay has been used in several studies to detail virus-host cell PPIs. A Y2H assay was used to screen the nonstructural protein 1 (NS1) of human RSV (respiratory syncytial virus) against a library of human genes (Dong et al. 2022). Human RSV, which infects the lower respiratory tract, is a major cause of hospitalization and mortality in children under 5 and adults over 65 years of age (Carvajal et al. 2019). Severe RSV infections have been linked to asthma and bronchiolitis onset in young children. Understanding how host cells respond to RSV infection would increase the likelihood of developing therapeutics that could be used as prophylactics or treatments to decrease the public health burden caused by this virus.

RSV-infected cells express low levels of type I interferons (IFNs) (Loveys et al. 2000), which are cytokines that serve to mobilize the host immune system to attack infected cells. The RSV genome contains two open reading frames (ORFs) that encode for NS1 and NS2 and two nonstructural proteins that have been implicated as IFN antagonists (Atreya et al. 1998). Consistent with its role as an IFN antagonist, cells infected with mutant forms of RSV that do not express NS1 and NS2 express higher levels of IFN-β mRNA levels compared to cells infected with

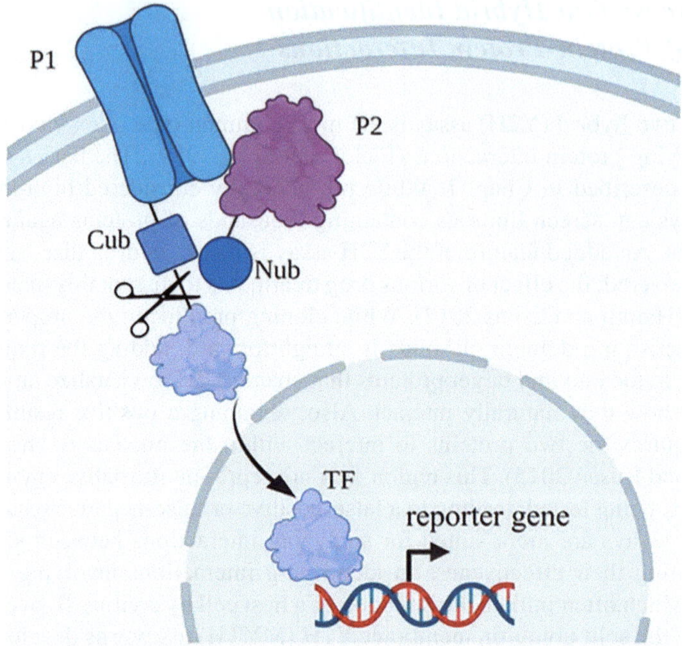


**Fig. 5.3** Split ubiquitin membrane yeast two-hybrid (MYTH) assay for identifying interactions between membrane proteins. In this system, two proteins (P1 and P2) are labeled with either the carboxy- (Cub) or amino-terminal regions of the protein ubiquitin. If an interaction between P1 and P2 occurs, a function ubiquitin protein is formed that activates a protease that releases a transcription factor (TF). The TF then transcribes a reporter gene indicating a positive interaction between P1 and P2

wild-type virus (Dong et al. 2022). Of these two proteins, NS1 is the more potent IFN suppressor, while NS2 serves to enhance the anti-IFN effect of NS1. NS1 inhibits RIG-1 (retinoic acid-inducible gene) activity and forms a viral degradosome within the host cell's mitochondria where it degrades several proteins involved in the type I IFN pathway. NS1 is also suspected of modulating the expression of various innate immune response genes through its association with gene promoter and enhancer regions. Unfortunately, the exact mechanism of how NS1 modulates host cell transcription is not fully understood.

A Y2H screen was performed by fusing the RSV NS1 protein with a GAL4 DNA-binding domain (GAL4-BD) and two different libraries of human prey proteins to GAL4 activation domains (GAL4-AD) (Dong et al. 2022). Co-expression of the plasmids in yeast cells resulted in a total of 156 positive colonies. The strategy followed by the investigators (as described below) shows how discovery-driven data identified in proteomic studies is converted into hypothesis-driven studies that can be tested. Of the 156 positive results, only those that were identified in at least 3 separate colonies were considered for further validation (see Table 5.1). This reduced the total number of potential PPIs between NS1 and host proteins to 15.

**Table 5.1** Proteins found to associate with respiratory syncytial virus (RSV) nonstructural protein 1 (NS1) via a yeast two-hybrid (Y2H) assay

| Gene name | Total no. of positive yeast colonies |
|---|---|
| *Phosphodiesterase 9A (PDE9A)* | 37 |
| *Kinesin light chain 4 (KLC4)* | 21 |
| *Golgin subfamily A member 2 (GOLGA2)* | 16 |
| *Mediator ofRNA polymerase II transcription subunit 25 (MED25)* | 10 |
| *Mediator ofRNA polymerase II transcription subunit 15 (MED15)* | 8 |
| *Signal-induced proliferation-associated 1-like protein 1 (SIPA1L1)* | 8 |
| *High mobility group 20A (HMG20A)* | 7 |
| *Inhibitor of nuclear factor kappa B kinase regulatory subunit gamma (IKBKG)* | 6 |
| *Centrosomal protein 290 (CEP290)* | 5 |
| *Thioredoxin domain containing 11 (TXNDC11)* | 5 |
| *Rabenosyn (RBSN)* | 5 |
| *Rabenosyn-5 (ZFYVE20)* | 4 |
| *Coiled-coil α-helical rod protein 1 (CCHCR1)* | 3 |
| *Nucleoside diphosphate kinase 3 (NME3)* | 3 |
| *Retinoic acid-induced 14 (RAI14)* | 3 |

Within these 15 potential PPIs were 2 subunits of the human Mediator complex: MED15 and MED25. Mediator proteins are part of a multi-subunit nuclear complex found that plays a role in RNA polymerase II transcription and regulates the expression of several immune response genes (Poss et al. 2013; Wang et al. 2016). In total, 8 and 10 of the positive yeast colonies expressed MED15 and MED25, respectively. Of the ten clones that expressed MED25, only six expressed full-length protein while the other four expressed truncated forms. All the expressed clones, however, contained the activator-interacting domain (ACID) of MED25. The ACID domain binds both viral and human transcription factors to regulate gene expression (Zhang et al. 2017). The number of positive yeast colonies containing the ACID domain of MED25 suggests this region of the protein interacts with the RSV NS1 protein. The investigators also focused on MED25 since it had been identified using a BioID proximity screen as interacting with NS1 (Van Royen et al. 2022).

As previously described, discovery-driven results lead to hypothesis-driven studies. To validate an interaction between NS1 and MED25, a split luciferase complementation assay was performed using various segments of both proteins to localize the specific regions of the proteins that interact (Dong et al. 2022). Three different MED25 constructs were co-expressed with full-length NS1: full-length MED25, truncated MED25 containing the ACID domain (MED25-ACID), and truncated MED25 containing the von Willebrand Factor domain (MED25-VWF). The RSV NS1 and MED25 proteins were cloned into the NanoLuc luciferase plasmid 114 and 11S vectors, respectively, and transfected into 293 T cells. After 24 hours, the cells were lysed, and the luciferase substrate furimazine was added. If the expressed NS1 and MED25 proteins (or fragments) are associated, the complete luciferase protein forms. The luciferase protein then decarboxylates furimazine to furimamide resulting in luminescence.

The luminescence observed when NS1 and full-length MED25 constructs were co-expressed in cells confirmed their interaction. A positive interaction was also observed between NS1 and MED25-ACID, while no interaction between NS1 and MED25-VWF was observed. These results showed not only that NS1 and MED25 interact but also that the interaction involved the ACID domain of MED25. The study also expressed NS1 in both NanoLuc luciferase plasmids and showed that this protein was able to homodimerize.

Once the key interaction domain within MED25 had been determined, it seemed logical to determine the key domain within NS1. Based on previous studies, the investigators created a series of NanoLuc luciferase 114 constructs containing various substitution mutations within the α3 helix region of NS1 along with a construct in which the entire α3 helix was deleted. These constructs were co-transfected into 293 T cells with the MED25-ACID plasmid. Some of the mutant NS1 and the α3-deletion mutant constructs failed to interact with MED25-ACID. These results not only confirmed the Y2H assay findings but localized the sites of interaction between NS1 and MED25 to the α3 helix and ACID domains of the respective proteins.

The interaction between NS1 and MED25 was also confirmed via a pull-down assay using various glutathione-S-transferase (GST)-NS1 and MED25-ACID recombinant proteins expressed in *E. coli*. The complex was extracted from the cell lysate using glutathione beads, separated using SDS-PAGE and stained using Coomassie blue. Two very prominent bands corresponding to NS1 and MED25--ACID were observed. To strengthen their conclusion that NS1 and MED25 interact, the investigators conducted a nuclear magnetic resonance (NMR) spectroscopy interaction experiment by titrating the NS1 α3 peptide into a solution of $^{15}$N-labeled MED25-ACID. As increasing amounts of NS1 α3 peptide were added, changes in the position of specific MED25-ACID amide signals were observed.

### *5.3.4 Pull-Down Assays*

As described in Chap. 1, pull-down assays in the form of immunoprecipitation (IP), co-IP, or affinity purification (AP) combined with mass spectrometry (MS) are invaluable tools for discovering potential interactions between virus and host proteins (Veenstra 2021b). Pull-down assays rely on isolating a specific bait protein using a molecule that has a high affinity for the bait protein and captures other proteins associated with the bait protein. It is analogous to fishing with a rod and reel. The hook (high-affinity molecule) has a high affinity for the worm (bait protein), which captures a fish (associated proteins). Continuing the theme of interaction between the RSV NS1 protein and host cell proteins, a group recently performed an AP study in which NS1 tagged with both streptavidin and hemagglutinin was expressed in HEK 293 T cells (Pei et al. 2021). After AP, proteins within the extracted complex were digested into peptides, which were identified using LC/MS. A total of 21 proteins were identified, with 13 of them being subunits of the Mediator complex, including MED25. Other proteins that bound RSV NS1 included proliferation-associated protein 2G4 (an RNA-binding protein involved in growth regulation), ribosomal protein L9 and L15 (which are part of the ribosomal 60S complex), ribosomal protein S3 (which is part of the 40S complex), and the human leukocyte antigen (HLA) subunits A, B, and C.

Done properly, the results of proteomic experiments lead to novel hypotheses that generate new experiments to further understand the data generated in the original proteomic experiment. Of particular interest in the previous study was the interaction between NS1 and MED25. To confirm this interaction, MED25 was tagged with FLAG peptide and expressed in A549 cells. The cells were subcellularly fractionated into cytosolic, nuclear, and chromatin fractions. Protein complexes were extracted from each fraction using an anti-FLAG antibody targeting the FLAG--tagged MED25 protein and identified using LC/MS. NS1 was found in all three, while MED25 was found primarily in the nuclear fraction. Performing the reverse AP using FLAG-tagged NS1 revealed MED25 (along with other members of the MED family) bound within this complex. Analysis of a chromatin fraction extracted from these cells revealed the presence of both these proteins and further substantiated and localized the NS1/MED25 PPI.

Now that there was substantial validated data showing that NS1 and MED25 interact within the nucleus, the next step was to demonstrate the extent of this interaction and its possible effect within infected cells. A chromatin immunoprecipitation (ChIP)-seq study was performed using NS1 and MED25 in RSV-infected A549 cells that were transfected with hemagglutinin-tagged NS1 plasmid (51). A ChIP--seq study identifies the interaction sites of DNA-binding proteins by cross-linking the DNA-protein complexes (Nal et al. 2001). The DNA-protein complexes are immunoprecipitated and fragmented, and excess unbound oligonucleotides are removed using an exonuclease. The sites at which the protein is bound are then determined via DNA sequencing. Comparison of the NS1 and MED25 ChIP-seq results showed significant overlap (Pei et al. 2021). A genome annotation analysis was performed to determine if the identified DNA sequences coincided with any functional genomic elements. Most of the NS1- and MED25-binding sequences corresponded to promoters and 5′ untranslated regions (UTRs). Most of the regions where NS1 bound are located within 10 kb of genes (64%), coincided with promoters (12%), and overlapped with enhancers (73%). Infection of the cells with RSV affected approximately 40% of the genes that contained NS1-bound promoters. Further analysis showed that almost all the enhancer regions where NS1 bound were found within lung tissue. Since enhancers regulate the temporal and quantitative transcription of genes in a highly tissue-specific manner, finding that many of the NS1-specific enhancers are found in lung tissue suggests that NS1 may control

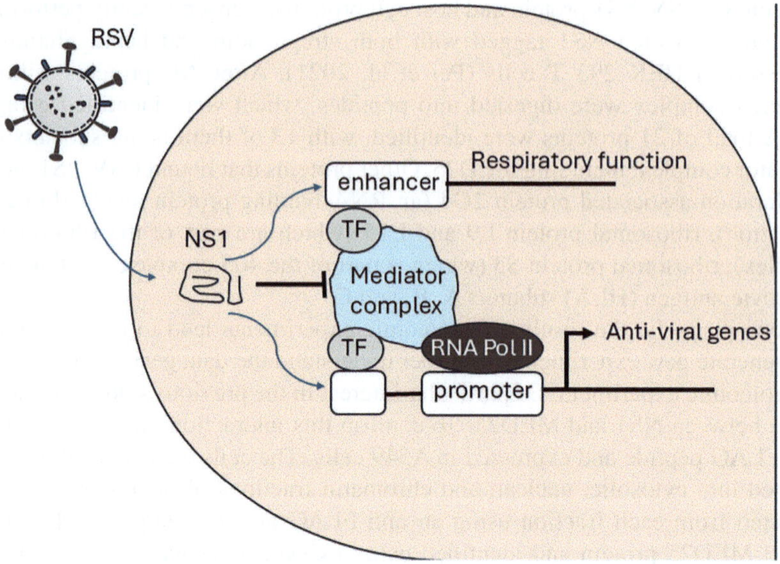


**Fig. 5.4** Model of how respiratory syncytial virus (RSV) nonstructural protein 1 (NS1) helps the virus evade the host immune system. When it enters the host cell, RSV releases NS1, which travels to the nucleus and binds to regulatory elements near genes (both enhancers and promoters) that are differentially expressed during RSV infection. NS1 binding to these sites inhibits the expression of many antiviral genes. These effects may be mediated by NS1 binding to regions that also bind Mediator, a key transcriptional regulatory complex that forms physical bridges between enhancers and promoters that are essential for transcriptional regulation

the expression of genes required for respiratory function, including their response to RSV infection.

The conclusions of this study corroborated those found using a Y2H assay; upon RSV infection, NS1 impacts host gene expression via interaction with MED25 and other Mediator proteins (see Fig. 5.4). The studies went to great lengths to characterize the interaction, even describing the domains within the proteins that interacted and measuring the affinity between the two proteins. Honestly, proteomic technology (i.e., Y2H and AP-MS) generated a very small amount of the results (probably less than 5%). This fact may make the reader wonder why is proteomics such a big deal. The answer lies in the discussion in Chap. 1 describing the differences between hypothesis- and discovery-driven science. While the functional details on how NS1 interferes with an infected RSV cell's immune response could only be determined using basic research, the fact that NS1 and MED25 interact was most efficiently discovered using proteomics. Using a trial-and-error approach to identify this interaction carries a high risk of failure and a potentially large waste of time and money. The proteomic approach not only identified the interaction between NS1 and MED25 but also generated additional data to develop further hypotheses that could be explored to further characterize the interaction between RSV and host cells using basic research methods.

## 5.4 Virus-Host Protein-Protein Interaction Databases

There has been a lot of effort expended over the past several decades detailing how viruses infect and propagate within host cells. Identifying PPIs has been a major part of this effort. In the early days, the rate of discovery of novel virus-host cell PPIs was such that researchers could stay abreast of new findings by reading the latest published literature. The development of proteomic technologies, however, has greatly increased the rate at which PPIs are discovered and make it difficult to stay current with all the results being generated. This increased rate of discovery (not just in proteomics but in every other "omics" field) has spawned the creation of virus-host cell protein interaction databases. Beyond simply reporting identified PPIs, some of these databases include computational tools to predict virus-host cell PPIs. In the following paragraphs, a description of some of the more comprehensive databases available for scientists interested in exploration of virus-host cell PPIs is provided.

The most popular generic virus-host cell PPI databases include EBI-GOA--nonIntAct, IntAct, BioGRID, VirHostNet, VirusMentha, Viruses.STRING, and HPIDB (Huntley et al. 2015; Del Toro et al. 2022; Oughtred et al. 2021; Guirimand et al. 2015; Calderone et al. 2015; Cook et al. 2018; Kumar and Nanduri 2010). These databases list PPIs for several different viruses and host cells. As of 2022, these seven databases listed a total of 343,866 interactions between virus and host cell proteins. Three other popular databases, VirusMINT, PHISTO, and HVIDB, focus specifically on interactions between viruses and human host proteins, while HCVpro is a resource for PPIs specifically between hepatitis C virus and human proteins (Chatr-aryamontri et al. 2009; Durmuş Tekir et al. 2013; Yang et al. 2021; Kwofie et al. 2011). EBI-GOA-nonIntAct, IntAct, BioGRID, VirHostNet, Viruses.MINT, and HPIDB contain experimentally validated PPIs that have been curated from literature. Virus.STRING, VirusMentha, PHISTO, HCVpro, and HVIDB contain PPIs that have been integrated from other databases. Parameters such as numbers of viruses, hosts, viral proteins, host proteins, and PPIs found within each database are given in Table 5.2. The interconnectivity between these databases that integrate PPIs from other databases is shown in Fig. 5.5. What is remarkable is that the overlap in information between the databases is not substantial, for example, as of 2022, the range of overlapping PPIs found in Viruses.STRING and seven of the other databases (i.e., EBI-GOA-nonIntAct, BioGRID, IntAct, VirHostNet, HPIDB, PHISTO, and HVIDB) range between 0.3% and 16.8% (Valiente 2022). Only VirusMentha had a significant overlap with Viruses.STRING (i.e., ~42%). Of the almost 300,000 PPIs listed in the EBI-GOA-nonIntAct, IntAct, BioGRID, VirHostNet, VirusMentha, Viruses.STRING, and HPIDB databases, there is not a *single* interaction that is found in all 7.

The types of interactions found within the databases could be classified as involving either central or peripheral proteins. Central interactions are between proteins that are cohesively connected to each other, while peripheral interactions involve proteins that are loosely connected to core members through links involving other proteins. Overlap between databases was much more likely to occur between

**Table 5.2** Parameters of various virus-host protein-protein interaction (PPI) datasets

| Database | Viruses | Hosts | Viral proteins | Host proteins | PPIs |
|---|---|---|---|---|---|
| Viruses.STRING | 186 | 61 | 1703 | 52,440 | 242,784 |
| HPIDB | 205 | 36 | 1387 | 7570 | 33,906 |
| VirHostNet | 128 | 6 | 984 | 7361 | 28,132 |
| IntAct | 197 | 68 | 1062 | 8102 | 22,727 |
| VirusMentha | 114 | 8 | 627 | 3624 | 10,626 |
| BioGRID | 13 | 6 | 50 | 2101 | 5157 |
| EBI-GOA-nonIntAct | 77 | 26 | 173 | 455 | 534 |
| HVIDB | 146 | 1 | 1313 | 7060 | 40,132 |
| PHISTO | 182 | 1 | 1700 | 6520 | 39,010 |
| VirusMINT | 28 | 1 | 287 | 287 | 372 |
| HCVPro | 1 | 1 | 7 | 138 | 140 |

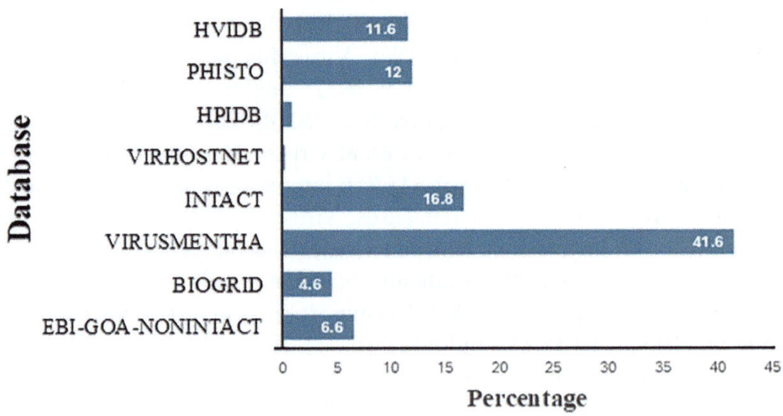


**Fig. 5.5** Percentage of protein-protein interactions within each of the individual virus/host protein databases listed that are also found within the Viruses.STRING database

peripheral proteins rather than central interactions. This overlap trend is not unexpected as discovery-driven studies make up a lot of data within databases. Discovery--driven studies are often polluted with proteins that bind to the central targeted protein with low affinity. These interactions are often nonspecific and associated with higher abundant cellular proteins (e.g., tubulin, actin, myosin, etc.). These results need to be carefully discerned or validated through more targeted studies.

## 5.4.1 *Analysis of PPIs Within Databases*

Databases may be the most underutilized resource in basic research today. While having comprehensive lists of PPIs involving viruses and host cell proteins is useful, unfortunately, they do not reveal the functional consequence of the interaction. A database called HVPPI (human-virus protein-protein interaction) was constructed to predict the function of the viral protein within the complex. Being able to predict the function of a viral protein enables the importance of the virus-host protein interaction to be discerned (Li et al. 2022a). The HVPPI database was populated with PPIs reported in over 200 published articles listed on PubMed. As of 2022, the HVPPI database included 27,293 virus-human PPIs involving 140 viral and 6059 human proteins. A total of eight virus families are included in the database. The greatest number of PPIs were collated for dengue (8410), hepatitis C (HCV, 5829), H1N1 (4094), and Zika (3588) viruses. There were 389 interactions included for SARS-CoV-2. The number of interacting human proteins was calculated for each viral protein. The results showed individual viral proteins associated with dengue, HCV, H1N1, and Zika viruses interacted with a higher number of individual human proteins. Zika, HPV, and Ebola virus proteins tended to interact with different numbers of human proteins, while dengue, HCV, and HIV proteins tended to interact with a similar number of human proteins. While most viral proteins interacted with several different human proteins, most human proteins interacted with only a single viral protein. Metadata including the experimental methods and the article's PubMed identifier were included. The HVPPI database can be queried by viral or human protein to obtain details of the virus-human protein interaction, predict the viral protein's function, as well as identify disease associations and candidate antiviral drugs.

The functions of the proteins that interacted with the virus proteins were identified using clusterProfiler and GO biological process categories. The functions most populated included "viral gene expression," "translational initiation," and "RNA catabolic processes." These results substantiate the well-established findings that viral proteins facilitate their replication in the host cell by co-regulating host cell RNA synthesis pathways. Viral proteins that affected cell immune response were identified by determining if they were enriched in human immune-related pathways.

The functions of the interacting human proteins were determined to understand the function of the viral proteins they interacted with. Analysis of the data showed a significant enrichment of human proteins involved in the regulation of the cell cycle and immune response, for example, nonstructural protein 1 (NSP1) of SARS-CoV-1 was found to interact with four proteins that were functionally annotated as involved in "DNA replication," "telomere maintenance," or "G1/S transition of mitotic cycle."

Significant numbers of human proteins that function within the immune response were also found to interact with viral proteins. For example, several human proteins involved in "regulation of type 1 interferon production" and "regulation of innate immune response" were found to interact with dengue/ZIKA and HVC/HPV viral proteins, respectively. Analysis of virus proteins showed that functional categories of "antigen processing and presentation," "antimicrobials," and "T-cell receptor signaling pathways" were commonly assigned to human proteins with which they formed an interaction. Proteins such as cluster of differentiation 8A (CD8A), CD4, CD74, and interferon gamma (IFNγ) were found to interact with viral proteins. Several HIV, dengue, and Zika viral proteins interacted with human proteins involved in antimicrobial pathways (e.g., C-C motif chemokine receptor 5, IL10, IL5, β2-microglobulin, peptidylprolyl isomerase, etc.). The obvious conclusion from examining the enriched functional categories of human proteins is that a general viral strategy is to control the cell cycle to stimulate its own reproduction while disrupting efforts by the host to eliminate it. Using information contained within databases of PPIs between viruses and human host cells can enable a more detailed view of the protein pathways targeted by the virus during infection.

### *5.4.2 Using Databases to Explain Symptoms of Viral Infections*

While their symptoms may be clear, there are several complications and comorbidities associated with viral infections that are still poorly understood. This fact is common for individuals suffering from the condition known as long COVID (Davis et al. 2023). These patients, who no longer test positive for SARS-CoV-2, continue to have lingering health problems long after the fever and cough associated with infection have passed. Examining PPIs in databases can start to provide clues to the pathogenesis behind various viral comorbidities. One of the tools integrated into the HVPPI database evaluates the association of the various PPIs with various diseases. Conditions found to be closely associated with long COVID included neurodegenerative, nervous system, and brain diseases (Song et al. 2021). Along with nervous system diseases, SARS-CoV-2 protein interactions were associated with metabolic and congenital diseases (Davis et al. 2023). Nervous system disease-related proteins that were found to interact with SARS-CoV-1 included peroxisomal biogenesis factor 14 (PEX14), PEX3, presenilin 1, and gelsolin. Aberrations in presenilin 1 levels and activity have been shown to be associated with Alzheimer's disease (AD). Presenilin 1 has been shown to alter the production of amyloid beta in neurons of patients with AD (Hampel et al. 2021). It is possible that an interaction with SARS--CoV-1 could alter the activity of the presenilin 1 resulting in a similar pathology.

Recently, a study of SARS-CoV-2-infected patients in China showed that 36% of these individuals suffered from neurological comorbidities (Mao et al. 2020). The HVPPI database revealed an interaction between the SARS-CoV-2 protein NSP16 and the human proteins tripartite motif-containing 32 (TRIM32) and ubiquitin protein ligase E3A (UBE3A) (Li et al. 2022a). While both proteins are associated with immune response, UBE3A is also associated with intellectual disabilities such as delayed development and severe speech impairment (Rotaru et al. 2020). This association suggests that interaction between UBE3A and NSP16 may play a role in central nervous system diseases observed in some patients. Collectively, there are results within the HVPPI database that could be explored as causing the symptoms associated with long COVID.

### 5.4.3 Using PPI Databases to Identify Potential New Drugs

One of the first steps in drug development is recognizing a druggable target. A druggable target must be a molecule that binds to a drug resulting in an alteration of its function resulting in a therapeutic benefit. Databases of PPIs between virus and host cell proteins are fantastic resources for identifying potential druggable targets that can disrupt the virus's ability to infect, propagate, and, in some cases, kill the host cell. While the search for novel drugs that disrupting virus-host cell interactions continues, using drugs that are intended to treat other human diseases is another potential source of finding therapeutics against viral diseases. The HVPPI database incorporated a network-based model that takes known druggable human proteins and determine if the protein interacts with a viral protein within the interactome database (Li et al. 2022a).

In the first stage of constructing this part of the database, FDA-approved drug--target interactions were downloaded from DrugBank. The result was 2170 drugs and 2694 targets. Candidate drugs were collated based on virus, for example, 45 small molecules and 1 biotech drug were identified for SARS-CoV-2. The number of viruses that an individual drug could potentially target was then determined. A small number of viruses were predicted to be treatable by most of the antiviral drugs. Encouragingly, 32 drugs that targeted 13 viruses were predicted to have broad-spectrum antiviral activity. Five of these drugs showed a significantly greater potential for having a broad-spectrum antiviral activity. Epirubicin, norfloxacin, and ribavirin were included within these five drugs consistent with empirical evidence of their broad-spectrum antiviral effects. While ribavirin is an antiviral known for its treatment of RSV and hepatitis C, neither epirubicin nor norfloxacin has been shown to possess antiviral properties. Epirubicin is an antineoplastic chemotherapy drug used in the treatment of breast cancers, while norfloxacin is a quinolone antibiotic used to kill or prevent the growth of bacteria. This comparison suggests that drug not conventionally believed to possess antiviral properties may be worth testing as virus therapies.

The PPIs involving SARS-CoV-1 and SARS-CoV-2 were studied in greater detail in hopes of predicting a viable drug candidate to combat these viruses. The top five candidate drugs identified against SARS-CoV-1 were ribavirin, mycophenolic acid, mycophenolate mofetil, omacetaxine mepesuccinate, and doxorubicin. Both ribavirin and mycophenolic acid targets inosine monophosphate dehydrogenase 2 (IMPDH2) (Naffouje et al. 2019). SARS-CoV-1 Nsp1 is known to bind to IMPDH2 and activate both NF-κB activity and cytokine induction, which promotes viral infection (Li et al. 2022b). Inhibiting IMPDH2 activity blocks coronavirus infection presumably by downregulating both these processes. Both ribavirin and mycophenolic acid have been shown to inhibit NF-κB activity and cytokine induction. In addition, the drugs doxorubicin, epirubicin, valrubicin, idarubicin, and teniposide, which all target DNA topoisomerase 2α (TOP2α) (Hevener et al. 2018), were predicted to be the most effective drugs for preventing SARS-CoV-2 infection. DNA methyltransferase 1 (DNMT1) was also identified as interacting with

SARS-CoV-2 Nsp1 protein. Studies have shown that SARS-CoV-2 infection causes a significant decrease in the levels of DNMT1, suggesting that blocking its interaction with Nsp1 could inhibit viral infection (Muhammad et al. 2021).

Testable hypotheses generated from this database provide a view of what discovery-driven research needs to produce usable results. While discovery-driven research has brought several interesting new discoveries, it has also produced an abundance of data that is often overlooked simply because of its sheer magnitude. If we are to be honest, only a very small amount of the data generated in proteomic studies is used to generate testable hypotheses. Most of the data ends up being archived. As of 2018, there were over 1700 molecular biology databases, and new ones were being created at a rate of >100 per year (Imker 2018). This plethora of databases contains all kinds of data; however, they are too often underutilized for a variety of reasons. These reasons include intimidation by database size, lack of time to learn how to interrogate database information, the complexity of the data, and finding relationships between different databases being difficult. In addition, many databases house algorithms to interrogate the data, but there is a learning curve required in their proper use. As the comfort level with discovery-driven research has continued to grow over the past 25 years, it is now time that the same evolution occurs with databases. The hope is that the next few years will see an influx of computer scientists collaborate with molecular biologists to mine the accumulated data for hypotheses that can be tested to develop such things as therapeutic drugs for viral infections. This collaboration may convert much of the already accumulated data into useful biological results.

## 5.5 Conclusions

The proteomic era has rapidly accelerated the discovery of how viruses and host cells interact. As late as the early 2000s, VOPBA was being used to detect, *not identify*, interactions between viruses and proteins separated on a gel (Martínez-Barragán and del Angel 2001). Soon after, MS was added to the VOPBA process to identify the protein(s) interacting with the virus (Thepparit and Smith 2004). Within the same time frame, Y2H was making strides in identifying virus/host cell interactions (MacFarlane and Uhrig 2008). While these methods yielded many important results, they were both binary methods—identifying single virus/host interactions per experiment. Proteomic technology in the form of LC/MS, combined with sample preparation methods such as AP and IP, rapidly increased not only the discovery rate of virus/host cell PPIs but also the number of proteins identified per experiment. Proteomics has enabled the intricate details of virus/host cell interactions to be revealed on a much greater scale than previously possible. This chapter, along with Chap. 1, provides examples of where proteomic technology has been used to discover virus/host cell PPIs that generate hypotheses that are tested using targeted experiments to provide increasing more detailed information about the interaction. In a crux, this progression is what has made proteomics so impactful. Proteomics' ability to rapidly discover PPIs provides the foundation by which testable hypotheses can be generated. Without this discovery ability, the pace of viral research would continue to be dominated by experiments that investigate a single hypothesis that have uncertain outcomes.

This increased data-gathering capability has led to another opportunity—the development of databases of experimentally measured PPIs between viruses and host cells. These databases are a necessity since the rate of discovery of PPIs is too much for a single lab to validate. Disseminating the data enables interested parties to either discover or validate novel interactions via analysis of the compiled results or their own experiments. Validation is simply done using targeted experiments on PPIs listed within the database. Discovery requires further computational analysis using software to determine some intrinsic characteristic of proteins found within the database. These characteristics can then be used to speculate on potential interactions that may occur between a virus and human protein. Turning database information into validatable results will require collaboration between molecular biologists, virologists, and computational scientists. This scenario is analogous to the early days of proteomics, which was dominated by analytical chemists who were well versed in LC/MS. Proteomics really started to make an impact in biology when molecular biologists started collaborating with analytical chemists. While the importance of this collaboration continues, the growth of proteomic databases will benefit from bringing more computational scientists onto the team.

## References

Abe M, Kobayashi T (2021) Imaging cholesterol depletion at the plasma membrane by methyl-β--cyclodextrin. J Lipid Res 62:100077

Arthos J, Cicala C, Martinelli E, Macleod K, Van Ryk D, Wei D et al (2008) HIV-1 envelope protein binds to and signals through integrin α4β7, the gut mucosal homing receptor for peripheral T cells. Nat Immunol 9:301–309

Atreya PL, Peeples ME, Collins PL (1998) The NS1 protein of human respiratory syncytial virus is a potent inhibitor of minigenome transcription and RNA replication. J Virol 72:1452–1461

Calderone A, Licata L, Cesareni G (2015) VirusMentha: a new resource for virus-host protein interactions. Nucleic Acids Res 43(Database issue):D588–D592

Carvajal JJ, Avellaneda AM, Salazar-Ardiles C, Maya JE, Kalergis AM, Lay MK (2019) Host components contributing to respiratory syncytial virus pathogenesis. Front Immunol 10:2152

Chang CK, Lin SM, Satange R, Lin SC, Sun SC, Wu HY et al (2021) Targeting protein-protein interaction interfaces in COVID-19 drug discovery. Comput Struct Biotechnol J 19:2246–2255

Chatr-aryamontri A, Ceol A, Peluso D, Nardozza A, Panni S, Sacco F et al (2009) VirusMINT: a viral protein interaction database. Nucleic Acids Res 37(Database issue):D669–D673

Cook HV, Doncheva NT, Szklarczyk D, von Mering C, Jensen LJ (2018) Viruses.STRING: a virus--host protein-protein interaction database. Viruses 10:519

d'Hérelle F (1926) The bacteriophage and its behavior. Williams and Wilkins, Baltimore

Davis HE, McCorkell L, Vogel JM, Topol EJ (2023) Long COVID: major findings, mechanisms and recommendations. Nat Rev Microbiol 21:133–146

Del Toro N, Shrivastava A, Ragueneau E, Meldal B, Combe C, Barrera E et al (2022) The IntAct database: efficient access to fine-grained molecular interaction data. Nucleic Acids Res 50(D1):D648–D653

Dimitrov DS (1997) How do viruses enter cells? The HIV-1 coreceptors teach us a lesson of complexity. Cell 91:721–730

Dimitrov D (2004) Virus entry: molecular mechanisms and biomedical applications. Nat Rev Microbiol 2:109–122

Dong J, Basse V, Bierre M, Peres de Oliveira A, Vidalain PO, Sibille P et al (2022) Respiratory syncytial virus NS1 protein targets the transactivator binding domain of MED25. J Mol Biol 434:167763

Drew K, Wallingford JB, Marcotte EM (2021) Hu.MAP 2.0: integration of over 15,000 proteomic experiments builds a global compendium of human multiprotein assemblies. Mol Syst Biol 17:e10016

Durmuş Tekir S, Çakır T, Ardiç E, Sayılırbaş AS, Konuk G, Konuk M et al (2013) PHISTO: a pathogen-host interactions search tool. Bioinformatics 29:1357–1358

Dutta S, Smith MD (2023) Detection of protein-protein interactions using the split-ubiquitin membrane-based yeast two-hybrid system. Methods Mol Biol 2690:37–57

Fields S, Song O (1989) A novel genetic system to detect protein-protein interactions. Nature 340:245–246

Foxman EF, Iwasaki A (2011) Genome-virome interactions: examining the role of common viral infections in complex disease. Nat Rev Microbiol 9:254–264

Galletta BJ, Rusan NM (2015) A yeast two-hybrid approach for probing protein-protein interactions at the centrosome. Methods Cell Biol 129:251–277

Giurgiu M, Reinhard J, Brauner B, Dunger-Kaltenbach I, Fobo G, Frishman G et al (2019) CORUM: the comprehensive resource of mammalian protein complexes-2019. Nucleic Acids Res 47:D559–D563

Gordon DE, Jang GM, Bouhaddou M, Xu J, Obernier K, White KM et al (2020) A SARS-CoV-2 protein interaction map reveals targets for drug repurposing. Nature 583:459–468

Guirimand T, Delmotte S, Navratil V (2015) VirHostNet 2.0: surfing on the web of virus/host molecular interactions data. Nucleic Acids Res 43(Database issue):D583–D587

Hamdi A, Colas P (2012) Yeast two-hybrid methods and their applications to drug discovery. Trends Pharmacol Sci 33:109–118

Hampel H, Hardy J, Blennow K, Chen C, Perry G, Kim SH et al (2021) The amyloid-β pathway in Alzheimer's disease. Mol Psychiatry 26:5481–5503

Havugimana PC, Goel RK, Phanse S, Youssef A, Padhorny D, Kotelnikov S et al (2022) Multiplex co-fractionation/mass spectrometry platform for accelerated protein interactome discovery. Nat Commun 13:4043

Hevener K, Verstak TA, Lutat KE, Riggsbee DL, Mooney JW (2018) Recent developments in topoisomerase-targeted cancer chemotherapy. Acta Pharm Sin B 8:844–861

Hu X, Zhu M, Liang Z, Kumar D, Chen F, Zhu L et al (2017) Proteomic analysis of BmN cell lipid rafts reveal role in Bombyx mori nucleopolyhedrovirus infection. Mol Gen Genomics 292:465–474

Huang Y, Song Y, Li J, Lv C, Chen ZS, Liu Z (2022) Receptors and ligands for herpes simplex viruses: novel insights for drug targeting. Drug Discov Today 27:185–195

Huntley RP, Sawford T, Mutowo-Meullenet P, Shypitsyna A, Bonilla C, Martin MJ et al (2015) The GOA database: gene ontology annotation updates for 2015. Nucl Acids Res 43:D1057–D1063

Imker HJ (2018) 25 years of molecular biology databases: a study of proliferation, impact, and maintenance. Front Res Metr Anal 3:18

Jaafar ZA, Kieft JS (2019) Viral RNA structure-based strategies to manipulate translation. Nat Rev Microbiol 17:110–123

Jones JE, Le Sage V, Lakdawala SS (2021) Viral and host heterogeneity and their effects on the viral life cycle. Nat Rev Microbiol 19:272–282

Kanehisa M, Furumichi M, Sato Y, Kawashima M, Ishiguro-Watanabe M (2023) KEGG for taxonomy-based analysis of pathways and genomes. Nucleic Acids Res 51:D587–D592

Karger A, Mettenleiter TC (1996) Identification of cell surface molecules that interact with pseudorabies vaccine. J Virol 70:2138–2145

Kumar R, Nanduri B (2010) HPIDB—a unified resource for host-pathogen interactions. BMC Bioinformatics 11 Suppl 6(Suppl 6):S16

Kumar CS, Dey D, Ghosh S, Banerjee M (2018) Breach: host membrane penetration and entry by nonenveloped viruses. Trends Microbiol 26:525–537

Kwofie SK, Schaefer U, Sundararajan VS, Bajic VB, Christoffels A (2011) HCVpro: hepatitis C virus protein interaction database. Infect Genet Evol 11:1971–1977

Labeau A, Fery-Simonian L, Lefevre-Utile A, Pourcelot M, Bonnet-Madin L, Soumelis V et al (2022) Characterization and functional interrogation of the SARS-CoV-2 RNA interactome. Cell Rep 39:110744

Li S, Zhou W, Li D, Pan T, Guo J, Zou H et al (2022a) Comprehensive characterization of human--virus protein-protein interactions reveals disease comorbidities and potential antiviral drugs. Comput Struct Biotechnol J 20:1244–1253

Li TW, Kenney AD, Park JG, Fiches GN, Liu H, Zhou D et al (2022b) SARS-CoV-2 Nsp14 protein associates with IMPDH2 and activates NF-κB signaling. Front Immunol 13:1007089

Loveys DA, Kulkarni S, Atreya PL (2000) Role of type I IFNs in the in vitro attenuation of live, temperature-sensitive vaccine strains of human respiratory syncytial virus. Virology 271:390–400

MacFarlane SA, Uhrig JF (2008) Yeast two-hybrid assay to identify host-virus interactions. Methods Mol Biol 451:649–672

Maddon PJ, Dalgleish AG, McDougal JS, Clapham PR, Weiss RA, Axel R (1986) The T4 gene encodes the AIDS virus receptor and is expressed in the immune system and the brain. Cell 47:333–348

Maginnis MS (2018) Virus-receptor interactions: the key to cellular invasion. J Mol Biol 430:2590–2611

Mao L, Jin H, Wang M, Hu Y, Chen S, He Q et al (2020) Neurologic manifestations of hospitalized patients with coronavirus disease 2019 in Wuhan, China. JAMA Neurol 77:683

Martínez-Barragán JJ, del Angel RM (2001) Identification of a putative coreceptor on Vero cells that participates in dengue 4 virus infection. J Virol 75:7818–7827

Matlin KS, Reggio H, Helenius A, Simons K (1981) Infectious entry pathway of influenza virus in a canine kidney cell line. J Cell Biol 91:601–613

May DG, Martin-Sancho L, Anschau V, Liu S, Chrisopulos RJ, Scott KL et al (2022) A BioID--derived proximity interactome for SARS-CoV-2 proteins. Viruses 14:611

Mørtz E, Vorm O, Mann M, Roepstorff P (1994) Identification of proteins in polyacrylamide gels by mass spectrometric peptide mapping combined with database search. Biol Mass Spectrom 23:249–261

Muhammad JS, Saheb Sharif-Askari N, Cui ZG, Hamad M, Halwani R (2021) SARS-CoV-2 infection-induced promoter hypomethylation as an epigenetic modulator of heat shock protein A1L (HSPA1L) gene. Front Genet 12:622271

Naffouje R, Grover P, Yu H, Sendilnathan A, Wolfe K, Majd N et al (2019) Anti-tumor potential of IMP dehydrogenase inhibitors: a century-long story. Cancers (Basel) 11:1346

Nal B, Mohr E, Ferrier P (2001) Location analysis of DNA-bound proteins at the whole-genome level: untangling transcriptional regulatory networks. BioEssays 23:473–476

Nomaguchi M, Fujita M, Miyazaki Y, Adachi A (2012) Viral tropism. Front Microbiol 3:281

Orchard S, Ammari M, Aranda B, Breuza L, Briganti L, Broackes-Carter F et al (2014) The MIntAct project—IntAct as a common curation platform for 11 molecular interaction databases. Nucleic Acids Res 42(Database issue):D358–D363

Oughtred R, Rust J, Chang C, Breitkreutz BJ, Stark C, Willems A et al (2021) The BioGRID database: a comprehensive biomedical resource of curated protein, genetic, and chemical interactions. Protein Sci 30:187–200

Paingankar MS, Arankalle VA (2014) Identification of chikungunya virus interacting proteins in mammalian cells. J Biosci 39:389–399

Pei J, Beri NR, Zou AJ, Hubel P, Dorando HK, Bergant V et al (2021) Nuclear-localized human respiratory syncytial virus NS1 protein modulates host gene transcription. Cell Rep 37:109803

Poss ZC, Ebmeier CC, Taatjes DJ (2013) The Mediator complex and transcription regulation. Crit Rev Biochem Mol Biol 48:575–608

Rotaru DC, Mientjes EJ, Elgersma Y (2020) Angelman syndrome: from mouse models to therapy. Neuroscience 445:172–189

Sharma A, Vasanthapuram R, M Venkataswamy M, Desai A (2020) Prohibitin 1/2 mediates Dengue-3 entry into human neuroblastoma (SH-SY5Y) and microglia (CHME-3) cells. J Biomed Sci 27:55

Sobhy H (2017) A comparative review of viral entry and attachment during large and giant dsDNA virus infections. Arch Virol 162:3567–3585

Song WJ, Hui CKM, Hull JH, Birring SS, McGarvey L, Mazzone SB et al (2021) Confronting COVID-19-associated cough and the post-COVID syndrome: role of viral neurotropism, neuroinflammation, and neuroimmune responses. Lancet Respir Med 9:533–544

Taharaguchi S, Kono Y, Ohta H, Takase K (2007) Putative host cell receptor for fowl adenovirus detected in gizzard. J Vet Med Sci 69:1203–1205

Thepparit C, Smith DR (2004) Serotype-specific entry of dengue virus into liver cells: identification of the 37-kilodalton/67-kilodalton high-affinity laminin receptor as a dengue virus serotype 1 receptor. J Virol 78:12647–12656

Thomas PD, Ebert D, Muruganujan A, Mushayahama T, Albou LP, Mi H (2022) PANTHER: making genome-scale phylogenetics accessible to all. Protein Sci 31:8–22

Valiente G (2022) The landscape of virus-host protein-protein interaction databases. Front Microbiol 13:827742

Van Royen T, Sedeyn K, Moschonas GD, Toussaint W, Vuylsteke M, Van Haver D et al (2022) An unexpected encounter: respiratory syncytial virus nonstructural protein 1 interacts with mediator subunit MED25. J Virol 96:e0129722

Veenstra TD (2021a) Omics in systems biology: current progress and future outlook. Proteomics 21:e2000235

Veenstra TD (2021b) Proteomic characterization of protein-protein interactions. Biology (Austin) 4:1027–1036

Walsh D, Mohr I (2011) Viral subversion of the host protein synthesis machinery. Nat Rev Microbiol 9:860–875

Wang J, Reuschel EL, Shackelford JM, Jeang L, Shivers DK, Diehl JA et al (2011) HIV-1 Vif promotes the G1-S-phase cell-cycle transition. Blood 117:1260–1269

Wang C, Du X, Mou Z (2016) The Mediator complex subunits MED14, MED15, and MED16 are involved in defense signaling crosstalk in *Arabidopsis*. Front Plant Sci 7:1947

Wang S, Wu R, Lu J, Jiang Y, Huang T, Cai YD (2022) Protein-protein interaction networks as miner of biological discovery. Proteomics 22(15–16):e2100190

Wilm M, Shevchenko A, Houthaeve T, Breit S, Schweigerer L, Fotsis T et al (1996) Femtomole sequencing of proteins from polyacrylamide gels by nano-electrospray mass spectrometry. Nature 379:466–469

Yamauchi Y, Greber UF (2016) Principles of virus uncoating: cues and the snooker ball. Traffic 17:569–592

Yang X, Lian X, Fu C, Wuchty S, Yang S, Zhang Z (2021) HVIDB: a comprehensive database for human-virus protein-protein interactions. Brief Bioinform 22:832–844

Zhang F, Ke J, Zhang L, Chen R, Sugimoto K, Howe GA et al (2017) Structural insights into alternative splicing-mediated desensitization of jasmonate signaling. Proc Natl Acad Sci USA 114:1720–1725

# Chapter 6
# The Pathology of Viral Infections

**Abstract** If you were to ask, "Biologically speaking, what are humans made of?", almost everyone would reply "Human cells and molecules, of course." This seemingly logical answer, however, does not truly capture the diversity of the human organism. Over the past couple of decades scientists have discovered that humans are a collective of cohabitating human, bacteria, and fungi cells along with countless numbers of viruses. Collectively, these microorganisms are referred to as the microbiome (Lederberg and McCray, The Scientist 15:8, 2001). The most recent estimates suggest the biological material from these microorganisms makes up as much as half of every human. Considering the size of the microbiome, it is not surprising that humans share an intimate relationship with viruses since they will be found wherever life exists.

**Keywords** Pathology • Host proteins • Human virome • Proximity labeling • Proteome of infected cells • iTRAQ • Gene ontology • SARS-CoV-2 receptor • Hepatitis A

## 6.1 Introduction

The study of the symbiotic relationship between the microbiome and humans started in the late seventeenth century when Antonie van Leeuwenhoek noted the differences in bacteria present in his saliva and feces (Lewandowska-Pietruszka et al. 2022). Fast forward over 300 years and even nonscientists are aware of the human microbiome. If you asked most people what the major component of the microbiome is, they would reply "bacteria." While there are an estimated 38 trillion bacteria within the microbiome (Sender et al. 2016), this number pales in comparison to that of viruses. With the advent of inexpensive genomic sequencing, scientists began to discover viruses throughout the human body. These viruses inhabit almost every body tissue and fluid including expected places such as the skin, mouth, respiratory tract, and blood. They also live in unexpected places such as cerebrospinal fluid, synovial fluid, and breast milk. The discovery of viruses in the central nervous

 
T. D. Veenstra, *Proteomic Applications in Virology*, Advances in Experimental Medicine and Biology 1511, https://doi.org/10.1007/978-3-032-22340-1_6

system was somewhat surprising as this system was thought to be a sterile environment (Kang et al. 2021). The number of viruses living on or inside an adult human is presently estimated to be near 400 trillion—approximately an order of magnitude greater than the estimated number of bacteria (Pride 2020).

While some cause illness, many viruses simply coexist within the human body and even provide health benefits to their host by infecting bacteria within our microbiomes. For example, bacteriophages (viruses that infect bacteria) found within mucosal surfaces are believed to provide antimicrobial defense (Barr et al. 2013). A metagenomic sequencing study of the global virome identified 19 different strains of redondovirus in the human respiratory and oropharyngeal sites (Abbas et al. 2019). While some were associated with lung and periodontal disease, many of these redondovirus strains are bacteriophages that provide antimicrobial protection against respiratory illness.

The human virome begins populating as soon as we are born. Shortly after birth, infants contain a diverse population of viruses in their gut, many ingested from breast milk (Taboada et al. 2021). An analysis of oropharyngeal and fecal samples collected from nine Mexican children starting at 2 weeks of age through to 12 months revealed the presence of primarily *Herpesviridae*, *Picornaviridae*, and *Reoviridae* viruses in the oropharynx with *Caliciviridae*, *Picornaviridae*, *Anelloviridae*, and *Reoviridae* viruses dominating fecal samples (Rivera-Gutiérrez et al. 2023). This study also showed that each child's virome is unique and is likely influenced by the environment, feeding, and age. As an infant ages, some viruses decline in number while the population of others increase as they are exposed to additional sources of air, water, food, and other people. These viruses continue to increase in diversity and number over the years and persist within cells that they infect. Once an individual reaches adulthood, their virome will become relatively stable. With the relative infancy of this field of study, it will be interesting to evaluate if there are any long-term health effects associated with specific virome compositions.

As mentioned previously, many viruses living in humans target the bacteria that make up our microbiomes (Tisza and Buck 2021). They are in the soil, water, air, living organisms, as well as extreme environments such as hot springs and acid mines. These bacteriophages infect and use the bacteria's molecular machinery to reproduce to an extent that they cause their host to burst, releasing more copies of itself to infect more bacteria. Bacteriophages are found everywhere bacteria exist because they hunt bacteria to ensure their survival and humans simply represent another hunting ground.

While knowing how many viruses live within our bodies may induce feelings of panic, most viruses do not cause any noticeable disease (Shaikh et al. 2023). For those that result in apparent symptoms, the severity of the disease depends on the virulence of the infecting virus and the susceptibility of the host (Glynn and Moss 2020). While pathogenicity refers to a virus's ability to cause disease, it does not indicate the measure of the disease's severity (Smith and Casadevall 2022). The term virulence is used to indicate the severity of a disease resulting from a virus's infection. Taken together, pathogenicity and virulence describe the capacity of a virus to cause disease in its host. These two factors, however, are not directly related to its transmissibility (or contagiousness) as a virus with a high virulence that results in the death of its host may have less opportunity to be transmitted (Mollentze et al. 2020).

## 6.2 Pathogenesis of Viruses

The pathogenesis of a virus is made up of three basic steps: (i) entry into host, (ii) replication, and (iii) exit from its primary host (see Table 6.1) (Baron et al. 1996). In the entry step, the virus must evade the host's innate and adaptive immune systems and interact with some component of a cell's surface to gain entry. In the replication step, the virus hijacks a portion of the host cell's transcriptional and translational machinery for its own replication processes. In the exit step, the virus leaves the host cell at a concentration sufficient to ensure infection of nearby cells. Throughout all these steps, the virus must evade the host's immune defense.

While each step in viral pathology is important, understanding the details of how a virus enters the cell and replicates is critical to developing treatments to prevent infection and spread. Identifying host proteins that are targeted by viral proteins during entry can help in the design of molecules that block subsequent infection. Identifying host proteins that are targeted by viral proteins during replication can help in the design of molecules to prevent local spread. Blocking local spread can prevent general spread and community transmission.

### *6.2.1 Identifying Host Proteins Involved in Viral Entry*

Viruses are very simple structures. They are genetic material housed within a protein coating, which is sometimes surrounded by a layer of lipids. This simple molecular formulation means that a virus can only thrive using the replication machinery of a host cell. This need requires the virus to enter a host cell. Regardless of whether the virus is enveloped or non-enveloped, it must bind to a cell surface receptor to gain entry. Once bound to the receptor, the virus enters either via an endocytic or non-endocytic route. In the endocytic route, the host cell's membrane forms a vesicle that surrounds the virus and brings it into the host's cytoplasm. In the

**Table 6.1** Essential steps in viral infection

| Step in infection process | Requirement for virus survival and progression of infection |
|---|---|
| Entry into host | Evade host's natural protective mechanisms<br>Bind to cell surface receptor |
| Replication | Hijack necessary host functions necessary to replicate<br>Evade host immune responses |
| Exit primary host | Exit host cell at concentration that ensures further infection<br>Evade host immune responses |

non-endocytic route, the virus fuses with the host cell's membrane and creates an opening that allows the virus to directly deposit its genome and associated proteins into the host's cytoplasm (see Fig. 6.1). Regardless of the method used, the necessary viral material gains access to the host machinery necessary to replicate itself. Identifying receptors on host cells is an important step in developing medications that can block the first step necessary for a virus to cause disease.

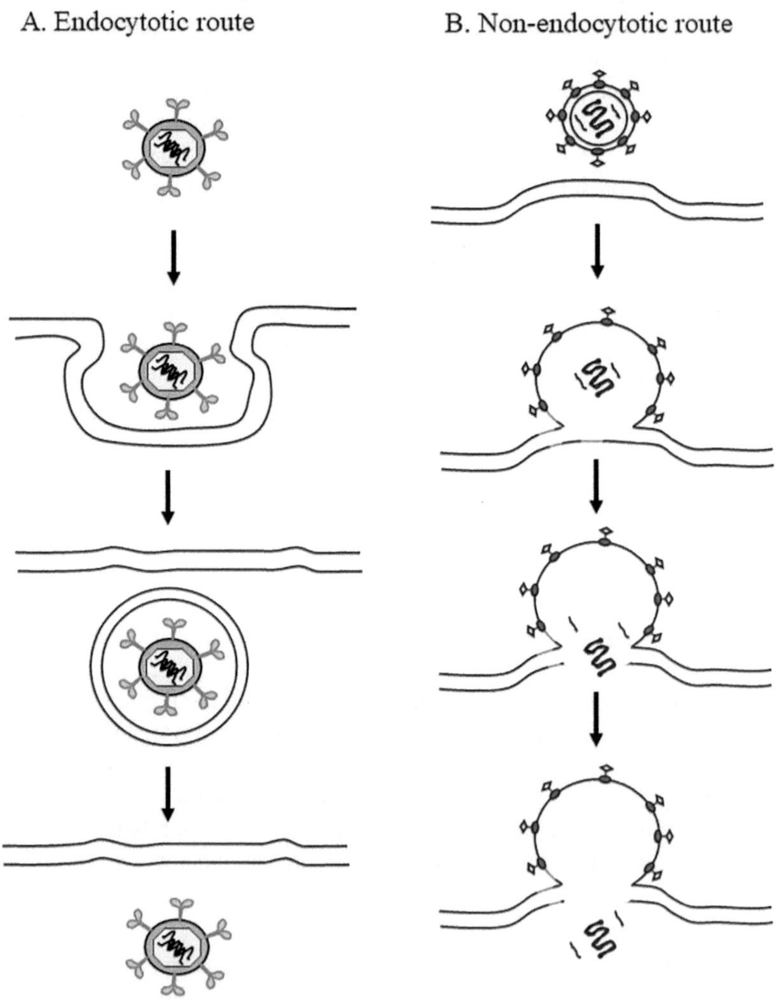


**Fig. 6.1** Endocytosis versus non-endocytosis pathways of viral entry into host cells. The chief difference is that in the endocytic route, the host cell's membrane forms a complete vesicle around the virus prior to bringing it into its cytoplasm. In the non-endocytotic route, the virus fuses with the host cell's membrane to create a passageway that allows the viral material necessary for its replication to enter the host

### 6.2.2 *Identifying Host Cell Receptors Via Immunoprecipitation*

Identifying host proteins that allow the virus to enter the cell requires focusing on the first barrier to entry: the cell membrane. Experiments that focus on this cellular region during viral entry is not trivial owing to solubility issues of membrane proteins. One of the most straightforward methods involves using a bait protein to extract a binding protein from a cell lysate and then determining if the interaction involved a cell membrane protein. The discovery that angiotensin-converting enzyme 2 (ACE2) is the cell surface receptor for severe acute respiratory syndrome coronavirus 2 (SARS-CoV-2) was published over two decades ago, during the infancy of the current proteomics era (Li et al. 2003). In this study, the African green monkey kidney cell line, Vero E6, was metabolically labeled with [$^{35}$S]cysteine and [$^{35}$S]methionine. A fusion protein made up of a human IgG1 fragment crystallizable (Fc) domain and residues 12-672 of the SARS-CoV-2 spike protein (S1-Ig) was incubated with these metabolically labeled Vero E6 or human embryonic kidney (HEK) 293T cells. While the fusion protein bound to Vero E6 cells, it did not bind to 293T cells. To characterize this interaction, two tagged versions of the S1 protein (S1-Ig and S1-c9; the S1 protein tagged with a nine residue peptide from the human complement component 9 protein) were separately mixed with Vero E6 cells and immunoprecipitated using an antibody against the Fc domain or the c9 peptide (see Fig. 6.2). As negative controls, three other constructs, c9-tagged glycoprotein-120 (gp-120-c9) and interferon alpha/beta receptor subunit 2 (IFNAR2-c9) along with IFNAR2-Ig, were incubated with Vero E6 cells. After immunoprecipitation using the appropriate antibodies, the immunoprecipitates were fractionated using sodium dodecyl sulfate-polyacrylamide gel electrophoresis (SDS-PAGE). Visualization of the gel resulted in the presence of a protein band with an apparent molecular weight of 110 kDa in the extract obtained from Vero E6 cells incubated with the S1-Ig construct. None of the other constructs showed evidence of a protein-protein interaction. In a separate experiment in which the samples extracted using the S1-Ig constructs were de-glycosylated using the PNGase F, two bands were observed at approximately 85 and 100 kDa demonstrating that the molecule captured from the Vero E6 was likely a glycosylated protein.

The original band at 110 kDa was excised from the gel, digested with trypsin, and identified using mass spectrometry (MS). As is often the case when analyzing a single gel band, multiple proteins were identified within it. These proteins were myosin 1b, major vault protein, and ACE2. Myosin 1b and major vault protein were eliminated as being the SARS-CoV-2 receptor since neither of these proteins are found on the cell's surface. ACE2, on the other hand, is an N-glycosylated type I transmembrane protein, so it fits the observed experimental data.

A critical step in any research project, especially proteomics studies in which multiple candidates are identified, is validation. In this study, the gene for ACE2 was cloned into an expression plasmid, which was transfected into 293T cells. The transfected cells bound to the SARS-CoV-2 spike protein/IgG1 fusion protein, while un-transfected cells did not. In addition, immunoprecipitation of the

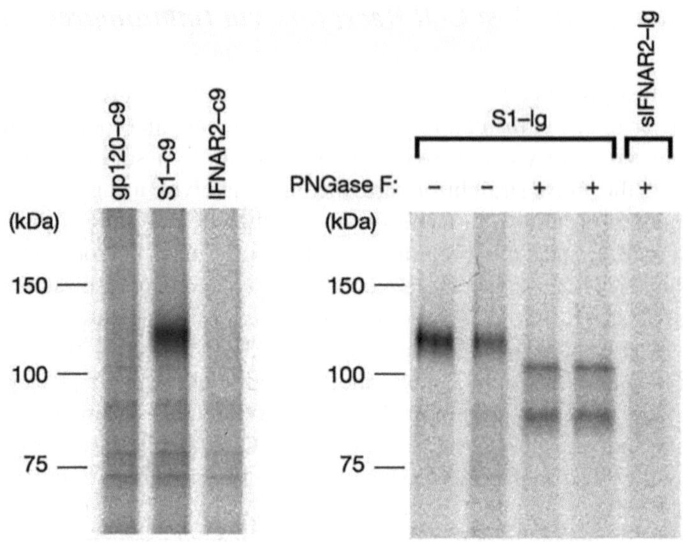


**Fig. 6.2** Metabolically labeled Vero E6 cell lysates were incubated with glycoprotein-120 (gp120), SARS-CoV spike protein (S1), or interferon alpha/beta receptor subunit 2 (IFNAR2), each containing a nine-residue peptide tag (c9) at its C terminus. The lysates were immunoprecipitated using an anti-c9 antibody and analyzed using SDS-PAGE (left panel). Metabolically labeled Vero E6 cell lysates incubated with S1-Ig or sIFNAR2-Ig were immunoprecipitated using anti-Ig antibody. Treatment of the extracts using PNGase F demonstrated that the protein extracted using the S1-Ig construct was likely glycosylated. (Reproduced with permission from Li et al. (2003); https://creativecommons.org/publicdomain/mark/1.0/)

transfected 293T cell lysate was recognized using an anti-ACE2 antibody. Finally, digestion of ACE2 extracted from these cells using PNGase F showed a single band at 85 kDa, representing the non-glycosylated form of the receptor. Taken together all the data shows a high affinity association between ACE2 and the SARS-CoV-2 spike protein.

Early studies of viruses generally identified a single (or couple of) key cell surface protein that mediated cell entry. For example, Kaposi's sarcoma-associated herpesvirus (KSHV) entry into host cells was originally associated with the glycosaminoglycan heparin sulfate (Birkmann et al. 2001). Over time, not only was KSHV cell entry shown to be more complex but was also found to be cell-type dependent. For example, the entry of KSHV into human microvascular dermal endothelial cells is initiated by the attachment of the virus to heparin sulfate. Following attachment, KSHV interacts with the integrins α3β1, αVβ3, and αVβ5, ephrin receptor A2 (EphA2), and the 12-transmembrane glutamate/cystine exchange transporter xCT (Veettil et al. 2008). These interactions induce phosphorylation of the multifunctional protein c-Cbl (along with additional signaling proteins), which then interacts with myosin IIA. Myosin IIA forms a bleb around the virus, which is then retracted into the cell along with EphA2 and calcium and integrin binding protein 1 (CIB1). The formed macropinosome then releases the KSHV capsid into the cytoplasm. The interactions involving KSHV were mostly identified using a hypothesis-driven approach utilizing antibodies against the putative host cell receptors.

Hypothesis-driven studies have enabled the identification of host cell receptor--virus interactions such as CCR-5/HIV (Deng et al. 1996; Dragic et al. 1996) and sialic acid/influenza A (Varghese et al. 1992); however, the proteomics era with its discovery-driven strategies has expanded our knowledge on the intricate details of viral entry. For example, a proteomics approach using immunoprecipitation followed by MS identification identified integrin $\alpha4\beta7$ as an HIV-1 cell surface receptor that also directs CD4+ T cells to migrate from circulation to the gut (Arthos et al. 2008). As a result, CD4+ T cells are massively depleted within the gut contributing to further HIV-1-induced immune function. HIV-1 gp120 binds to integrin $\alpha4\beta7$ through a peptide motif that mimics the natural ligands of the receptor. Once gp120 binds to integrin $\alpha4\beta7$, lymphocyte function-associated antigen-1 (LFA-1) is rapidly activated. As LFA-1 is the central integrin involved in the formation of virological synapses, its activation facilitates the spreading of HIV-1 between cells. The discovery of the interaction between integrin and gp120 provides the first clue as to why HIV-1 patients can suffer from a range of gastrointestinal symptoms ranging from swallowing pain and difficulties to nausea, vomiting, abdominal pain, diarrhea, and rectal tenesmus.

### *6.2.3 Identifying Host Cell Receptors Via Proximity Labeling*

Antibodies targeted at specific receptors are an invaluable tool for identifying protein interactions between viruses and host cells, and the reader will see their utility throughout this book. There is, however, as the proverb says, "More than one way to cook an egg." A simple fundamental characteristic of interacting proteins is that they must be near one another. The enzyme-mediated activation of radical source (EMARS) method takes advantage of this fact. EMARS uses horseradish peroxidase (HRP) to form covalent bonds between aryl azide or tyramide molecules (Kotani et al. 2022). Using aryl azide enables molecules within 200–300 nm of the HRP-labeled target molecule to be labeled, while tyramide will only cause molecules within about 20 nm of each other to be labeled. The aryl azide and tyramide molecules can be conjugated to other molecules that can be visualized via fluorescence, affinity purified using an antibody, etc.

In an application of EMARS (see Fig. 6.3), the SARS-CoV-2 spike protein S1 was conjugated to HRP via a mouse IgG constant domain (S1-Fc-HRP). The hypothesis was that S1-Fc-HRP would bind to the S1-specific cell surface receptors and HRP would catalyze the addition of fluorescein to other proteins within the vicinity of this receptor. An anti-fluorescein antibody could then be used to immunoprecipitated fluorescein-labeled proteins from the proteome sample.

Caco-2 and A549 cells were incubated with the S1-Fc-HRP molecule followed by the addition of fluorescein-conjugated tyramide. Since A549 cells do not contain the ACE2 receptor, this cell line served as a negative control. After lysing both cell

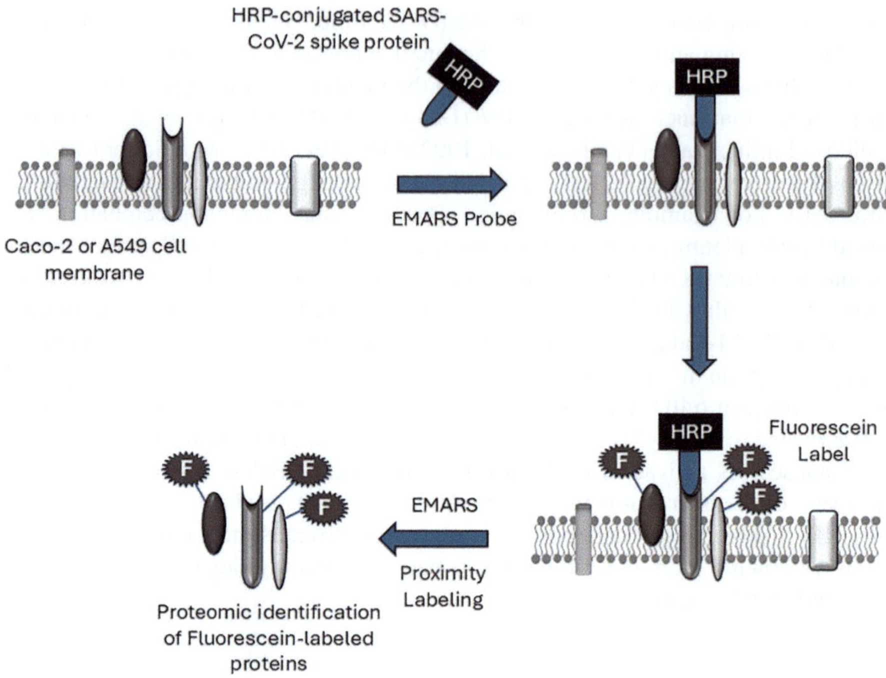


**Fig. 6.3** Process of identifying candidate membrane proteins involved in SARS-CoV-2 entry using enzyme-mediated activation of radical source (EMARS). Cells are incubated with a construct containing horseradish peroxidase (HRP) coupled to SARS-CoV-2 spike protein using a mouse antibody constant domain (S1-Fc-HRP). The cells are then incubated with fluorescein--conjugated tyramide molecules that will form covalent linkages with proteins in the vicinity of S1-Fc-HRP-labeled proteins. The fluorescein-labeled proteins are then immunoprecipitated using an anti-fluorescein antibody and identified using liquid chromatography coupled with a mass spectrometer operating in a tandem mass spectrometry mode (LC/MS$^2$). (Reproduced with permission from Li et al. (2003); https://creativecommons.org/licenses/by/4.0/)

types separately, plasma membranes were isolated via centrifugation and resolubilized using an NP-40 containing buffer since the goal is to identify cell surface proteins responsible for viral entry. Fluorescein-labeled proteins were then isolated using an anti-fluorescein antibody coupled to Sepharose beads. After washing the antibody-conjugated beads several times to remove nonspecifically bound proteins, fluorescein-labeled proteins were eluted by heating at 95 °C for 5 min.

The eluted proteins were digested into peptides using trypsin and identified using LC/MS with an LTQ Orbitrap XL mass spectrometer. The identification strategy did not require identification of the fluorescein-labeled peptides; simple identification of peptides originating from a protein was considered sufficient to consider the protein as potentially interacting with the SARS-CoV-2 spike protein. For a protein to be considered as a potential interactor, it had to be identified in Caco-2 cell but not in A549 cells. Other commonly known contaminants such as keratin, immunoglobulins, histones, albumin, trypsin, or actin were also considered as nonspecific background. In the analysis of Caco-2 cells, 38 different membrane proteins were

**Table 6.2** Selected Caco-2 membrane proteins identified using the enzyme-mediated activation of radical source (EMARS) method to identify potential membrane proteins associated with SARS-CoV-2spike protein binding to the cell surface

| Protein | Number of unique peptides identified |
|---|---|
| Cadherin 17 | 11 |
| Prominin-1 (CD133) | 7 |
| Cadherin-related family member 5 | 4 |
| CD276 antigen | 4 |
| Voltage-dependent anion-selective channel protein 1 | 4 |
| Cadherin-related family member 2 | 4 |
| Leucine-rich repeat transmembrane protein FLRT3 | 5 |
| CD59 glycoprotein | 2 |
| Basal cell adhesion molecule | 2 |
| Podocalyxin | 3 |
| Sodium/potassium-transporting ATPase subunit beta-1 | 2 |
| Cell adhesion molecule 1 | 2 |
| Adhesion G-protein coupled receptor G6 | 2 |
| Complement decay-accelerating factor | 2 |
| 4F2 cell surface antigen heavy chain | 2 |
| Plastin-2 | 2 |

selected as possibly interacting with the SARS-CoV-2 spike protein, while for A549 cells this number was only 8. These eight proteins were considered false positives since this cell type does not possess the ACE2 receptor. A list of proteins identified by at least two peptides in the immunoprecipitate extracted from the Caco-2 cells is provided in Table 6.2.

It is at this stage that discovery proteomics requires the development of a hypothesis since, like any scientific finding, proteomic results need to be validated. This mantra is especially true for many types of discovery proteomics methods because of the high chance of obtaining a false positive. In this study, several proteins were selected for further testing including ACE2, prominin-1 (CD133), cadherin 17 (CDH17), and voltage-dependent anion-selective channel protein 1 (VAPA). Western blotting was performed to demonstrate expression of ACE2 and the candidate proteins in transfected cells. Immunofluorescence using labeled antibodies was performed to demonstrate colocalization of SARS-CoV-2 S1 with CDH17, VAPA, and CD133 proteins. Colocalization of these proteins was also demonstrated using transmission microscopy with antibodies conjugated with gold particles.

To test the involvement of each candidate in viral infection, ACE2-expressing HEK293T cells were transfected with expression vectors containing the various potential binding partners. ACE2-expressing HEK293T cells transfected individually with either CD133, CDH17, or VAPA showed higher spike protein binding than HEK293T cells that only expressed ACE2. HEK293T cells that only expressed CD133, CDH17, or VAPA without co-expression of ACE2 did not bind spike protein, showing that while these candidates enhanced spike protein binding, none of them were the primary SARS-CoV-2 receptor.

## 6.3 Identifying Proteins Involved in Viral Replication

Once a virus gets into host cells, the action really begins to ramp up. Owing to their limited genome, viruses do not possess the biological components necessary to propagate themselves. Most scientists are well versed in the central dogma that many (but not all) viruses integrate their genetic material into the host cell's genome and then allow the cell's transcriptional and translational proteins to reproduce the molecules required for the virus to reassemble. This simplistic dogma does not honor the intricacies of how viruses manipulate host cell proteins to ensure their own survival. Not only must the virus be replicated, it must also evade a range of host immune defenses designed to stop it. While the SARS-CoV-2 pandemic provided the perfect storm for applying proteomics technology to viral pathology, its utility for demonstrating what happens when a virus finally enters the cell had been known for years.

### *6.3.1 Host Proteins Within Virions*

Many of the details of what happens within an infected cell have been made possible by the high-throughput data gathering capabilities of proteomics. While a specific virus-host cell interaction study will be discussed later, the speed of MS and database searching has enabled viruses to be viewed from many different angles. Most studies characterize what happens to proteins within infected host cells, but other investigations sought to identify what host cell proteins are found in virions (Lippé 2012; Chen et al. 2023). Many different virions, including vaccinia, influenza, HIV, vesicular stomatitis, avian leukosis, and herpes, have been examined to identify their host cell protein content. These studies rely on a straightforward proteomics approach of isolating the virions, extracting their protein content, enzymatically digesting the proteins into peptides, and identifying the peptides using LC/$MS^2$. A study of eight different extracellular herpes virions strains identified host proteins in all the samples (Lippé 2012). A total of 173 unique host proteins were found across all 8 strains, with the number of host proteins found within each individual virion strain ranging from 49 to 4. Nine host proteins were identified in at least four of the different types of herpes virions. These proteins included various Rab GTPases, translational factors, pyruvate kinase M2, and glyceraldehyde-3--phosphate dehydrogenase (GAPDH), along with high abundance proteins such as actin, myosin, and heat shock proteins.

The results show that it is common for different herpes virion strains to contain similar host proteins and that host-pathogen interactions take place both within the virus and host cell. Since these virions are purified from host cell cultures, the argument can be made that the cellular proteins are simply contaminants. The consistency of some proteins being found in related and unrelated virions, however, suggests that they have some biological relevance. Classification of these proteins into functional categories shows that the most populated categories included proteins involved in cytoskeletal formation, trafficking, and metabolism, as well as cell proliferation, migration, and death.

The fact that virions incorporate host proteins is consistent with the parasitic nature of viruses, which rely on host cell proteins rather than encoding them in their genomes. But what advantage does incorporating host cell proteins infer upon the virus? There are several possibilities. The presence of fully functional trafficking proteins may allow the virus to immediately initiate transport to sites of replication within the cell as soon as it enters. Previous studies have shown that viruses utilize many of the same transport proteins that are used to traffic cellular proteins (Mercer et al. 2010). A particularly diverse group of Rab proteins were identified as being present within herpes simplex virus type 1 (HSV1) and pseudorabies virions (Lippé 2012). This family of small monomeric GTPases is part of the RAS superfamily and functions in vesical sorting, transport, tethering, docking, and fusion. It is possible that these proteins regulate the final envelopment step of virus formation via their fusion and tethering function. Rab proteins are known to be involved in autophagy via the formation of autophagosomes (Chua et al. 2011). It is possible that Rab proteins found within virions become incorporated during formation of the viral envelope. While the exact roles of these cell proteins within virions remain uncertain, the high-throughput, discovery-driven approach of MS-based proteomics enables the generation of multiple, testable hypotheses. It is also possible as more data is discovered that these cell proteins may function as druggable targets to inhibit viral infection or replication.

## 6.4 Characterizing Changes in the Proteome of Virus-Infected Cells

As mentioned above, once a virus enters a cell, the action really begins. Viruses start manipulating the cell to produce the parts necessary for its own replication and survival. While it seems to behave like a cooperative hostage, the host cell starts targeting the virus through a series of antiviral factors (Niu et al. 2023; Denz et al. 2023). In turn, the virus enacts countermeasures to subvert the host cell defenses (Saxena et al. 2023). The challenge in proteomics is capturing data on these dynamic events that are occurring simultaneously. We are often trapped into thinking we need to obtain all this data in a single experiment. Accurately capturing what is occurring within a virally infected cell requires collating data captured from many different types of experiments.

### 6.4.1 *Dynamic Changes in the Proteome of Adenovirus-Infected Cells*

One of the most fundamental, yet effective, experiments for capturing effects on a cell's proteome is measuring changes in protein abundances in perturbed and control cells. This strategy has been employed in several viral infectivity studies. For example, a study analyzed the entire cell proteome of human bone osteosarcoma epithelial U2OS cells infected with adenovirus serotype 5 (Ad5) (Dybas et al. 2021). Ad5 is a non-enveloped virus with stable double-stranded linear DNA genome (see Fig. 6.4). Ad5 is used extensively in gene transfer experiments since it can infect a

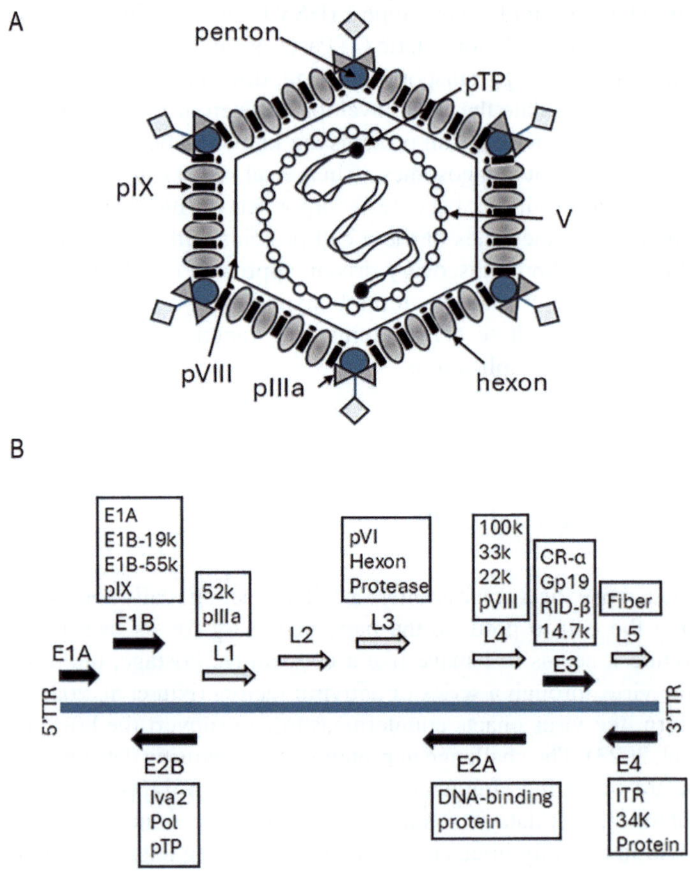


**Fig. 6.4** Adenovirus structure and genome organization. (**a**) Graphical representation of adenovirus structure and various proteins. pTP, preterminal protein: hexon; a major capsid protein: pIIIa; structural protein involved in DNA packing: pVIII; capsid-stabilizing protein: V; core protein: IX; capsid cement protein: penton; role in viral attachment and internalization. (**b**) Adenovirus genome organization showing various early (E) and late (L) transcripts and proteins encoded by each transcript

wide range of different cell types. Scientists are also able to integrate large genes within its genome using various homologous recombination methods (Douglas 2007).

The genome of Ad5 is approximately 36 kb and encodes over 40 proteins (Saha et al. 2014). Its coding regions are designated early (E) or late (L) depending on whether they are expressed before or after DNA replication within the host cell. It is fascinating how viruses have evolved processes that organize the translation of their genomic material in an "as needed, time-dependent manner." As you read through the following paragraph, I encourage the reader to consider the role of each gene product and the order in which they are produced. The first genes transcribed when Ad5 infects cells are contained within the early regions E1A, E1B, E2A, E2B, E3, and E4. These genes produce proteins involved in transcription of other viral regions as well as promoting viral production by the cell (Saha and Parks 2017). The E1A proteins induce host cell mitogenic activity and promote expression of other viral genes (Flint 1999). Viral DNA replication is stimulated by the E2 proteins, while E3 and E4 proteins alter host immune responses and cell signaling (Horwitz 2004; Weitzman 2005). Four intermediate/late expressing genes produce proteins that assist in packaging viral DNA into immature virions. Activation of the major late promoter (MLP) following the start of virus DNA synthesis allows expression of the late genes encoding primarily virion structural proteins (Saha et al. 2014). The late gene regions L1–L5 are transcribed from an alternatively spliced transcript and expressed following the initiation of viral DNA synthesis. These genes primarily express virion structural proteins. Other late coding regions express RNA products that inhibit activation of the cell's interferon response and antagonize host cell micro-RNA processing as a mechanism to influence host gene expression. Of particular interest are six viral proteins expressed from genes in the E4 region (E4orf1, E4orf2, E4orf3, E4orf4, E4orf6, E4orf6/7) that counteract antiviral proteins produced by the host cell during infection (Dybas et al. 2021).

Just to summarize the sequence of virus gene translation described above, it shows that the virus first expresses genes necessary to transcribe other genes within the virus and promote cell mitosis, enabling the virus to be more readily replicated. The second phase of gene expression affects the host immune response to prevent the viral products from being eliminated early in its replication process. The third phase expresses genes that produce proteins necessary for the initial packaging of the virus. In the fourth phase, the final group of structural proteins is expressed to complete the structural organization of the virus. Finally, the virus directs the expression of RNA molecules that inhibit the host cell from producing interferons that would signal the immune system to come to its defense. Although they are nonliving, the way that viruses can coordinate replication with their host cell is astonishing!

To identify global quantitative changes to the host proteome, human alveolar basal epithelial A549 cells were infected with Ad5 (Dybas et al. 2021). Since every proteome is dynamic, the cells were harvested 16, 24, and 48 h postinfection. The proteome was extracted from cells at each time point along with mocked infected cells to serve as a negative control. Each proteome was tryptically digested into peptides. The peptides were prefractionated using a mini-reversed phase column into eight fractions using sequential elution with increasing concentrations of acetonitrile in ammonium formate buffer. These fractions were combined into four aliquots that were individually analyzed using LC/$MS^2$. Each replicate (infected cells at each time point and mock-infected cells) was analyzed four times. The $MS^2$ data was processed to identify the peptides present within each sample.

To identify changes in relative protein abundances between the different proteome samples, protein quantities for each infection time point were compared to those found within the mock-infected A549 cells. A protein's abundance was considered to be altered if it fulfilled one of the three following conditions: (i) the protein had to be quantified in a minimum of two biological replicates of infected and mock cells, and the $\log_2$-fold change in its abundance had to be at least 0.585 ($P < 0.05$); (ii) the protein had to be quantified in a minimum of three biological replicates of the infected or mock cells, but only one replicate for the comparative replicative and the $\log_2$-fold change in its abundance had to be at least 0.585; or (iii) the protein had to quantified in a minimum of three biological replicates of infected or mock cells but not observed in any of the other replicates. In this case the protein was considered as being uniquely identified in whatever replicate it was detected. Quantification was performed using a label-free approach in which the areas of unique peptide peaks were calculated. In cases where multiple peptides were detected for a specific protein, the unique peptide abundances would be summed up to provide a single protein abundance measurement. To explain it using simplistic terms, quantitation between proteins in different samples was essentially determined using the sum of the areas of peptides that correspond to a specific protein.

Between 6000 and 7000 proteins were quantified in each biological replicate. The Pearson correlation coefficient within each of the four replicates was 0.96, demonstrating the analysis was highly reproducible. General differences in each sample were identified within each biological replicate using principal component analysis (PCA). While PCA provides an indication that the overall proteomes of the replicates are different, they do not really provide much molecular information. Analysis of individual samples showed that within the infected cells, 214, 659, and 2103 proteins decreased in abundance at 16, 24, and 48 h postinfection, respectively, compared to mock-infected cells. An increase in abundance was observed for 533, 433, and 887 proteins at 16, 24, and 48 h postinfection, respectively. A total of 126 proteins decreased, and 165 proteins increased in abundance at all three measured time points. Along with host proteins, 45 viral proteins were also quantitated. These proteins were transcribed from both early (i.e., E1A, E1B, E2, E3, and E4) and late-expressed (L1–L5) genes. The abundance of these proteins predictably increased over time, confirming the virus was progressing normally through the infectious cycle.

So what types of host cell proteins were affected during infection? The DNA repair proteins MRE11, RAD50, and LIG4 were detected only in mock-infected cells. While the inability to detect them in infected cells prevented their $\log_2$-fold change from being calculated, nevertheless it suggested that viral infection inhibits their production in host cells. A decrease in abundance of these proteins would prevent them from being available to repair potential sites of integration of viral genes into the host genome. The levels of DAXX, a protein that interacts with integrase-viral DNA complexes to repress viral gene expression (Tang et al. 2015), were also lower within infected cells. Not only were the abundances of these proteins decreased, but evidence of their ubiquitination was detected suggesting their degradation was mediated by the E4orf6-E1B55k E3 ligase complex: a complex known for its capacity to degrade host cell proteins that function to inhibit viral replication (Querido et al. 2001). The abundance of other proteins known to be targeted by E4orf3, namely, TRIM24 and TRIM33 and GTF2I, were also found to be lower within infected host cells. Both TRIM24 and TRIM33 are involved in silencing retrovirus via epigenetic silencing (Margalit et al. 2020). TRIM24 and TRIM33 have also been shown to reduce expression of the Epstein Barr virus (EBV) lytic switch gene *BZLF1* (De La Cruz-Herrera et al. 2023). In contrast, activation of EBV in latent infected cells showed an increase in the SUMOylation of TRIM24 and TRIM33, causing a disruption in the complex necessary to inhibit *BZLF1* expression (Dybas et al. 2021). The *general transcription factor I* (*GTF2I*) gene encodes for two proteins: TFII-I and Bruton tyrosine kinase-associated protein-135 (BAP-135) (Pérez Jurado et al. 1998). TFII-I is a transcription factor that regulates the expression of genes that assist in cell growth and division and calcium influx (Caraveo et al. 2006). BAP-135 activity is associated with B cells, the antibody producers of the immune system (Yang and Desiderio 1997). BAP-135 protein function is activated when B cells sense pathogens. In this situation, BAP-135 helps coordinate B cells to mature, divide, and produce antibodies.

#### 6.4.1.1 Classification of Host Proteins Affected by Adenovirus Infection

Considering the number of direct E4 region targets identified within the whole cell proteome dataset, it was hypothesized that many of the remaining proteins whose abundance level decreased in response to viral infection were also antiviral factors. The abundance of these proteins varied over the 0-, 16-, 24-, and 48-h time points showing the viral infection causes a dynamic change in the host cell proteome. Since such a large number (i.e., 2976) of proteins decreased in abundance due to viral infection, instead of analyzing them individually, they were grouped to identify cellular processes that were affected at each time point (see Table 6.3). These proteins (and their $\log_2$-fold change) were analyzed using Gene Ontology (GO), classifying the genes based on their annotated biological process.

The GO analysis provides an interesting overview of early, mid, and late events that occur during infection of host cells. The top ten populated GO categories at the 16-h postinfection time point were heavily populated with metabolism-related proteins. Many viral infections induce high glycolytic rates within their host, which results in the downregulation of other metabolic pathways as resources are shifted (Thai et al. 2014). Adenoviruses are known to upregulate glycolysis in infected cells, so the downregulation of other metabolic pathways is consistent with this observation. While some of the cell functions downregulated 24 h postinfection overlap with those seen at 16 h (i.e., cell organization, metabolic processes, etc.),

**Table 6.3** Heatmap showing the most enriched Gene Ontology (GO) biological process terms identified within the host cell's proteome that are significantly decreased at 16, 24, and 48 h postinfection (compared to mock-infected cells)

| Time post-infection (hours) | | | |
|---|---|---|---|
| 16 | 24 | 48 | Cell Process |
| | | | Negative regulation of nitrogen compound metabolic processes |
| | | | Alpha-linolenic acid metabolic processes |
| | | | Drug metabolic processes |
| | | | Small molecule metabolic processes |
| | | | Lipid catabolic processes |
| | | | Organelle organization |
| | | | Positive regulation of cellular component organization |
| | | | Membranous septum morphogenesis |
| | | | Regulation of transcription by RNApolII |
| | | | Positive regulation of endocytosis |
| | | | Chromosome organization |
| | | | Cell component organization or biogenesis |
| | | | Cell macromolecule metabolic processes |
| | | | Organonitrogen compound metabolic processes |
| | | | Protein-DNA complex subunit organization |
| | | | Protein-containing complex subunit organization |
| | | | Long-chain fatty acid metabolic processes |
| | | | Organic cyclic compound biosynthetic processes |
| | | | Oxidation-reduction processes |
| | | | RNA processing |
| | | | Gene expression |
| | | | ncRNA metabolic processes |
| | | | Ribonucleoprotein complex biogenesis |
| | | | RNA splicing |
| | | | Mitochondrial gene expression |

Heatmap colors correspond to lower abundance observed at specific time point (i.e., darker color equals lower abundance). GO terms with colors at multiple time points indicate that changes in proteins related to the specific biological process were observed at each time point

many new functions become downregulated at later time points (Dybas et al. 2021). These include various protein-DNA complexes as well as chromosome organization suggesting that the virus has begun to influence genetic factors within the host. As the postinfection time reaches 48 h, there appears to be a dramatic shift in the function of proteins that are downregulated. While many of the same protein categories involved in protein-DNA complexes and chromosome regulation that were downregulated at 24 h were further downregulated 48 h postinfection, significant numbers of downregulated proteins at 48 h are involved in processes related to RNA. The categories included RNA processing and splicing, noncoding RNA metabolism, and formation of ribonucleoprotein complexes.

What can sometimes get lost in the scientific detail is the speed at which a virus acts upon its host. Within 16 h, Ad5 has already affected the host cell's metabolism, by downregulating unnecessary energy-expending pathways to increase energy reserves within the host. Within 24 h, the virus has "hijacked" many genomic processes to prepare the host cell to start producing proteins needed for the virus to replicate. By 48 h, the virus had impacted the host cell's ability to process RNA and prevents it from generating antiviral signals, such as interferons. Taken collectively, this data shows that Ad5 dynamically affects specific host cell functions at different stages of infection in an orderly manner that reflects the various stages of viral replication.

### 6.4.2 Validation of Host Proteins Affected by Adenovirus Infection

A consistent need for these types of whole cell proteomic analyses is validation of the results. Unfortunately repeating the same experiment is not considered validation in proteomics, rather orthogonal techniques such as Western blotting or ELISAs are required. The main challenge with these techniques is they measure a single protein per experiment. Considering many whole cell proteomic studies identify thousands of proteins showing an increase or decrease in abundance, it is impossible to validate every finding. This stage is where logic and reason in selecting proteins (and the number of proteins) to validate must be exercised. In this study, proteins within the GO category "Chromosome Organization" were selected since this function was the most highly enriched at 24 h postinfection and remained highly enriched through 48 h (Dybas et al. 2021). Several targets of the E4 region of Ad5, which counteracts antiviral proteins, were represented in this dataset. MRE11, RAD50, and LIG4 were found within this category; however, these proteins were not selected for further validation since other studies have shown them to be targets of Ad5 (Schwartz et al. 2008). Lymphoid-specific helicase (HELLS/SMARCA6/LSH), ATP-dependent DNA helicase Q4 (RECQL4), sequestosome-1 (SQSTMI/p62), and DDB1- and CUL4-associated factor 1 (DCAF1) were selected for validation owing to their specific cellular functions. HELLS/SMARCA6/LSH functions in nucleosome remodeling through regulation of DNA methylation (Jenness et al. 2018). RECQL4 participates in response to DNA damage (Schurman et al. 2009). SQSTMI/p62 plays a role in the immune response to viral infection by promoting autophagy (Schwob et al. 2019). The kinase DCAF1, which was downregulated only at the 48-h time point, is known to inactivate genes via phosphorylation of histone 2A (Shin et al. 2023). All these proteins showed an Ad5-dependent decrease in abundance during infection as measured via immunoblotting. While these experiments only validated approximately 0.1% of the results of the whole cell proteome study, extrapolating the limited immunoblot results onto the proteome results provides an additional layer of confidence in the whole cell proteome results.

## 6.5 Proteomic Analysis of Virus Infection Using Protein Isotopic Labeling

The Ad5 infection study described above utilized a label-free method in which the area of peaks associated with unique peptides is compared across samples. Label-free methods are often used to analyze clinical samples since they do not require additional sample preparation steps and are relatively inexpensive to perform. Unfortunately, this method analyzes samples separately, increasing the rate of error due to LC/MS issues such as ion suppression, peptide selection, chromatography variation, etc. Owing to the mass spectrometer's data-dependent selection of peptides and the complexity of the samples being analyzed, comparisons between samples are difficult since peptides analyzed in one sample have a reasonable chance of being overlooked in another.

Recent years have seen an influx of protein labeling methods that employ stable isotope tags to quantitate changes in protein abundances. In contrast to label-free methods, labeling methods allow multiple samples to be compared within the same LC/MS experiment. This ability serves to increase the overlap between samples and minimizes instrumental and sample variation. Some of the most popular methods for isotopically labeling proteins are isobaric tandem mass tags (TMTs), isobaric tags for relative and absolute quantitation (iTRAQ), and stable isotope labeling of amino acids in cell culture (SILAC). While each method has been used to study the infection of host cells by viruses, TMTs were used in a proteomic technique called quantitative temporal viromics (QTV) (Fletcher-Etherington and Weekes 2021). The QTV method has been used to study many human viruses including EBV, BK polyomavirus, vaccinia virus, HSV-1, and influenza A virus (IAV) (Ersing et al. 2017; Caller et al. 2019; Soday et al. 2019; Soh et al. 2020; Bogdanow et al. 2023).

Isobaric tags (i.e., TMT, iTRAQ, etc.) have an identical mass; however, the heavy isotopes incorporated into their structures are placed at different atomic locations enabling the portions of the tags liberated using $MS^2$ to be distinguished from one another (see Fig. 6.5) (Chen et al. 2007). After proteolytic digestion of the individual proteome samples, the peptides are labeled with different isobaric tags, combined, and analyzed using a single LC/$MS^2$ experiment. During the initial MS survey scan, peptide ions are isolated and fragmented. Fragmentation cleaves the isobaric tags off the peptides as well as provides the necessary information to identify the peptide. The quantity of each peptide is measured by comparing the intensity of the reporter ion cleaved from the individual tags, which have a distinct mass. The variety of tags available allows up to 8 or 16 samples to be directly compared using iTRAQ or TMT, respectively.

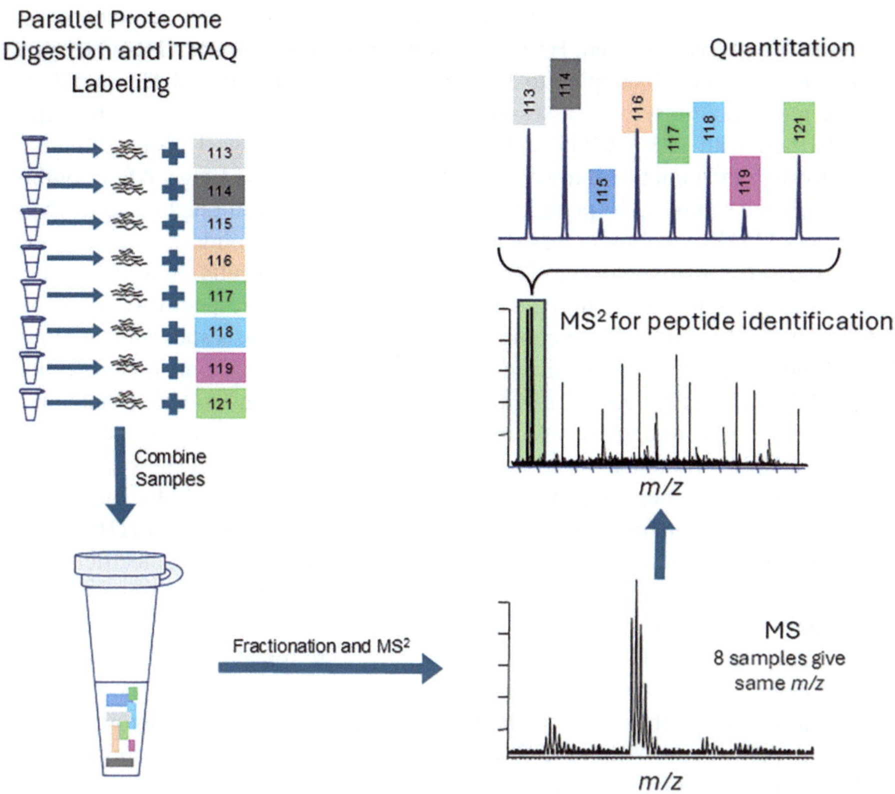


**Fig. 6.5** Process of identifying proteomic changes in cells using quantitative temporal viromics (QTV). In QTV protein samples are digested into peptides that are identified and quantified using liquid chromatography combined with tandem mass spectrometry (LC/MS$^2$). Host cells are infected, lysed at selected times postinfection, and processed for MS analysis. QTV can utilize a variety of different quantitation methods including metabolic prior to cell lysis and chemical labeling after cell lysis. Multiple samples can be compared within a single LC/MS$^2$ analysis. Abbreviations: iTRAQ isobaric tag for relative and absolute quantitation

### *6.5.1 Proteomic Analysis of H1N1- and H5N1-Infected Mast Cells*

A study used iTRAQ to measure quantitative changes in proteins that occurred within human mast cells separately infected with H1N1 (swine) and H5N1 (avian) influenza viruses (Wu et al. 2019). Mast cells are found in all tissues and serve as immune modulators. They release cytokines and chemokines to recruit neutrophils and dendritic cells to eliminate pathogenic infections (Bradding et al. 1993). While their role in innate immunity against parasites and bacteria is well known, their function in viral immunity is unclear. It is known that mast cells recruit natural killer and cytotoxic T cells in response to dengue virus (St John et al. 2011). In addition, they have been shown to play a role in pathogenesis of influenza A viral infection in mouse lung cells (Hu et al. 2012). While only three different cell treatments were analyzed (i.e., H1N1 infected, H5N1 infected, and control), 8-plex iTRAQ reagents were used enabling 8 samples to be concurrently compared and the sample-to-sample reproducibility to be assessed. Three H1N1- and H5N1-infected samples were labeled with 115/116/117 and 118/119/121 iTRAQ tags 12 h postinfection, respectively, while control samples were labeled with 113/114 tags. After combining the labeled samples, the peptides were identified and quantitated using $LC/MS^2$.

Almost 11,000 peptides, corresponding to 3266 unique peptides, were quantified in this study. Comparison of the H1N1-infected and control cells showed 12 and 21 proteins to be significantly upregulated or downregulated, respectively, in H1N1-infected cells. The GO cellular component enrichment analysis showed that almost half of these proteins were localized to the nucleus. These proteins included Ski2-like RNA helicase (SKIV2L), CCR4-NOT transcription complex subunit 4 (CNOT4), interferon regulatory factor 2 binding protein-like, and DExH-box helicase 29 (DHX29) proteins. Surprisingly, many of the proteins that showed a change in abundance were localized to the Golgi apparatus and intracellular membranes suggesting these structures may play important roles with respect to H1N1 infection. These proteins also participated in processes such as transport, electrical stimulus response, RNA processing, mRNA splicing, signal transduction, and transcription regulation. Other proteins were associated with molecular functions such as poly(A) and GTP binding. Five pathways were found to be highly enriched with proteins found to be differentially abundant in H1N1-infected cells (i.e., RNA degradation, thyroid hormone signaling pathway, spliceosome, phagosome, and metabolic pathways). Within the RNA degradation pathway, the proteins CNOT4, SKIV2L, and U6 small nuclear RNA and mRNA degradation associated protein (LSM7) were significantly upregulated. Protein-protein interaction analysis showed that SKIV2L and LSM7 interact within the same protein complex. The results suggest that H1N1 infection affects the degradation of various RNAs within mast cells.

A total of 30 proteins were upregulated, and 45 were downregulated in cells infected with H5N1. Two major classes of binding proteins were highly represented within these groups. Many of these proteins were involved in binding DNA, poly(A)-RNA, ATP, and metals such as zinc. Three groups (i.e., RNA, GTP, and calcium ion binding) were found to be upregulated, while eukaryotic translation initiation factor 4E (eIF4), cell division cycle 20 (cdc20), and protein inhibitor of activated STAT1 proteins were downregulated. Other functions that were affected by H5N1 infection compared to control cells included transcription coactivator activity, cell-cell adhesion, and protein transport.

The results provide an overview of changes in protein abundances that occur within specific regions of infected mast cells. Of these 41 total proteins shown to be over- or underexpressed in H5N1-infected cells, 10 were localized to the membrane. Almost half were cytoplasmic and nuclear proteins including mTOR, the apoptotic factors Bcl2-associated X, and Bax. Continuing this theme of proteomic changes in various subcellular locations, other proteins were localized to the mitochondria, nucleosome, meiotic spindle, and even extracellular exosomes. More than half (i.e., 26) of the changes occurred to proteins associated with protein binding, primarily poly(A)-RNA and ATP binding. Other unique functions such as GTPase activity, hetero- and homo-dimerization activity, ATPase activity, ribosome binding, serine/threonine kinase activity, and kinase binding differed between H1N1- and H5N1-infected cells.

Gene ontology comparison of H1N1- and H5N1-infected cells showed an increase in microtubule (i.e., tubulin, etc.), translation and transcription, cell cycle arrest, cell differentiation, and signal transduction proteins in H5N1-infected cells. The downregulated proteins were involved in different biological processes including maintenance of transmembrane electrochemical gradient (i.e., Bax and ATP1A3 proteins), fertilization (Bax and Y-box binding protein), receptor internalization (arrestin 2 and transferrin receptor), protein transport, and negative regulation of protein binding (Bax and aurora kinase A). KEGG pathway analysis showed that many proteins that are part of the hypoxia-inducible factor-1 (HIF-1) signaling pathway were either up- or downregulated in H5N1-infected cells compared to those infected with H1N1. Other proteins that are not directly part of the HIF-1 pathway but interact with HIF-1 signaling proteins (e.g., mTOR, 6-phosphofructo-2-kinase/fructose-2,6-biphosphatase 3, glucose transporter 1, and transferrin receptor) showed differences in abundance depending on the specific viral infection.

Proteomics has been used in several studies to evaluate the effect of different influenza A strains on various cell types (Bogdanow et al. 2019; Wang et al. 2019; Nyman and Matikainen 2018; Yang et al. 2021). These studies have also used many different techniques (i.e., 2D-PAGE, SILAC, iTRAQ, etc.). These studies show that different viruses can have grossly different effects on the same cell line. While some of these differences are likely due to experimental variation, much of it is due to the unique nature of the virus-host cell interaction. The previous study shows that the uniqueness of this interaction is very discrete as different strains of the same virus can affect the proteome of the same cell type in profoundly different ways (Wu et al. 2019). The subtlety is most obvious in the activation of the HIF-1 signaling pathway. The HIF-1 signaling pathways control the activity of various transcription factors that express genes to coordinate the cell's response to hypoxia (O'Rourke et al. 1997). In the present study, various members of this signaling pathway were differentially abundant in mast cells infected with H1N1 or H5N1. This result shows that the HIF-1 signaling pathway is more strongly activated in H5N1-infected mast cells.

What is remarkable about these studies is the vast and intricate nature of the interactions between viruses and their host cells. The interaction is often viewed simply as the virus entering the cell and using the host's transcriptional and translational machinery to make the parts sufficient to make copies of themselves. The large amount of data gathered in proteomic studies has demonstrated there is much more to this interaction. Not only do viruses have mechanisms to hijack cell machinery, but they have also evolved mechanisms to combat attempts by the host cell to eliminate their presence. Temporal studies also show that the virus commandeers certain proteins within the host cell in a time-dependent, organized fashion.

Understanding the dynamic interplay between viruses and their host will aid in the development of effective treatments, such as vaccines and antiviral therapies, designed to limit viral spread both within an individual and the community.

## 6.6 Viral Egress

An often-underappreciated action performed by viruses is how they leave their host cell. It would be a fool's errand for a virus to enter a cell and replicate without being able to leave and propagate in other cells. Without the ability to leave, a virus would quickly exhaust its host's resources. Not only would a lack of resources stop its continued replication but also make the existing virions susceptible to destruction by the host's immune response. While most drugs aim to prevent viral entry and replication within host cells, there are notable studies to identify medications to prevent viruses from exiting host cells. For example, Marburg virus (MARV) relies on an interaction between its matrix protein (mVP40) and the host E3 ubiquitin ligase Nedd4 protein to promote its budding and release (Harty et al. 2000; Liang et al. 2017; Han et al. 2016). A compound termed, FC-10696 was designed to target the interaction between mVP50 and Nedd4, effectively inhibiting the egress of MARV from its host (Han et al. 2021). Marburg-infected mice treated with this drug showed delayed onset of clinical symptoms and lower viral loads compared to controls. In addition, 14% of the animals given a lethal dose of MARV survived after being treated with FC-10696.

FC-10696-treated mice displayed delayed onset of weight loss and clinical signs and significantly lower viral loads compared to controls, with 14% of animals surviving 21 days following a lethal MARV challenge. While a drug like FC-10696 that targets viral exit would unlikely be a first-line medication, it could be used in cases where a diagnosis of infection was delayed or in combination with an antiviral that targets viral entry or replication.

### *6.6.1 Proteomic Analysis of Viral Egress*

Proteomic studies that describe the mechanism(s) of how viruses exit cells are much less developed than those that measure viral entry and propagation. The difficulty mainly resides in the inability to capture populations of cells in the process of viral egress without having a substantial background of viral replication occurring. This deficiency makes using global proteomic studies for measuring proteomic changes that specifically reflect the virus exiting the cell almost impossible. Enveloped viruses exit the cell by acquiring the membrane of the host cell they are exiting (Welsch et al. 2007). The membrane, however, can come from the plasma membrane or an internal compartment such as the endoplasmic reticulum (ER). If the virus egresses by acquiring its coat from the plasma membrane, changes in plasma membrane proteins would be expected to occur, and the virus finds itself immediately outside of the cell. If the virus acquires its coat from an internal compartment, protein changes would occur within this compartment membrane as well as with proteins involved in the secretion pathway that moves the virion from the lumen to the plasma membrane.

Some viruses are released into the extracellular space simply through the lysis of the host cell (Dimitrov 2004). In this scenario, specialized viral proteins are produced that weaken the cell's wall. As the number of replicated virions increases, the cell's internal osmotic pressure reaches a point that the water pressure can no longer be contained by the cell. The bursting of the host cell releases virions into the surrounding environment, and they go on to infect nearby cells. Some viruses are released from cells enclosed in membranous vesicles, known as exosomes, without the need for cell lysis (van der Grein et al. 2019). By exiting in this manner, the host cell is preserved and continues to be used by the virus for replication. The vesicle also shields the virus from the immune system. While often used to describe any membrane-bound vesicle, the term exosome properly refers to extracellular vesicles that are between 40 and 150 nm in diameter and form through the inward budding of endosomal membranes. Since endosomes often contain multiple intraluminal vesicles, they are sometimes referred to as multivesicular endosomes or multivesicular bodies (MVBs). While MVBs generally transport materials to the lysosome, some move to and fuse with the plasma membrane to release their contents into the extracellular space (see Fig. 6.6). While their existence and functions are well known, the specific signals that control their formation and movements are not well understood.

### 6.6.2 *Proteomic Analysis of Non-lytic Release*

The non-lytic release of viruses is an important function that drives disease pathogenesis, but the mechanism of this process is not well understood primarily due to the limitations of static experimental methods trying to detail a dynamic process. A study of the non-lytic release of quasi-enveloped hepatitis A virus (eHAV) using a quantitative proteomics approach showed that the process is highly specific and controlled (McKnight et al. 2017). In this study, stable isotope labeling by amino acids in cell culture (SILAC) was used to quantify proteins from infected (heavy stable isotope labeled) and noninfected (control; light stable isotope labeled) human hepatomic cells (Huh-7.5). The SILAC method is very straightforward and was one of the earliest stable isotope labeling methods developed for quantitative proteomics (see Fig. 6.7) (Ong et al. 2002; Veenstra et al. 2000). In SILAC, cells to be compared are cultured in either normal isotopic abundance medium or a medium containing an amino acid that contains specific heavy isotopes (e.g., Lys-$^{13}C_6$, Arg-$^{13}C_6$, etc.). After being cultured in the light or heavy isotopic media, the cells are treated as per experimental protocol. After treatment, equal numbers of cells from the comparative cultures are combined, and the proteins are extracted and digested into peptides

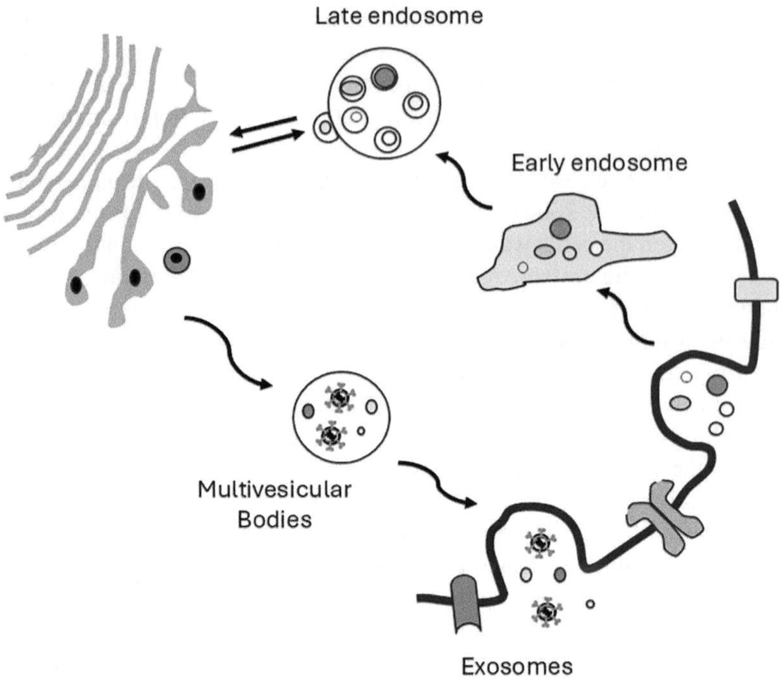


**Fig. 6.6** Exosome production. Exosomes originate from the inward budding of the cell membrane that encapsulates extracellular materials, which can include virions. These early endosomes mature into late endosomes that transport materials to the Golgi apparatus. After processing and sorting, cellular material is packaged into multivesicular bodies that fuse with the plasma membrane allowing molecules, including viruses, to exit the cell

(if a bottom-up proteomic strategy is to be used). The peptides are analyzed using LC/MS$^2$. The origin of each peptide can be determined by their relative masses as those extracted from the cells cultured in the heavy isotopic medium will have a predictable higher mass than those cultured in normal isotopic medium. The relative abundance of the peptides in the different cultures is measured by comparing their peak heights or area.

Quasi-enveloped hepatitis A virus proteins associated with the virion containing exosomes were compared to exosomes from control cells by using a series of centrifugation steps to isolate particles that were of similar size and density from the two cell samples. After confirming the presence of HAV in the heavy isotope-labeled preparation, it was mixed with the parallel fraction isolated from the normal isotope-labeled cells. The combined sample was digested into peptides, which were identified and quantitated using LC/MS$^2$. A separate experiment was conducted in which infected cells were labeled with the normal isotopes and the cells being compared were combined prior to enrichment of the virions and parallel exosomes.

Overall, the study identified 638 proteins, with 294 of these being found in both heavy and light isotope-labeled samples. Peptides corresponding to 105 host proteins that were at least twofold enriched in the eHAV vesicles were identified in at least 2 of 4 replicate samples. This number dropped to 26 for proteins enriched by

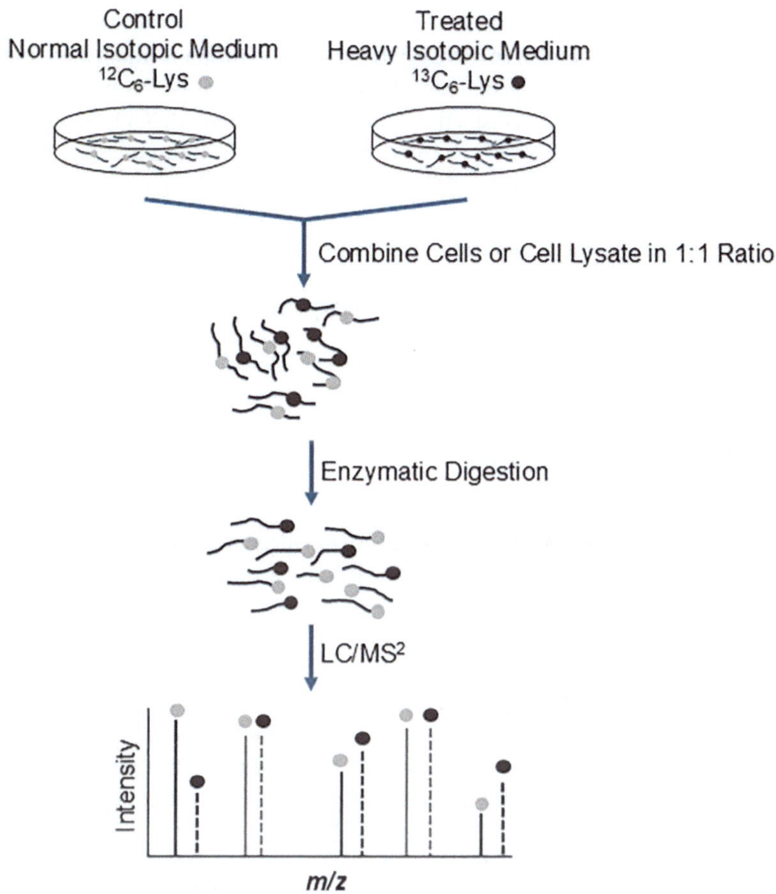


**Fig. 6.7** Stable isotope labeling by amino acids in cell culture (SILAC). In SILAC, cell populations are cultured in normal isotopic abundance medium or medium containing a heavy isotope-labeled version of a specific amino acid (e.g., $^{13}C_6$-Lys as shown above). After treating one of the cultures, equal amounts of each are combined and analyzed using bottom-up proteomics. Differences in protein abundance are determined by comparing the heights or areas of the peptide peak pairs originating from the different cell cultures

at least twofold in at least 3 eHAV vesicle replicates. Over 90% of the proteins had been previously identified in exosomes. Specifically, several lysosome-associated proteins were identified in all four virion samples. Of the 26 proteins observed in at least 3 replicates, 15 were human, and 11 were of indeterminate origin and may have originated from the bovine serum that was added as part of the cell culture media. Of the 26 proteins, 5 are known to be functionally associated with the endosomal sorting complexes required for transport III (ESCRT-III) complex. This complex is made up of cytoplasmic proteins that cause remodeling so that membranes can bend and/or bud away from the cytoplasm (Vietri et al. 2020). The ESCRT-III complex functions in all ESCRT-mediated processes. During membrane detachment (i.e., abscission) and viral budding, this complex forms long filaments that wrap around the site of membrane constriction to initiate membrane cleavage (Pfitzner et al. 2020). The five ESCRT-III complex proteins identified included charged multivesicular body proteins (CHMP) 1A, 1B, and 4B as well as syntenin (SDCBP) and programmed cell death 6-interacting *protein* (PDCD6IP, also known as ALIX) (McKnight et al. 2017). Syntenin and ALIX form a pathway along with syndecan to affect intraluminal budding of endosomes resulting in the generation of MVB that often forms into exosomes. Two Ras-associated binding (Rab) GTPases, RAB5C and RAB7A, were also found to be upregulated within HAV-infected cells. These proteins function in endosomal trafficking. Taken together, these proteins show that eHAV egress is conducted via an MVB-like endosomal export pathway.

### *6.6.3 Validation of Quasi-Enveloped Hepatitis A Exosome Proteins*

As discussed throughout this book, any discovery-driven proteomics study requires hypothesis-driven experiments to confirm at least a portion of the results. In this study, eHAV was immunoprecipitated from serum collected from humans in the early stage of eHAV infection. The patients were acquired during a nationwide outbreak within Korea and confirmed to only contain eHAV, with undetectable levels of non-enveloped virus. Immunoprecipitation using anti-ALIX and DPP4 antibodies precipitated eHAV from sera confirming the inclusion of both proteins within the virions. To assess the functional relevance of these proteins to viral egress, eHAV-infected cells were transfected with small interfering RNAs targeting the ESCRT-III-associated proteins ALIX, CHMP2A, CHMP2B, and IST1 (McKnight et al. 2017). While CHMP1A and CHMP1B knockdowns were not as efficient as those seen for the other proteins, the partial knockdown of CHMP1B enhanced eHAV egress. Knockdown of CHMP1B caused the levels of eHAV found in the supernatant to increase 68%. Knockdown of CHMP2A alone or a combination of CHMP2A and CHMP2B together resulted in a 63% and 75% reduction in eHAV egress, respectively. Knockdown of CHMP2B alone only resulted in 18% reduction. Knockdown of DPP4 and RAB7A had no effect on eHAV egress, suggesting that while these proteins are part of the eHAV MVB, they do not play a role in viral egress.

The proteomic data shows that the packaging of eHAV virions into MVBs is a highly selective and specific sorting process. All four structural major capsid proteins of eHAV (i.e., VP1–VP4) were identified with high coverage; however, no peptides from larger nonstructural viral proteins were detected (McKnight et al. 2017). While nonstructural eHAV proteins are likely present in high abundance within the cell, they are not co-packaged as part of the eHAV vesicles. While the specific signals that dictate what proteins are ultimately packaged within the eHAV capsid are still unknown, it is likely that ALIX-VP2 interactions play a role in this sorting (Bissig and Gruenberg 2014).

An interesting finding in the above study was that while ESCRT-III-associated proteins played a role in eHAV egress, no proteins associated with ESCRT-I or II were identified. This result is surprising considering that ESCRT-I and ESCRT-II proteins play an important role in exosome biogenesis (Ju et al. 2021). Tumor susceptibility gene 101 (TSG101) is associated with ESCRT-I and is usually present in exosomes derived from MVBs (Nabhan et al. 2012). The finding that ESCRT-III-associated proteins, but not those linked with ESCRT-I or ESCRT-II, play a role in MVB formation in eHAV infection and exosome biogenesis contradicts existing paradigms. This result shows that eHAV uses only part of the known exosome biogenesis pathway. In a bigger sense, it points to the fact that individual viruses may have evolved customized methods for exiting infected cells.

Customized egress mechanisms for viruses are epitomized in the comparison of eHAV with poliovirus and other enteroviruses. Poliovirus exits from infected cells non-lytically in autophagosome-derived vesicles that contain LC3-II (microtubule-associated proteins 1A/1B light chain 3B), a protein that is localized to autophagic structures throughout their life cycle (Taylor et al. 2009). These poliovirus vesicles are larger than those associated with eHAV and contain many more viral capsids. While autophagy may play a role in HAV egress, LC3-II was not detected in any HAV viral preparations, nor were these virions immunoprecipitated using an anti-LC3-II antibody. In addition, the knockdown of beclin 1, another protein associated with poliovirus autophagy, had no effect on the exit of eHAV from the cell (McKnight et al. 2017). Hepatitis E virus (HEV) provides another example of the potential uniqueness of viral exit strategies. HEV is phylogenetically distinct from picornaviruses, being classified within the *Hepeviridae* family. HEV, along with bluetongue virus (family *Reoviridae*) and HAV, exits cells via non-lytic processes using ESCRT-associated proteins (Roy 2020). Unlike HAV, however, both HEV and bluetongue virus rely on TSG101 for cell exit (Kumar et al. 2024). It is likely that the non-lytically released, non-enveloped viruses (including HAV, HEV, and bluetongue) exit cells using a variety of heterogeneous mechanisms.

## 6.7 Conclusions

There are approximately 270 known viruses that infect humans (Forni et al. 2022). This number will continue to increase as technology identifies additional viruses that infect humans and viruses evolve to form new strains. While viruses can be classified based on their genetic material (DNA or RNA), structure, and mode of replication, the more we learn about them, the more diverse they seem to be. Their diversity is exemplified by possessing either RNA or DNA genomes that can be single or double stranded. They infect a variety of different hosts, encompassing all organisms presently living on Earth. Owing to differences in their genetic material, viruses express a variety of structural proteins that lead to morphological diversity. These differences at the genetic and structural levels have led to a variety of viral functions that continue to increase as they continue to evolve.

As the various host cell processes affected by the presence of the virus become better understood, researchers are finding that virus-host interactions have their own individual signatures. These processes include viral entry, proliferation, and egress. Different viruses use unique methods for entering cells. While access begins with attachment of the virus to a cell receptor, entry is gained either through the penetration of or fusion with the cellular membrane. The molecular interactions made at the cell surface that enable viral entry are very specific and highly detailed. This specificity is a major reason why humans can survive within an environment in which we are surrounded by viruses. If viruses were simply able to enter our cells through passive diffusion, it is unlikely humans would thrive. Once inside, viruses proliferate by hijacking the cell's replication machinery. After reassembling their required parts, virions exit the cell to infect others. Much like cell entry, the processes by which viruses replicate are highly diverse and vary depending on the genomic and proteome content of the virus. Non-enveloped viruses are generally assumed to exit cells via cell lysis, while enveloped viruses exit via budding through cell membranes. Release via cell lysis requires no specific mechanism because the cell membrane that encases the virions simply degrades.

The past few years have seen the impact that genomic technology has had on viral research. New strains of viruses are quickly identified, and diagnosis of infected individuals is extremely fast and accurate. What proteomics is teaching us is that viruses have unique pathways of cell entry, replication, and egress. Not only is proteomics enabling the data-gathering throughput necessary to detail the processes used by viruses to thrive, but they are demonstrating that the complexity of virus-host cell interactions is much greater than we ever imagined. The molecular processes involved in viral propagation are very complex. In addition, the coordination of molecular replication and virion formation is coordinated at higher levels than previously understood. The ability of proteomics technology to unravel all these processes will continue to lead to greater opportunities to develop medications that can prevent viral infection and community spread. These medications will hopefully usher in a time when pandemics are no longer a threat to human existence.

## References

Abbas AA, Taylor LJ, Dothard MI, Leiby JS, Fitzgerald AS, Khatib LA et al (2019) Redondoviridae, a family of small, circular DNA viruses of the human oro-respiratory tract associated with periodontitis and critical illness. Cell Host Microbe 26:297

Arthos J, Cicala C, Martinelli E, Macleod K, Van Ryk D, Wei D et al (2008) HIV-1 envelope protein binds to and signals through integrin alpha4beta7, the gut mucosal homing receptor for peripheral T cells. Nat Immunol 9:301–309

Baron S, Fons M, Albrecht T (1996) Viral pathogenesis. In: Baron S (ed) Medical Microbiology, 4th edn. University of Texas Medical Branch at Galveston, Galveston. Chapter 45

Barr JJ, Auro R, Furlan M, Whiteson KL, Erb ML, Pogliano J et al (2013) Bacteriophage adhering to mucus provide a non-host-derived immunity. Proc Natl Acad Sci USA 110:10771–10776

Birkmann A, Mahr K, Ensser A, Yağuboğlu S, Titgemeyer F, Fleckenstein B et al (2001) Cell surface heparan sulfate is a receptor for human herpesvirus 8 and interacts with envelope glycoprotein K8.1. J Virol 75:11583–11593

Bissig C, Gruenberg J (2014) ALIX and the multivesicular endosome: ALIX in Wonderland. Trends Cell Biol 24:19–25

Bogdanow B, Wang X, Eichelbaum K, Sadewasser A, Husic I, Paki K et al (2019) The dynamic proteome of influenza A virus infection identifies M segment splicing as a host range determinant. Nat Commun 10:5518

Bogdanow B, Katsimani N, Liu F, Selbach M (2023) Combining metabolic pulse labeling and quantitative proteomics to monitor protein synthesis upon viral infection. Methods Mol Biol 2610:149–165

Bradding P, Feather IH, Wilson S, Bardin PG, Heusser CH, Holgate ST et al (1993) Immunolocalization of cytokines in the nasal mucosa of normal and perennial rhinitic subjects. The mast cell as a source of IL-4, IL-5, and IL-6 in human mucosal inflammation. J Immunol 151:3853–3865

Caller LG, Davies CTR, Antrobus R, Lehner PJ, Weekes MP, Crump CM (2019) Temporal proteomic analysis of BK polyomavirus infection reveals virus-induced G2 arret and highly effective evasion of innate immune sensing. J Virol 93:e00595–e00519

Caraveo G, van Rossum DB, Patterson RL, Snyder SH, Desiderio S (2006) Action of TFII-I outside the nucleus as an inhibitor of agonist-induced calcium entry. Science 314:122–125

Chen X, Sun L, Yu Y, Xue Y, Yang P (2007) Amino acid-coded tagging approaches in quantitative proteomics. Expert Rev Proteomics 4:25–37

Chen S, Yan Y, Gao L, Gao S, Feng K, Li H et al (2023) Proteomic profiling of purified avian leukosis virus subgroup J particles. Vet Microbiol 284:109821

Chua CE, Gan BQ, Tang BL (2011) Involvement of members of the Rab family and related small GTPases in autophagosome formation and maturation. Cell Mol Life Sci 68:3349–3358

De La Cruz-Herrera CF, Tatham MH, Siddiqi UZ, Shire K, Marcon E, Greenblatt JF et al (2023) Changes in SUMO-modified proteins in Epstein-Barr virus infection identifies reciprocal regulation of TRIM24/28/33 complexes and the lytic switch BZLF1. PLoS Pathog 19:e1011477

Deng H, Liu R, Ellmeier W, Choe S, Unutmaz D, Burkhart M et al (1996) Identification of a major co-receptor for primary isolates of HIV-1. Nature 381:661–666

Denz PJ, Speaks S, Kenney AD, Eddy AC, Papa JL, Roettger J, et al (2023) Innate immune control of influenza virus interspecies adaptation. bioRxiv 23:2023.08.23.554491

Dimitrov DS (2004) Virus entry: molecular mechanisms and biomedical applications. Nat Rev Microbiol 2:109–122

Douglas JT (2007) Adenoviral vectors for gene therapy. Mol Biotechnol 36:71–80

Dragic T, Litwin V, Allaway GP, Martin SR, Huang Y, Nagashima KA et al (1996) HIV-1 entry into $CD4^{+}$ cells is mediated by the chemokine receptor CC-CKR-5. Nature 381:667–673

Dybas JM, Lum KK, Kulej K, Reyes ED, Lauman R, Charman M et al (2021) Adenovirus remodeling of the host proteome and host factors associated with viral genomes. mSystems 6:e0046821

Ersing I, Nobre L, Wang LW, Soday L, Ma Y, Paulo JA et al (2017) A temporal proteomic map of Epstein-Barr virus lytic replication in B cells. Cell Rep 19:1479–1493

Fletcher-Etherington A, Weekes MP (2021) Quantitative temporal viromics. Ann Rev Virol 8:159–181

Flint J (1999) Organization of the adenoviral genome. In: Seth P (ed) Adenoviruses: basic biology to gene therapy. R.G. Landes Company, Austin, pp 17–30

Forni D, Cagliani R, Clerici M, Sironi M (2022) Disease-causing human viruses: novelty and legacy. Trends Microbiol 30:1232–1242

Glynn JR, Moss PAH (2020) Systematic analysis of infectious disease outcomes by age shows lowest severity in school-age children. Sci Data 7:329

Han Z, Sagum CA, Bedford MT, Sidhu SS, Sudol M, Harty RN (2016) ITCH E3 ubiquitin ligase interacts with Ebola virus VP40 to regulate budding. J Virol 90:9163–9171

Han Z, Ye H, Liang J, Shepley-McTaggart A, Wrobel JE, Reitz AB et al (2021) Compound FC-10696 inhibits Egress of Marburg Virus. Antimicrob Agents Chemother 65:e0008621

Harty RN, Brown ME, Wang G, Huibregtse J, Hayes FP (2000) A PPxY motif within the VP40 protein of Ebola virus interacts physically and functionally with a ubiquitin ligase: implications for filovirus budding. Proc Natl Acad Sci USA 97:13871–13876

Horwitz MS (2004) Function of adenovirus E3 proteins and their interactions with immunoregulatory cell proteins. J Gene Med 6:S172–S183

Hu Y, Jin Y, Han D, Zhang G, Cao S, Xie J et al (2012) Mast cell-induced lung injury in mice infected with H5N1 influenza virus. J Virol 86:3347–3356

Jenness C, Giunta S, Müller MM, Kimura H, Muir TW, Funabiki H (2018) HELLS and CDCA7 comprise a bipartite nucleosome remodeling complex defective in ICF syndrome. Proc Natl Acad Sci USA 115:E876–E885

Ju Y, Bai H, Ren L, Zhang L (2021) The role of exosome and the ESCRT pathway on enveloped virus infection. Int J Mol Sci 22:9060–9071

Kang Y, Ji X, Guo L, Xia H, Yang X, Xie Z et al (2021) Cerebrospinal fluid from healthy pregnant women does not harbor a detectable microbial community. Microbiol Spectrum 9:e0076921

Kotani N, Nakano T, Kuwahara R (2022) Host cell membrane proteins located near SARS-CoV-2 spike protein attachment sites are identified using proximity labeling and proteomic analysis. J Biol Chem 298:102500

Kumar MA, Baba SK, Sadida HQ, Marzooqi SA, Jerobin J, Altemani FH et al (2024) Extracellular vesicles as tools and targets in therapy for diseases. Signal Transduct Target Ther 9:27–67

Lederberg J, McCray AT (2001) Ome Sweet'Omics – a genealogical treasury of words. The Scientist 15:8

Lewandowska-Pietruszka Z, Figlerowicz M, Mazur-Melewska K (2022) The history of the intestinal microbiota and the gut-brain axis. Pathogens 11:1540

Li W, Moore MJ, Vasilieva N, Sui J, Wong SK, Berne MA et al (2003) Angiotensin-converting enzyme 2 is a functional receptor for the SARS coronavirus. Nature 426:450–454

Liang J, Sagum CA, Bedford MT, Sidhu SS, Sudol M, Han Z et al (2017) Chaperone-mediated autophagy protein BAG3 negatively regulates Ebola and Marburg VP40-mediated Egress. PLoS Pathog 13:e1006132

Lippé R (2012) Deciphering novel host-herpesvirus interactions by virion proteomics. Front Microbiol 28(3):181

Margalit L, Strauss C, Tal A, Schlesinger S (2020) Trim24 and Trim33 play a role in epigenetic silencing of retroviruses in embryonic stem cells. Viruses 12:1015

McKnight KL, Xie L, González-López O, Rivera-Serrano EE, Chen X, Lemon SM (2017) Protein composition of the hepatitis A virus quasi-envelope. Proc Natl Acad Sci USA 114:6587–6592

Mercer J, Schelhass M, Helenius A (2010) Virus entry by endocytosis. Annu Rev Biochem 79:803–833

Mollentze N, Streicker DG, Murcia PR, Hampson K, Biek R (2020) Virulence mismatches in index hosts shape the outcome of cross-species transmission. Proc Natl Acad Sci USA 117:28859–28866

Nabhan JF, Hu R, Oh RS, Cohen SN, Lu Q (2012) Formation and release of arrestin domain-containing protein 1-mediated microvesicles (ARMMs) at plasma membrane by recruitment of TSG101 protein. Proc Natl Acad Sci USA 109:4146–4151

Niu S, Guo Y, Wang X, Wang Z, Sun L, Dai H et al (2023) Innate immune escape and adaptive immune evasion of African swine fever virus: a review. Virology 587:109878

Nyman TA, Matikainen S (2018) Proteomics to study macrophage response to viral infection. J Proteome 180:99–107

Ong SE, Blagoev B, Kratchmarova I, Kristensen DB, Steen H, Pandey A et al (2002) Stable isotope labeling by amino acids in cell culture, SILAC, as a simple and accurate approach to expression proteomics. Mol Cell Proteomics 1:376–386

O'Rourke JF, Dachs GU, Gleadle JM, Maxwell PH, Pugh CW, Stratford IJ et al (1997) Hypoxia response elements. Oncol Res 9:327–332

Pérez Jurado LA, Wang YK, Peoples R, Coloma A, Cruces J, Francke U (1998) A duplicated gene in the breakpoint regions of the 7q11.23 Williams-Beuren syndrome deletion encodes the initiator binding protein TFII-I and BAP-135, a phosphorylation target of BTK. Hum Mol Genet 7:325–334

Pfitzner AK, Mercier V, Jiang X, Moser von Filseck J, Baum B, Šarić A, Roux A (2020) An ESCRT-III polymerization sequence drives membrane deformation and fission. Cell 182:1140–1155.e18

Pride D (2020) The viruses inside you. Sci Am 6:46–53

Querido E, Morrison MR, Chu-Pham-Dang H, Thirlwell SW, Boivin D, Branton PE (2001) Identification of three functions of the adenovirus e4orf6 protein that mediate p53 degradation by the E4orf6-E1B55K complex. J Virol 75:699–709

Rivera-Gutiérrez X, Morán P, Taboada B, Serrano-Vázquez A, Isa P, Rojas-Velázquez L et al (2023) The fecal and oropharyngeal eukaryotic viromes of healthy infants during the first year of life are personal. Sci Rep 13:938

Roy P (2020) Bluetongue virus assembly and exit pathways. Adv Virus Res 108:249–273

Saha B, Parks RJ (2017) Human adenovirus type 5 vectors deleted of early region 1 (E1) undergo limited expression of early replicative E2 proteins and DNA replication in non-permissive cells. PLoS One 12:e0181012

Saha B, Wong CM, Parks RJ (2014) The adenovirus genome contributes to the structural stability of the virion. Viruses 6:3563–3583

Saxena R, Sharma P, Kumar S, Agrawal N, Sharma SK, Awasthi A (2023) Modulation of mitochondria by viral proteins. Life Sci 313:121271

Schurman SH, Hedayati M, Wang Z, Singh DK, Speina E, Zhang Y et al (2009) Direct and indirect roles of RECQL4 in modulating base excision repair capacity. Hum Mol Genet 18:3470–3483

Schwartz RA, Lakdawala SS, Eshleman HD, Russell MR, Carson CT, Weitzman MD (2008) Distinct requirement of adenovirus E1b55K protein for degradation of cellular substrates. J Virol 82:9043–9055

Schwob A, Teruel E, Dubuisson L, Lormières F, Verlhac P, Abudu YP et al (2019) SQSRTM-1/p62 potentiates HTLV-1 Tax-mediated NF-κB activation through its ubiquitin binding function. Sci Rep 9:16014

Sender R, Fuchs S, Milo R (2016) Revised estimates for the number of human and bacterial cells in the body. PLoS Biol 14:e1002533

Shaikh N, Swali P, Houben RMGJ (2023) Asymptomatic but infectious – the silent drive of pathogen transmission. A pragmatic review. Epidemics 44:100704

Shin Y, Kim S, Liang G, Ulmer TS, An W (2023) VprBP/DCAF1 triggers melanomagenic gene silencing through histone H2A phosphorylation. Biomedicine 11:2552

Smith DFQ, Casadevall A (2022) On the relationship between pathogenic potential and infective inoculum. PLoS Pathog 18:e1010484

Soday L, Lu Y, Albarnaz JD, Davies CTR, Antrobus R, Smith GL et al (2019) Quantitative temporal proteomic analysis of vaccinia virus infection reveals regulation of histone deacetylases by an interferon antagonist. Cell Rep 27:1920–1933.e7

Soh TK, Davies CTR, Muenzner J, Hunter LM, Barrow HG, Connor V et al (2020) Temporal proteomic analysis of herpes simplex virus 1 infection reveals cell-surface remodeling via pUL56-mediated GOPC degradation. Cell Rep 33:108235

St John AL, Rathore AP, Yap H, Ng ML, Metcalfe DD, Vasudevan SG et al (2011) Immune surveillance by mast cells during dengue infection promotes natural killer (NK) and NKT-cell recruitment and viral clearance. Proc Natl Acad Sci USA 108:9190–9195

Taboada B, Morán P, Serrano-Vázquez A, Iša P, Rojas-Velázquez L, Pérez-Juárez H et al (2021) The gut virome of healthy children during the first year of life is diverse and dynamic. PLoS One 16:e0240958

Tang SY, Wan YP, Wu YM (2015) Death domain associated protein (Daxx), a multi-functional protein. Cell Mol Biol Lett 20:788–797

Taylor MP, Burgon TB, Kirkegaard K, Jackson WT (2009) Role of microtubules in extracellular release of poliovirus. J Virol 83:6599–6609

Thai M, Graham NA, Braas D, Nehil M, Komisopoulou E, Kurdistani SK et al (2014) Adenovirus E4ORF1-induced MTC activation promotes host cell anabolic glucose metabolism and virus replication. Cell Metab 19:694–701

Tisza MJ, Buck CB (2021) A catalog of tens of thousands of viruses from human metagenomes reveals hidden associations with chronic diseases. Proc Natl Acad Sci USA 118:e2023202118

van der Grein SG, Defourny KAY, Rabouw HH, Galiveti CR, Langereis MA, Wauben MHM et al (2019) Picornavirus infection induces temporal release of multiple extracellular vesicle subsets that differ in molecular composition and infectious potential. PLoS Pathog 15:e1007594

Varghese JN, McKimm-Breschkin JL, Caldwell JB, Kortt AA, Colman PM (1992) The structure of the complex between influenza virus neuraminidase and sialic acid, the viral receptor. Proteins 14:327–332

Veenstra TD, Martinović S, Anderson GA, Pasa-Tolić L, Smith RD (2000) Proteome analysis using selective incorporation of isotopically labeled amino acids. J Am Soc Mass Spectrom 11:78–82

Veettil MV, Sadagopan S, Sharma-Walia N, Wang FZ, Raghu H, Varga L et al (2008) Kaposi's sarcoma-associated herpesvirus forms a multimolecular complex of integrins (αVβ5, αVβ3, and αβ1) and CD98-xCT during infection of human dermal microvascular endothelial cells, and CD98-xCT is essential for the postentry stage of infection. J Virol 82:12126–12144

Vietri M, Radulovic M, Stenmark H (2020) The many functions of ESCRTs. Nat Rev Mol Cell Biol 21:25–42

Wang Q, Liu R, Li Q, Wang F, Zhu B, Zheng M et al (2019) Host cell interactome of PB1 N40 protein of H5N1 influenza A virus chicken cells. J Proteome 197:34–41

Weitzman MD (2005) Functions of the adenovirus E4 proteins and their impact on viral vectors. Front Biosci 10:1106–1117

Welsch S, Müller B, Kräusslich HG (2007) More than one door – budding of enveloped viruses through cellular membranes. FEBS Lett 581:2089–2097

Wu H, Zhang S, Huo C, Zou S, Lian Z, Hu Y (2019) iTRAQ-based proteomic and bioinformatic characterization of human mast cells upon infection by the influenza A virus strains H1N1 and H5N1. FEBS Lett 593:2612–2627

Yang W, Desiderio S (1997) BAP-135, a target for Bruton's tyrosine kinase in response to B cell receptor engagement. Proc Natl Acad Sci USA 94:604–609

Yang Y, Zhang Y, Yang C, Fang F, Wang Y, Chang H et al (2021) Differential mitochondrial proteomic analysis of A549 cells infected with avian influenza virus subtypes H5 and H9. Virol J 18:39

# Chapter 7
# Proteomic Analysis of COVID-19 Infection

**Abstract** There are things that I remember that seemed innocuous at the time but ended up being major events. As a young boy growing up in Canada, I remember listening to a radio broadcast about a young man who was planning to run across the country to raise money for cancer research. I recollect obscure details such as him filling a bottle with water taken from the Atlantic Ocean. The broadcast received little fanfare, and honestly, I cannot remember the date. I was listening to it only because I was waiting for the broadcast of a hockey game to begin. Attention to this endeavor grew over the next weeks and months as the young man, Terry Fox, ran across Canada. Terry captivated the entire nation and was able to raise $1.7 million dollars until his journey ended on September 2, 1980, after 3339 miles near Thunder Bay, Ontario (https://www.tfri.ca/about/terry-fox). A nationally televised telethon was held a week after his run ended, which raised $10.5 million. Over the next several months over $23 million was raised. His fame was such that letters simply addressed "Terry Fox" were successfully delivered to a specific location. Although Terry passed away on June 28, 1981, his legacy continued. He was the youngest person ever to receive the Order of Canada. A total of 32 roads and streets, 14 schools, the Terry Fox Research Institute, and countless athletic venues are all named in honor of this courageous young man (https://www.twinkl.com/search?q=terry+fox&c=226&r=teacher&fl=12&fco=4213). The Terry Fox Run is still held in communities across Canada every year to raise money for cancer research. Almost 20 years after his run, a national survey named him as the second greatest Canadian of all time (https://en.wikipedia.org/wiki/The_Greatest_Canadian).

Why do I bring up this story? It reminded me of the beginnings of the coronavirus disease 2019 (COVID-19) pandemic. I can remember early in 2020 hearing news stories about a virus spreading throughout Wuhan, China. I must admit I was not concerned by the report and, having lived through other warnings such as bird and swine flu, did not consider this virus as a global threat. Wow, I was wrong! Within a few days the first confirmed case was reported in Washington state, and the entire city of Wuhan (population 11 million) was placed on lockdown (https://www.cdc.gov/museum/timeline/covid19.html). Owing to its genetic similarity to the virus responsible for the severe acute respiratory syndrome (SARS) outbreak of 2003, this novel coronavirus was called SARS-CoV-2 (Coronaviridae Study Group

 
T. D. Veenstra, *Proteomic Applications in Virology*, Advances in Experimental Medicine and Biology 1511, https://doi.org/10.1007/978-3-032-22340-1_7

of the International Committee on Taxonomy of Viruses, Nat Microbiol 5:536–544, 2020). By mid-February 2020, over 1000 people had died because of SARS-CoV-2 infection. On March 11, 2020, the World Health Organization (WHO) declared COVID-19 a pandemic (https://www.cdc.gov/museum/timeline/covid19.html). By the end of March, Drs. Deborah Brix and Anthony Fauci predicted that up to 200,000 Americans would die from COVID-19 even if precautions (social distancing, mask wearing, etc.) were taken (https://theweek.com/speedreads/905623/dr-birx-faces-200000-coronavirus-deaths-even-everything-almost-perfectly). As of April 2024, the number of deaths due to COVID-19 were approximately 1.2 and 7.0 million within the USA and worldwide, respectively (https://www.worldometers.info/coronavirus/coronavirus-death-toll/). After more than 3 years, COVID-19 was no longer considered a public health emergency of international concern. While this declaration means the pandemic is over, COVID-19 remains a disease that will be part of the world community for years to come.

**Keywords** SARS-CoV-2 • COVID-19 • Proteomic diagnostic COVID-19 test • Serum proteomics • Acute respiratory distress syndrome (ADRS) • Tissue proteomics • Proteomics of cell surface proteins • SARS-CoV-2/human interaction map • Drug repurposing • COVID database

## 7.1 Introduction

Although this book is focused more on proteomics than viruses, it is always useful to have some knowledge on the basic structure and function of the entity being studied. SARS-CoV-2 is a positive-sense single-stranded RNA (+ssRNA) with a genome just under 30 kb in size (see Fig. 7.1) (Cao et al. 2021). The genome contains two major open reading frames, ORF1a and ORF1b, which encompass two--thirds of SARS-CoV-2's genome (Pereira 2020). The ORF1a and ORF1b genes encode for polyprotein 1a (pp1a) and pp1b, respectively, which are proteolytically processed to produce 16 nonstructural proteins (NSPs). Both pp1a and pp1b undergo autoproteolytic cleavage by the papain-like (i.e., NSP3) and 3C-like cysteine protease (i.e., NSP5) (Moustaqil et al. 2021). The RNA-dependent RNA polymerase (RdRp) of SARS-CoV-2, which is critical for the replication and transcription of SARS-CoV-2 since it synthesizes RNA from an RNA template, is comprised of NSPs 7, 8, and 12 and an RNA template primer (Malone et al. 2022). The spike (S), envelope (E), membrane (M), and nucleocapsid (N) proteins are expressed from sequences at the 3′-end of the viral genome. The virus genome also encodes for nine accessory proteins (Liu et al. 2014). While these accessory proteins are not required for the virus to replicate and many of their functions are not known, some play a role in mediating antiviral responses by the host. A complete list of the proteins expressed by SARS-CoV-2 and their inferred functions (based on homology to other SARS--CoV viruses) are listed in Table 7.1.

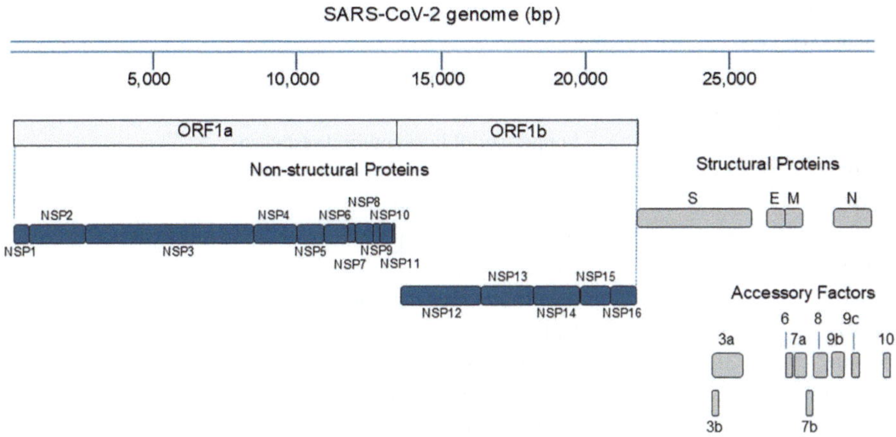


**Fig. 7.1** Genome of severe acute respiratory syndrome coronavirus 2 (SARS-CoV-2). The SARS--CoV-2 genome is approximately 30 kilobases in size and encodes for 16 nonstructural proteins (NSPs), 4 structural proteins, and 9 accessory proteins

SAS-CoV-2 is an enveloped virus in which a lipid bilayer surrounds its capsid (see Fig. 7.2). The four major structural SARS-CoV-2 proteins are encoded by the remaining one-third of the genome. These include the spike glycoprotein (S), nucleocapsid protein (N), membrane protein (M), and envelope protein (E). The S glycoprotein interacts with host cell receptors. The M glycoprotein is comprised of the cytoplasmic, the transmembrane, and N hydrophilic domains and is responsible for assembly of the replicated viral progeny and promotes installation of the S glycoprotein (de Haan et al. 1999). The E protein is essential for viral assembly and release from the host cell. The N protein is the most abundant protein in SARS-CoV-2 and is expressed in the early stage of infection (Huang et al. 2024). It forms a core ribonucleoprotein through its binding to viral RNA. This enables the viral RNA to enter and interact with biomolecules within the host cell. The N protein carries out a diverse group of functions including (i) decreasing host immune response, (ii) promoting inflammation, and (iii) inhibiting the host's cell cycle (Zhang et al. 2022).

## 7.2 The SARS-CoV-2 Life Cycle

The components necessary for the replication of SARS-CoV-2 enter the host cells through an endocytosis pathway that is initiated by the binding of the S protein to the angiotensin-converting enzyme 2 (ACE2) on the surface of the host cell (see Fig. 7.3) (Wan et al. 2020; V'kovski et al. 2021). SARS-CoV-2 can infect ACE2--deficient cells using the alternative receptor CD147 (Wang et al. 2020). The transmembrane protease, serine 2 (TMPRSS2), is also a key factor in the endocytosis

**Table 7.1** Proteins expressed by SARS-CoV-2 genome along with their predicted molecular weight (MW) and inferred function

| Protein | MW (kDa) | Inferred function |
|---|---|---|
| *Nonstructural proteins* | | |
| NSP1 | 19.8 | Suppressor of host antiviral response |
| NSP2 | 70.5 | Unknown |
| NSP3 | 217.3 | Papain-like protease |
| NSP4 | 56.2 | Involved in viral replication with NSP3 and NSP6 |
| NSP5 | 33.8 | Cysteine protease |
| NSP6 | 33.0 | Involved in viral replication with NSP3 and NSP4 |
| NSP7 | 9.2 | Part of RNA polymerase with NSP8 |
| NSP8 | 21.9 | Part of RNA polymerase with NSP7 |
| NSP9 | 12.4 | ssRNA binding protein |
| NSP10 | 14.8 | Essential for NSP16 methyltransferase activity |
| NSP11 | 1.3 | Short peptide |
| NSP12 | 106.7 | RNA polymerase |
| NSP13 | 66.9 | Helicase/triphosphatase activity |
| NSP14 | 59.8 | 3′-5’exonuclease |
| NSP15 | 38.8 | Uridine-specific endoribonuclease |
| NSP16 | 33.3 | RNA-cap methyltransferase |
| *Structural proteins* | | |
| S | 141.2 | Spike glycoprotein |
| E | 8.4 | Envelope protein involved in viral assembly |
| M | 25.1 | Membrane glycoprotein that is major component of envelope |
| N | | Binds to viral RNA to form helical capsid |
| *Accessory proteins* | | |
| Orf3a | 31.1 | Mediates viral release via lysosomal exocytosis pathway |
| Orf3b | 6.5 | Unknown |
| Orf6 | 7.3 | Type 1 IFN antagonist |
| Orf7a | 13.7 | Unknown |
| Orf7b | 5.2 | Modulates IFN pathways and cell adhesion |
| Orf8 | 13.8 | Unknown |
| Orf9b | 10.8 | Suppressor of host antiviral response |
| Orf9c | 8.0 | Unknown |
| Orf10 | 4.4 | Unknown |

*S* spike glycoprotein, *E* envelope protein, *M* membrane glycoprotein, *N*nucleocapsid protein, *Orf* open reading frame, *NSP* nonstructural protein

pathway. This protease cleaves the S glycoprotein, enabling the membranes of SARS-CoV-2 and its host to fuse (Evans and Liu 2021). During the process of endocytosis, the virus's + ssRNA enters the cytoplasm, and ORF1a and ORF1b are directly translated by the host ribosome into pp1a and pp1b. Both pp1a and pp1b are cleaved by the viral protease into 16 NSPs, which assemble to form the replication and transcription complex. The RdRp part of this complex catalyzes the formation of genomic RNA from the original SARS-CoV-2 +ssRNA. Subgenomic viral

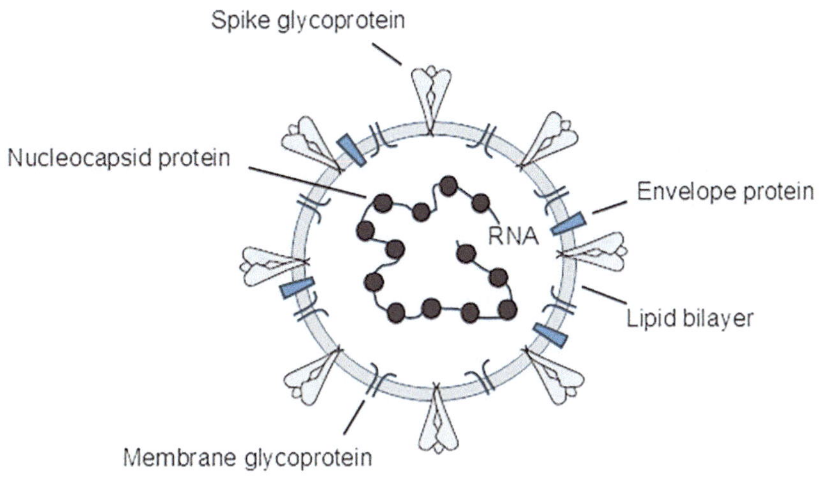


**Fig. 7.2** Structure of severe acute respiratory syndrome coronavirus 2 (SARS-CoV-2)

mRNAs are 5′ capped and translated into other necessary viral proteins. These viral proteins along with the genomic RNA travel to the endoplasmic reticulum where the progeny virions are assembled. After they undergo final processing (i.e., glycosylation) and maturation in the Golgi, the progeny viruses are exocytosed out of the host cell (V'kovski et al. 2021).

## 7.3 SARS-CoV-2 Infects Proteomic Research

The COVID-19 pandemic created a frenzy of new research into SARS-CoV-2 causing major changes of direction in established research areas. Sequencing the genome of the virus became a priority so that its behavior, evolution, and spread could be measured (Zhu et al. 2020). SARS-CoV-2 diagnostic tests, many based on existing technology, were quickly customized and mass produced. The race to develop vaccines that offered protection from SARS-CoV-2 became a major effort for both public and private laboratories. While not a new technology, the public quickly became aware of mRNA vaccines to fight COVID-19 through the efforts of both Pfizer-BioNTech and Moderna. Various treatments—some legitimate, others unproven—were quickly recommended and advertised. People were encouraged to wear masks, avoid large groups, and work from home, while many businesses closed, and the world's economy was thrown into turmoil.

Proteomics also quickly turned its attention to SARS-CoV-2. This focus is illustrated by a bar graph of the number of manuscripts listed on PubMed containing the term "proteomics" along with other specific virus types (see Fig. 7.4). The term "proteomics" along with one of either "influenza," "HIV," "herpes," or "measles" showed a relatively constant number of published manuscripts over the past decade.

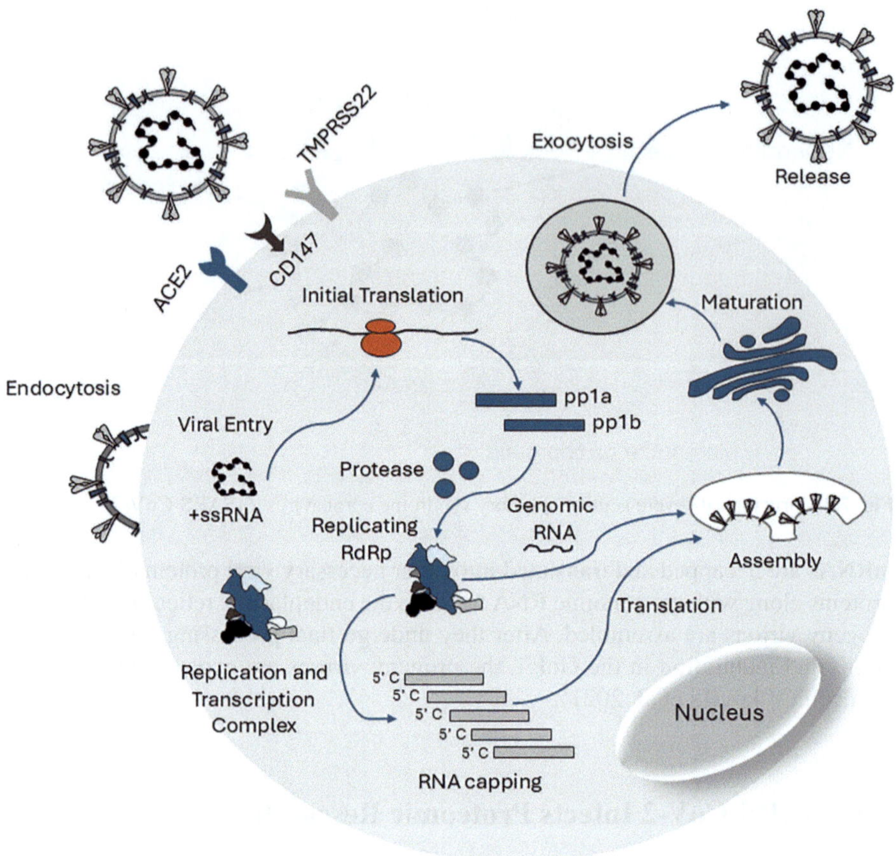


**Fig. 7.3** The SARS-CoV-2 life cycle includes viral entry, replication and transcription, assembly, and release. SARS-CoV-2 enters host cells through an endocytosis pathway mediated with the angiotensin-converting enzyme 2 (ACE2). Once in the cytoplasm, the viral RNA (ORF1a and ORF1b) is translated by the host ribosome into pp1a and pp1b. The viral polyproteins are cleaved into NSPs and assemble into the necessary replication and transcription complexes. After 5′-capping, the subgenomic viral mRNAs serve as templates for viral protein translation. The viral proteins travel to the endoplasmic reticulum and Golgi body where they are assembled into progeny virions. Once assembled, the virions are exocytosed to complete the life cycle

A dramatic increase, however, was observed in the number of manuscripts that included the words "SARS" and "proteomics." Since the origin of the SARS-CoV-2 pandemic began in late 2019, there were no publications that included both "SARS--CoV-2" and "proteomics" prior to 2020. Demonstrating science's rapid response, 295 papers were published containing the terms "SARS-CoV-2" and "proteomics" in 2020, and this number increased to 586 the following year. While the graph shows a greater number of manuscripts that include the term "SARS," most of these are specific to SARS-CoV-2 but used other names to describe this virus such as 2019-nCoV.

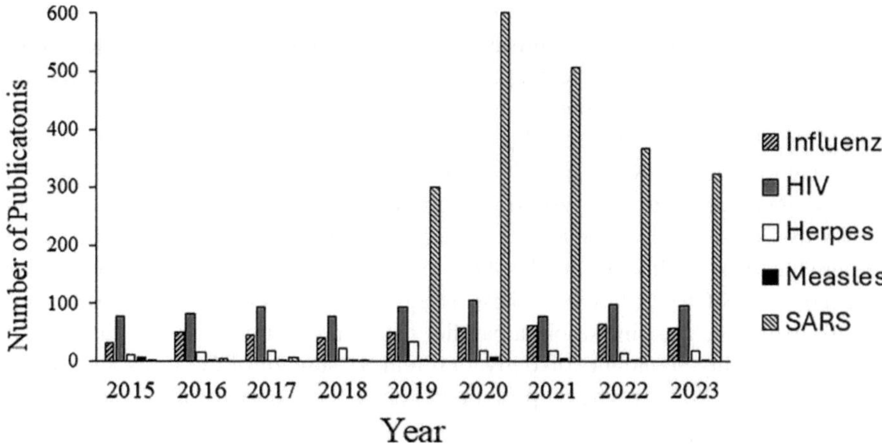


**Fig. 7.4** Graph showing the number of manuscripts listed on PubMed whose abstract contains the term "proteomics" along with one of the following: "influenza," "HIV," "herpes," "measles," or "SARS"

## 7.4 Novel Proteomic Tests to Diagnose COVID-19

One of the major initiatives when COVID-19 was first recognized was developing diagnostic tests. The first tests were based on real-time reverse transcription polymerase chain reaction (RT-PCR). These diagnostic tests are considered the gold standard owing to their high sensitivity and accuracy (Tang et al. 2020). Unfortunately, they require specialized equipment and reagents that limit its accessibility. Point-of-care (POC) immunoassay-based tests were also developed as these provided good detection characteristics, but probably more importantly they are easy to use and inexpensive (Younes et al. 2020). Since antibody recognition is not absolute, these immunoassays can suffer from interference from the presence of other molecules in the clinical sample or a lack of sensitivity (Lai et al. 2021).

Since it is an invaluable tool for analyzing proteins, mass spectrometry (MS) has continued to increase its impact in many different areas of life sciences. One of these impacted areas is clinical diagnostics (Wenk et al. 2024; Meng and Veenstra 2011). Targeted MS assays have been developed to measure clinical biomarkers such as troponin I, apolipoproteins, thyroglobulin, lactate dehydrogenase, hepatocyte growth factor receptor, human epidermal growth factor receptor-2, and glycated hemoglobin. Before describing the development of a targeted SARS-CoV-2 assay, the use of these tests with respect to other diagnostic tests needs to be put into proper perspective. While targeted MS assays are highly specific, they require specialized equipment and skill that limits their use and increases their cost relative to immunoassays. Targeted MS assays are analogous to RT-PCR in that patient samples are acquired at a clinic and sent to specialized laboratories that generate results that are sent back to the clinic. While developing a targeted MS test may require less time and expense than generating an antibody, they still require significant time, labor, and expensive equipment. I do not want to overhype targeted MS assays—they do not have the widespread utility of immunoassays—but they provide a complementary platform to other testing platforms. They are also useful in specialized tests when there is not sufficient patient demand to develop an antibody-based test.

In Chap. 3, a targeted MS test that analyzed 985 respiratory tract samples previously diagnosed using RT-PCR (540 positive and 445 negative for SARS-CoV-2) was described (Cardozo et al. 2020). To summarize, two peptides originating from the SARS-CoV-2 nucleoprotein were selected as the final surrogate peptides to measure the virus in clinical samples. Other similar studies also targeted the nucleoprotein of SARS-CoV-2; however, these assays could not detect low viral load samples. To overcome this limitation, a parallel reaction monitoring (PRM) assay that targeted the highly abundant spike glycoprotein S1 subunit of the virus was developed (Zhou et al. 2023a). The S1 protein was chosen not only for its abundance but also because its sequence varies depending on the variant of the virus allowing the assay to be easily modified to identify different mutant strains (Candido et al. 2022).

A PRM assay works just like a multiple reaction monitoring (MRM) assay, which detects specific peptides and their fragments as they are analyzed using the mass spectrometer. To identify the peptides that gave the best overall performance, samples containing 20 or 2 ng of the S1 protein were digested into tryptic peptides and analyzed using liquid chromatography-tandem MS (LC/MS$^2$). From the 20- and 2-ng samples, 29 and 17 peptides were identified, respectively. These peptides were evaluated based on four criteria: (i) high signal intensity, (ii) no missed tryptic cleavages, (iii) uniqueness, and (iv) glycosylation state. The uniqueness of the peptide selected for monitoring the virus is critical as homology searching comparing SARS-CoV, MERS-CoV, HKU1, and OC43 using the UniProtKB/Swiss-Prot database showed that the S1 proteins of these viruses are 66.3%, 31.4%, 30.1%, and 30.0% similar to the SARS-CoV-2 S1 protein. Glycosylated peptides were eliminated as the structures of these posttranslational modifications vary and cause uncertainty in MS detection. Ultimately eight peptides successfully passed all the criteria and were identified in both the 20- and 2-ng samples. None of these eight peptides were found in a search of the proteomes of other pathogens, coronaviruses, microbes, or humans.

To test the assay's ability to measure these eight peptides, various amounts of the S1 protein were mixed with concentrations of the virus's nucleocapsid, membrane, and envelope proteins. The proteins were mixed in ratios that reflected their relative abundances in the wild-type virus. After tryptic digestion, PRM analysis of the peptide mixture was conducted using LC/MS$^2$ with an Easy-nLC 1200 system and an Orbitrap Exploris 480 mass spectrometer. As little as 0.01 picogram of the SARS-CoV-2 S1 protein could be identified in mixtures containing the other structural proteins. No non-S1 peptides were identified demonstrating the assay's high selectivity.

Using this or any type of targeted MS method for diagnosing COVID-19 requires significant development. The major deficiency is throughput (Maxwell et al. 2023). To make them competitive with current viral diagnostic methods, sample treatment and MS analysis times must be reduced (or at least performed on a massively parallel automated processing platform). While a recent targeted MS method demonstrated the ability to analyze 400 samples per day (Drouin et al. 2024), this throughput pales in comparison to the thousands of samples that can be diagnosed using RT-PCR (Bustin and Nolan 2020). Coupling ion mobility MS could significantly reduce the required LC separation time and minimize any chemical background detectable by the spectrometer. But even shortening the LC/$MS^2$ time does not obviate the fact that only a single sample can be analyzed per instrument. The most obvious means of increasing throughput further would require adding more MS instruments, which would require a huge increase in cost to achieve the throughput comparable to RT-PCR and immunoassays. While tremendous progress has been made, the goal of using targeted MS assays as a routine diagnostic tool will require significant technological advances to achieve the low-cost throughput presently available using other methods.

## 7.5 Host Response to COVID-19

While developing clinical assays is important, it is presently not the optimal use of proteomics for furthering our knowledge of SARS-CoV-2. The reason is that genetic-based and immunological methods are much faster and cheaper at diagnosing infected patients while providing high levels of sensitivity and accuracy. The major impact proteomics can have is evaluating the effects of the virus on cell function and how it affects host cells and the entire organism. In the next section, I will focus on studies that characterize the effect of SARS-CoV-2 on the cell's proteome and circulating proteins.

### *7.5.1 Proteomic Analysis of Blood Proteins*

SARS-CoV-2, the virus that causes COVID-19, enters the body via droplets that originate from the cough, sneeze, or breath of an infected individual. These droplets could be airborne or acquired from touching an infected surface and then touching your own eyes, nose, or mouth. The virus then travels to the mucous membranes in your throat, and within a couple of weeks symptoms such as fever, cough, shortness of breath, fatigue, sore throat, runny nose, etc., develop. Many individuals report loss of smell and taste (Lamers and Haagmans 2022). Just like all viruses, SARS--CoV-2 exerts its main effects by entering host cells via a receptor. Like other SARS--CoV viruses, SARS-CoV-2 enters host cells using ACE2 as its receptor. The entry of SARS-CoV-2 through binding to ACE2 is aided by transmembrane serine protease 2 (TMPRSS2) (Hoffmann et al. 2020). This protease cleaves the S1 protein of the virus into a form that enables it to bind ACE2. Normally, ACE2 helps regulate blood pressure and electrolyte levels via its role in the renin-angiotensin-aldosterone system (RAAS) where it converts angiotensin II into angiotensin 1–7 or converts angiotensin I to angiotensin 1–9, which is further converted to angiotensin 1–7 by ACE or neutral endopeptidase (NEP). The expression patterns of ACE2 in human tissues are similar to known SARS-CoV-2 target cells (Hikmet et al. 2020). While the highest level of ACE2 expression is within cells of the intestine, colon, kidney, or heart muscle, cells within the upper respiratory tract also contain high levels of ACE2. These areas track with the upper respiratory system being the predominant location of viral replication during the early stages of infection and some of the organ complications observed in patients that suffer long-term COVID infection.

The expression of ACE2 in the lung is limited to type II alveolar (ATII) cells (Gerard et al. 2021). ATII cells serve two primary roles in the lungs; they are the progenitors of type I alveolar (ATI) cells (the cells that facilitate gas exchange between the blood and alveoli) and secrete surfactants that prevent alveolar collapse. ATII cells also play a role in innate immune response through the secretion of the surfactant proteins A and D. Both these surfactants bind bacteria, fungi, and viruses to facilitate their removal by macrophages. When SARS-CoV-2 infects these cells, it rapidly replicates, which can lead to their death along with generating both innate and adaptive immune responses. These immune responses lead to severe inflammatory reactions that damage tissue, first within the lung and then systemically. The inflammatory response leads to the release of cytokines that mediate acute respiratory distress syndrome (ARDS). ARDS is characterized by a dysregulated production of cytokines that causes hyperinflammation and damages the lung tissue (Leisman et al. 2020). This damage causes fluid to collect in the lungs and a loss of surfactant for ATII cells, which negatively impacts the ability of ATII cells to facilitate gas exchange. The lungs are thus unable to supply the organs with enough oxygen, which leads to organ (e.g., kidneys, lungs, liver, etc.) failure.

Another cause of organ failure is the hyperproduction of cytokines—known as a cytokine storm (Fajgenbaum and June 2020). These cytokines enter circulation and recruit immune cells to attack noninfected cells throughout the body. This attack continues to feed the production of cytokines, which exacerbates the storm. Clinical studies have identified several interleukins (i.e., 6, 1b, 10, 17, and 1ra), along with tumor necrosis factor (TNF), granulocyte macrophage-colony stimulating factor (GM-CSF), interferon-induced protein 10 (IFN-10), and monocyte chemotactic protein-3 (MCP-3) as the primary cytokines released in circulation during a cytokine storm.

Since these cytokines circulate, it would make sense to analyze serum or plasma to identify pathological mechanisms resulting from COVID-19 as well as identify infection-specific biomarkers. A study published in 2023 compared the proteomes of serum samples obtained from 15 men infected with SARS-CoV-2 and 5 healthy men (Babačić et al. 2023). The serum samples were profiled by prefractionation using high-resolution isoelectric focusing (HiRIEF) followed by LC/MS$^2$. The result was the identification of 15,425 serum peptides originating from 2037 proteins. Over 50% of these proteins had previously been identified using a SOMAscan approach in which aptamers (single-stranded DNA or RNA molecules that behave like antibodies in that they bind to specific target molecules with high affinity and specificity) are used to identify proteins in complex samples (Gold et al. 2012).

Comparison of proteins identified in the samples taken from COVID-19 and healthy patients using principal component analysis (PCA) found that of the over 2000 proteins identified, 1779 could be clearly assigned to unique proteins (Babačić et al. 2023). Of these, about 35% (i.e., 619) were altered in abundance in COVID-19 patients compared to samples taken from healthy controls. The largest protein abundance increases in COVID-19 patients were seen for IFN-induced proteins including IFN-stimulated 15-kDa protein (ISG15), interleukin enhancer-binding factor 2 (ILF2), myxovirus resistance protein 1 (MX1), interferon-stimulated gene 20 (ISG20), leucine aminopeptidase 3 (LAP3), and ubiquitin-conjugating enzyme E2 L6 (UBE2L6). The proteasomal protein proteasome activator subunit 1 (PSME1), along with proteasome subunits β type-3, and α types-3, 4, 5, and 7 (PSMB3, PSMA3, PSMA4, PSMA5, and PSMA7) were all upregulated in COVID-19 patients, while 26S proteasome non-ATPase regulatory subunit 2 (PSMD2) was the only downregulated proteasomal protein. Several acute phase proteins such as lipopolysaccharide binding protein (LBP), orosomucoid 1 and 2 (ORM1 and 2; also known as α-1-acid glycoprotein 1 and 2), and C-reactive protein (CRP) were elevated in COVID-19 serum samples. Grouping the proteins into functional clusters identified proteins involved in innate and adaptive immunity, cytokine signaling, neutrophil degranulation, and posttranslational modification being upregulated in COVID-19 samples. Functional clusters downregulated in the serum collected from COVID-19 infected individuals included extracellular matrix organization, hemostasis, and biological development. Many proteins involved in fatty acid metabolism, including low-density lipoprotein receptor-related protein 2 (LRP2) and other apolipoproteins, exhibited the largest decrease in abundance in COVID-19 serum samples. Along with the large decrease in albumin levels, these results could reflect the impaired nutrition that occurs during systemic inflammation.

### 7.5.2 Determining the Tissue Origin of SARS-CoV-2-Affected Proteins

Since no human cell is more than four cell units away from the circulatory system, proteins found in the blood originate from every possible location in the human body. It is well established that viral infections affect the expression of proteins outside of their specific site of infection. While SARS-CoV-2 specifically targets respiratory and intestinal cells that express ACE2 and TMPRSS2 (Hoffmann et al. 2020), it also causes immune and secondary responses in other cell types (Sievers et al. 2024; Veluswamy et al. 2021; Beltrami et al. 2023). Therefore, it is important to determine which protein abundance changes observed in serum are directly derived from the site of infection or are a result of an induced immune or secondary response. To identify the tissues of origin, two studies were performed. One study utilized an infected cell line to determine if proteins identified in the serum samples originated from lung tissue, while a second study performed a meta-analysis to correlate serum protein changes to a database of mRNA expression levels of specific tissues.

### 7.5.3 Serum Proteins Traceable to SARS-CoV-2 Infection Site

To identify proteins that originate from the site of infection, human lung adenocarcinoma airway epithelial cells (Calu-3) were infected with SARS-CoV-2 and samples collected 3 h, as well as 1, 3, and 7 days postinfection (Babačić et al. 2023). Samples were also collected at the same time points from noninfected (i.e., negative control) and SARS-CoV-2-infected Calu-3 cells treated with UV radiation to inactivate the virus (i.e., mock-infected cells). Thirteen peptides originating from SARS-CoV-2 itself were quantified in the infected Calu-3 cells. The abundance levels of these viral proteins peaked at 3 days postinfection and declined by day 7. Six, five, and two of the peptides originated from the virus's nucleocapsid phosphoprotein, S1, and membrane glycoprotein, respectively.

While no changes were found 3 h and 1 day postinfection, 532 and 893 proteins were differentially abundant between infected and noninfected cells at days 3 and 7 postinfection, respectively. The proteins that exhibited the greatest increase in abundance within infected cells were, not surprisingly, IFN-α- and IFN-γ-responsive proteins such as interferon-induced protein with tetratricopeptide repeats (IFIT) 1, 2, 3, and 5. All these proteins are involved in antiviral defense (Diamond and Farzan 2013). Other IFN-stimulated proteins include the cytokines interferon-stimulated gene 15 (ISG15), which helps to modulate natural killer (NK) and T cell activity (Kang et al. 2022), along with the RNases ISG20 and 2′-5′-oligoadenylate synthetase (OAS) 1 and 2, which inhibit viral replication by degrading both viral and cellular RNA (Hoffmann et al. 2020; Hikmet et al. 2020). As anticipated, the class 1 antigen-presenting molecules β-2 microglobulin, HLA-A, HLA-B, and HLA-E were upregulated in infected cells.

Forty-four of the proteins that were differentially abundant in the Calu-3 cells 3 days postinfection were also found to be altered in serum taken from infected patients (Babačić et al. 2023). Seventy-seven differentially abundant proteins in the cell line were also altered in infected patients 7 days postinfection. Demonstrating the challenge in comparing in vivo and in vitro results, only 77% and 47% of the overlapping proteins at day 3 and 7, respectively, had the same directional abundance change. Comparison of the proteomes extracted from COVID-19 patient sera and Calu-3 cells infected for 3 and 7 days revealed 15 proteins that were consistently upregulated and a single protein that was consistently downregulated. The upregulated proteins were either IFN activated or associated with the proteosome (see Table 7.2). The single downregulated protein, aldolase C, catalyzes the conversion of fructose-1,6-bisphosphate to glyceraldehyde-3-phosphate (G3P) and dihydroxyacetone phosphate (DHAP) during glycolysis.

The upregulation of ISG15, MX1, STAT1, and LGALS3BP in COVID-19 patient sera was corroborated in a previous study using a SOMAscan aptamers (Sullivan et al. 2021), while another study had observed upregulation of LAP3 peptidase, NAMPT, and THOP1 in similarly treated samples (Filbin et al. 2021). While these studies also detected UBE2L6, TYMP, SERPINB1, NAMPT, and aldolase C, they did not observe any change in their abundance. While only a small number of

**Table 7.2** List of proteins identified as being up- or downregulated in sera taken from COVID-19 patients and Calu-3 cells 3 and 7 days postinfection with SARS-CoV-2

| Upregulated | | Downregulated | |
|---|---|---|---|
| Function | Protein | Function | Protein |
| IFN activated | ISG15, MX1, UBE2L6, ISG20, STAT1, LAP3 peptidase | Glycolysis | Aldolase C |
| Proteosome | PSME1, PSMB10, PSMB8, TYMP, LGALS3BP, SERPINB1, WARS1, NAMPT, THOP1 | | |

proteins showed consistent changes within sera and infected cells, this result demonstrates that viral effects that occur within cells can be detected within blood samples. An interesting finding in this study was that the proteosome-related proteins PSME1, PSMB10, PSMA7, and PSMB8 were only identified in patient serum but not COVID-19-infected cells in culture (Babačić et al. 2023). This result suggests that these proteins were released because of systemic rather than localized effects of host cell infection. This previously unreported finding has implications on deciphering the overall mechanism of how SARS-CoV-2 affects the entire body.

### *7.5.4 Tissue Origin of Blood Proteins Identified in SARS-CoV-2-Infected Patients*

While several acute phase response and coagulation factor proteins identified in the serum of COVID-19 patients are likely secreted by the liver during systemic inflammation, it is more challenging to assign the origin of other differentially abundant proteins. It is impossible to quantify the level of each protein in every tissue, and even if this was possible, it still does not absolutely determine which tissue a serum-based protein originated from. To generate testable hypotheses, the list of 1779 differentially abundant serum proteins was analyzed against their tissue of origin based on their mRNA expression as listed within the Human Protein Atlas (HPA) (Karlsson et al. 2023). Eighty-nine of the proteins that showed a change in abundance in serum could be traced back to a specific tissue or organ. Of these, 57 likely originated from the liver. Of the five that were predicted to originate from the intestine, four (ZG16, NLRP6, APOA4, and VIL1) showed a decrease in abundance in serum, and only the microvillar protein, CDHR2, was increased in abundance. Both zymogen granule protein 16 (ZG16) and nucleotide-oligomerization domain-like receptor 6 (NLRP6) function as immune modulators to protect the gut from various pathogens. Cadherin-related family member 2 (CDHR2) plays a role in cell adhesion particularly within the brush border of the intestines. Its downregulation could contribute to the malnutrition observed in many COVID-19 patients or increase SARS-CoV-2's access to host cells. The lymphoid organ proteins, STAB2, immunoglobulin-like receptor B1 (LILRB1), and CR2 showed a decreased level in serum taken from COVID-19 patients. LILRB1 binds to class I MHC molecules causing a downregulation in the immune response. Its decrease in expression in serum taken from COVID-19 patients may suggest that the virus releases factors that prevent it from being presented to immune effector cells. Since LILRB1 is downregulated in infected patients, it is unable to inhibit the hyperinflammation observed in these individuals. The lung-enriched surfactant B protein (SFTPB) was increased in abundance in infected patient sera, likely because of its leakage from damaged ATII cells, a major target of SARS-CoV-2.

Just because the abundance of a circulating protein is found to be up- or downregulated in COVID-19 patients does not mean it is specifically related to the site of SARS-CoV-2 infection (i.e., lung, intestines, or lymphoid tissue). The differential abundance of proteins originating from healthy tissues can also be reflected in circulation. Tracing a protein's origin would require some type of tagging technique such as cell-specific isotope labeling. In lieu of this capability, the best that can be done is to compare results from various studies to accumulate as much circumstantial evidence of a protein's origin as possible. To acquire such evidence, the results of this study (Babačić et al. 2023) were compared to two other datasets that contained lists of differentially abundant proteins identified in tissues from patients who died from COVID-19 (Nie et al. 2021; Katze et al. 2002). Filtering out common circulating proteins and immunoglobulins, many proteins identified in these two studies showed a similar directional change as observed in this study (Babačić et al. 2023). Of the core set of 15 proteins that were consistently elevated in serum at days 3 and 7 post-SARS-CoV-2 infection of Calu-3 cells, 11 were also elevated in at least 1 human organ obtained from the deceased individuals (i.e., NAMPT, ISG15, MX1, STAT1, TYMP, LAP3, ISG20, UBE2L6, the proteasomal proteins PSMB8 and PSMB10, and LGALS3BP). An additional 70 proteins that were upregulated on days 3 and 7 within infected Calu-3 cells were also upregulated in at least 1 patient's organ. Most of these proteins, however, did not show consistent organ specificity between patients. Therefore, it is likely that their change in abundance was related to the host systemic response and not from direct viral infection.

## 7.6 Value in Identifying Serum Proteins Affected by SARS-CoV-2

Although the pandemic is over, COVID-19 is here to stay and will continue to cause long-term health effects or death in many individuals. Since it is going to be around for a while, identifying diagnostic biomarkers, beyond the virus itself, would be useful in minimizing its spread, while prognostic biomarkers would be invaluable for treating patients. Blood represents an attractive sample for detecting these markers, but it is also the most challenging proteome to characterize, because of its high dynamic range of protein concentrations. The analysis of serum and other tissues from SARS-CoV-2-infected patients provides a model for characterizing the organism-level effects of a disease. Since it is impossible for a single study to cover its entirety, this study collated the results from a variety of studies to gain a highly comprehensive view of what SARS-CoV-2 does to a human's proteome. Overall, several novel alterations in the proteome of COVID-19 patients were identified along with several others being confirmed. Many of these proteins were identified in only one or two of the combined studies, confirming the complementary nature of different proteome platforms.

This study demonstrated several important findings. The abundances of SARS-CoV-2 viral proteins are the highest 3 days postinfection, which would explain why some studies fail to detect proteomic changes in COVID-19 patients. As far as specific findings are concerned, it showed that changes in the abundance of proteins involved in IFN signaling, proteasome formation, and ubiquitination were most similar between studies of in vivo and in vitro systems (Babačić et al. 2023). Large increases in the abundance of proteins associated with the NF-kB, IFN-α, and IFN-γ pathways were observed in the serum proteome from individuals with severe COVID-19. These findings are consistent with the expected innate immune response that accompanies viral infection (Sijts and Kloetzel 2011). The abundance of two other IFN-activated proteins, ISG15 and MX1, could be correlated in both serum and Calu-3 cells. Other proteins highly abundant in serum of COVID-19 patients were those associated with proteosomes. Proteosomes are crucial for the degradation of ubiquitinated proteins and proteolyze viral proteins so that they can be presented via MHC class I proteins to immune cells (Jackson et al. 2022). The discovery of proteosome proteins being identified demonstrates an important feature of proteomic analysis: its specificity. Not only did the study identify that proteosome proteins were upregulated, but it also identified that it was the constitutional β1 (PSMB6) and β5 (PSMB5) and immunoproteasome subunits β2i (PSMB10) and β5i (PSMB8) that were more abundant in COVID-19 serum.

A meta-analysis comparing serum, in vitro (Calu-3 cells), and in vivo (organs from COVID-19 deceased patients) proteomes found a core set of 11 proteins that were consistently upregulated (Babačić et al. 2023). These proteins could serve as a panel for diagnosing patients in the future or following the effectiveness of various treatments. Although this study analyzed serum, in vitro, and in vivo samples, it is important to note that the results do not provide absolute proof that the observed changes in protein abundances directly derive from SARS-CoV-2 infection or a specific organ. At this stage of proteomics technology, however, it suggests some of the changes observed in the serum proteome of COVID-19 patients originate from alterations in the proteomes of cells and organs during SARS-CoV-2 infection. Other changes undoubtedly originate from changes in multiple cell types and organs as the virus induces a systemic host response. Just like good science, however, these results demonstrate that soluble blood proteome alterations can be traced to SARS-CoV-2-infected cells and can be used to develop further hypotheses to test the overall systemic effect of the virus as well as identify blood-based markers of the disease.

## 7.7 Proteomic Analysis of SARS-CoV-2 Cell Surface Interactions

Understanding the complete repertoire of interactions that occur at the host cell surface is important for developing therapeutics to prevent COVID-19. Besides personal hygiene, the first line of defense against a virus is preventing it from interacting with its host cells. The entry of SARS-CoV-2 into its host is initiated via low-affinity interactions with ubiquitous components of the host cell membrane (Partridge et al. 2021). Once the virus is loosely tethered, it binds with higher affinity to ACE2. With the assistance of TMPRSS2, ACE2 binds to the SARS-CoV-2 spike protein, causing a change in the conformation of the viral protein (Hoffmann et al. 2020; Partridge et al. 2021; Brito and Andrews 2011). While the involvement of ACE2 and TMPRSS2 was quickly discovered, it is likely that there are additional proteins involved in SARS-CoV-2 binding to host cells. For example, as mentioned previously, SARS-CoV-2 can infect ACE2-deficient cells using the alternative receptor CD147 (Wang et al. 2020). The challenge in identifying these additional proteins is that they are likely lower-affinity interactions. Nevertheless, they are important in recognizing how the virus invades cells and provide potential drug targets for inhibiting viral infection.

Affinity methods using immunoprecipitation (IP) or affinity tagging/purification (AP) combined with liquid chromatography-mass spectrometry (LC/MS)-based detection have been invaluable in identifying protein-protein interactions (PPIs). Unfortunately, these methods preferentially identify the highest affinity and abundant PPIs and are biased against lower-affinity interactions with membrane proteins due to solubility limitations (Guo et al. 2023). To capture lower-affinity interactions in their native conformation, proximity labeling (e.g., BioID, APEX, etc.) has been used to identify cell surface proteins that are in proximity to SARS-CoV-2 spike protein during cell infection. As discussed in more detail in Chap. 1, proximity labeling methods use a modified bait protein that can be activated to label nearby proteins with a biotin tag (Veenstra and Veenstra 2024; May et al. 2022). Biotinylated proteins are extracted from the mixture and identified. While it has been used to characterize PPI between SARS-CoV-2 spike protein and host proteins, the spike-BioID enzyme fusion protein could not be stably expressed resulting in confounding results (Datta et al. 2023). Another study used a different strategy. It covalently modified SARS-CoV-2 spike protein with a small (i.e., <1000 Da) photocatalytic molecule (PC) that when triggered using light activates a biotin-diazine probe resulting in the attachment of biotin to nearby proteins (Marquez et al. 2024). Human embryonic kidney (HEK) 293 T cells that expressed ACE2 on their surface were incubated with the PC and biotin-diazirine probe and irradiated with blue light. Biotin-labeled proteins are then extracted using streptavidin magnetic beads and identified using LC/MS.

As normal for discovery-driven proteomic studies, many proteins were found to be biotin labeled. It is important to remember, however, that proximity does not necessarily equate to functional relevance. Of the 67 biotin-labeled proteins, 30 were chosen for further scrutiny. Each of the 30 candidate proteins was individually depleted from ACE2-expressing HEK293T cells using the CRISPR/Cas9 system, and the ability for SARS-CoV-2 to infect each cell line was evaluated using a virus pseudo-particle that contained a firefly luciferase reporter gene. The extent of pseudo-viral entry into any of the cell lines was monitored using a luciferase assay. Depletion of two proteins, prostaglandin F2 receptor inhibitor (PTGFRN) and ephrin B1 (EFNB1), resulted in the higher effect on pseudo-viral entry causing a 71% and 40% reduction, respectively.

The role of PTGFRN and EFNB1 was further tested using actual SARS-CoV-2 particles in HK-2 cells that express ACE2 endogenously and are susceptible to SARS-CoV-2 infection. A 70–80% reduction in SARS-CoV-2 infection was observed when either of the proteins were knocked down. Interestingly, EFNB1, but not PTGFRN, knockdown resulted in an 80% reduction in the level of ACE2, suggesting that this protein may stimulate viral infection by regulating ACE2 expression.

Having established PTGFRN and EFNB1 as interacting with SARS-CoV-2 spike protein, it is important to ask if this result makes sense. PTGFRN is known to interact with a binding partner of CD9 and CD81 (Fénéant et al. 2014). Both HCV entry and influenza virus envelope formation have been shown to be modulated via interactions between PTGFRN and CD81 (Earnest et al. 2017). CD9 is known to promote the entry of MERS-CoV (Luo et al. 2011). EFNB1 is involved in IL-6 signaling during T-cell-mediated immune reactions and enhanced virus entry (Vasta 2009). Other PPIs with the spike protein that were discovered but not further validated include integrin β1 (ITGB1) and galectins 1 and 3. Both galectins have been shown to interact with viral glycans and other host cell receptor to enhance viral attachment and entry (Machala et al. 2019). Galectin 9, however, which inhibits human cytomegalovirus infection, was not identified in this study (Yeung et al. 2022). It is not yet understood whether galectin 9 enhances or inhibits SARS-CoV-2 infection (Du et al. 2023; Ismailia et al. 2012). Integrin β1 binds to viral surface proteins, facilitating viral entry in host cells via endocytosis (Simons et al. 2021; Zhou et al. 2023b). While there are still several candidates to test, this study increases the number of cell surface proteins that can be targeted using antagonistic drugs to inhibit SARS-CoV-2 entry.

## 7.8 A SARS-CoV-2/Human Cell Interaction Map

Proteomics' two biggest contributions to science have arguably been the ability to rapidly identify proteins and PPIs. Identifying proteins has been used to discover proteins whose abundance is altered when a cell or organism is introduced to an external perturbation. Being obligate parasites, PPIs with its host cell is critical for its survival. Like all other proteins, viral proteins do not act in isolation but carry out their function through binding to other proteins. Understanding these PPIs reveal much about the mechanisms of how a virus infects and affects its host's response. Understanding these mechanisms can lead to ways to disrupt a virus's pathology.

### *7.8.1 Global Mapping of SARS-CoV-2/Human Protein-Protein Interactions*

As described in Chap. 1, the two main methods for generating global PPI maps are affinity purification followed by mass spectrometry (AP-MS) and yeast two-hybrid (Y2H) assays. Each method has its pros and cons and is generally considered to capture different but complementary PPI data. Being binary, Y2H captures interactions that are key connections between different protein pathways, while AP-MS captures greater number of interactions that can be either primary (i.e., bind directly to the bait protein) or secondary (i.e., bind to a protein that is binding to the bait protein). While studies have most often focused on one technique or the other, one group undertook the monumental task of generating a SARS-CoV-2/human PPI map using both AP-MS and Y2H (Chen et al. 2023).

Pairwise combinations of 28 SARS-CoV-2 ORFs and approximately 16,000 human ORFs were cloned into pDEST-AD (activation domain) and/or pDEST-DB (DNA-binding domain) vectors against ~16,000 human proteins (hORFeome V8.1) (see Fig. 7.5) (Chen et al. 2023). NSP11 was not tested owing to its small size (i.e., 1.3 kb). To minimize the number of false positives that needed validation, each ORF was cloned into both vectors so that all SARS-CoV-2 and human proteins would serve as both bait and prey in the Y2H assay. This comparison resulted in 896,000 (i.e., 28 SARS-CoV-2 proteins x 16,000 human proteins x 2) possible PPIs being tested. A total of 299 SARS-CoV-2-human PPIs were identified in the Y2H screen. Of these, 267 had not been previously described.

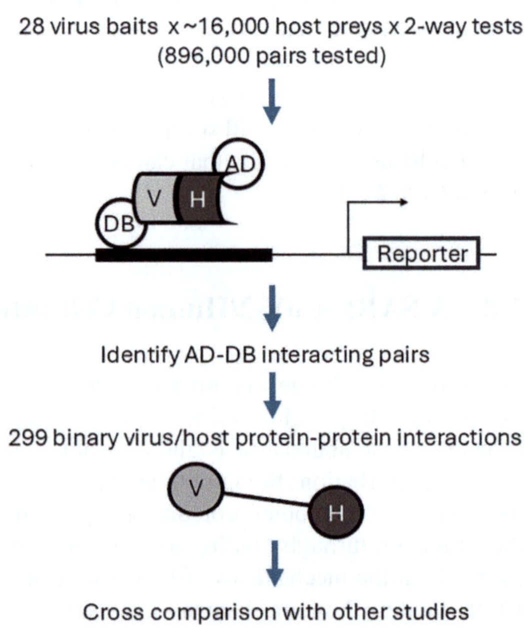


**Fig. 7.5** Schematic showing yeast two-hybrid (Y2H) strategy identify SARS-CoV-2-human protein-protein interactions

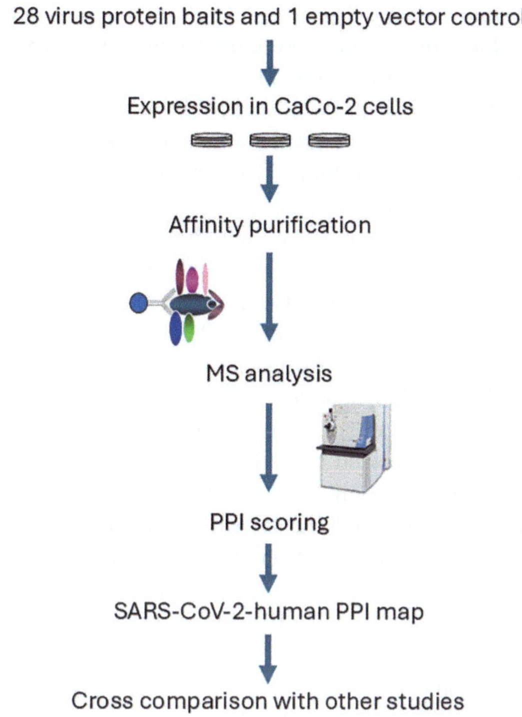


**Fig. 7.6** Schematic showing affinity purification-mass spectrometry (AP-MS) strategy to identify SARS-CoV-2-human protein-protein interactions

To complement the identified Y2H binary interactions, interactions involving 28 SARS-CoV-2 proteins and host proteins were studied using AP-MS (see Fig. 7.6). The SARS-CoV-2 genes were cloned into mammalian expression vectors that contained streptavidin, Myc, or FLAG affinity tags. Each of the vectors was individually expressed in the human intestinal epithelial cell line Caco-2. Caco-2 cells express both the S glycoprotein and TMPRSS2, which are required for SARS-CoV-2 entry into the cell (Hoffmann et al. 2020). In addition, this cell line has been used extensively to study SARS-CoV-2 infection, is easy to transfect, and propagates quickly (Hoshijima 2006). Consistent with the time point where SARS-CoV-2 abundance is at its peak, the proteomes were extracted from the cells 72 h postinfection. The supernatants were incubated with beads that bound streptavidin, Myc, or FLAG peptide depending on the vector that was used to clone each SARS-CoV-2 gene. The beads bind directly to the tagged SARS-CoV-2 bait protein expressed within the Caco-2 cells. By using gentle, non-denaturing washing buffers (i.e., neutral pH, low salt and detergent concentrations, etc.), proteins that bind to the bait protein can also be extracted. The challenge is achieving the optimal buffer condition to preserve as many specific PPIs while minimizing the number of nonspecific interactions (i.e., false positives). The extracted protein complexes were identified using LC/MS. After filtering for interactions that met specific thresholds, a total of 472 SARS-CoV-2-human PPIs were identified. Combining the AP-MS and Y2H results revealed a network of 739 interactions between the 28 SARS-CoV-2 proteins and 579 unique host proteins (Chen et al. 2023).

While most of the human proteins (i.e., 81%) only interacted with a single SARS-CoV-2 protein, α-actinin-4, α-actinin-1, tripartite motif containing 27 (TRIM27), and integrin subunit β1 binding protein 2 (ITGB1BP2) interacted with multiple viral proteins. These results are not surprising as ITGB1BP2, also known as melusin, acts as a sensor of mechanical stress, which can be caused by simply overexpressing any protein within a cell (Zhang et al. 2018). TRIM27 is involved in various cellular processes, including cell proliferation, apoptosis, and immune response, and would therefore be expected to interact with many different viral proteins (Oikonomou et al. 2011). The α-actinins are cytoskeletal proteins that interact with many different proteins owing to their high abundance (Li et al. 2021).

The viral protein N, which interacts with viral RNA to form a helical capsid, had the highest number of interactions within its network (Chen et al. 2023). This protein was followed by ORF7b and ORF9b. Orf7b is thought to interfere with IFN pathways and cell adhesion, while ORF9b suppresses host cell immune response. The E protein had the fewest interactions within its network. This result could be due to the native position of this protein within the envelope of the virus. Outside of its native environment, the E protein may lose many of its normal PPIs. NPS7, NPS1, and NPS4 also had fewer interactions compared to most SARS-CoV-2 proteins. Since NPS7 and NPS4 form complexes with other SARS-CoV-2 proteins, the expression of each of these proteins individually may hinder their ability to form PPIs. NPS1 is believed to interfere with antiviral response of the host. Its low number of PPIs may suggest its mechanism of interference is highly specific. Overall, the interacting partners for each viral protein did not significantly overlap, demonstrating the unique role of each SARS-CoV-2 protein.

While it is impossible to describe each PPI in this format, a summary analysis of those proteins from specific functional groups is provided. The biological processes most highly represented in the PPI data included protein translation, transcription, and neutrophil-mediated immunity. Pathway enrichment analysis found that proteins involved in processing within the endoplasmic reticulum, the formation of tight junctions, glycolysis, ribosomes, and protein export were highly represented in the data. Obviously this study was not the only large-scale PPI study conducted during the height of the COVID-19 pandemic. Like every scientific result, reproducibility is a key metric in determining its accuracy. The results of this AP/MS-Y2H study were compared to three previously published SARS-CoV-2-human PPI studies that exclusively used AP-MS (see Fig. 7.7) (Gordon et al. 2020; Stukalov et al. 2021; Xavier et al. 2009). While each study identified several unique factors, there was significant overlap between all four studies. While it may not seem like a big deal, this result is quite encouraging as many large-scale proteomic PPI studies often do not show much overlap between them. Differences in each study would be expected since different cells lines were used, and each had their own distinct sample preparation methods and computational analyses. A total of 218 PPIs found in this study were validated in at least one other of the published SARS-CoV-2-human PPI studies. Just over 62% (i.e., 361) of the identified human proteins had never

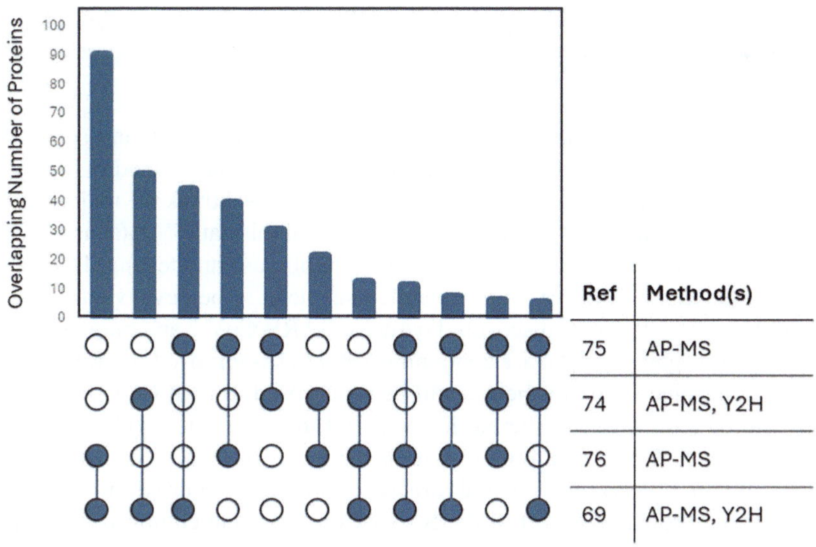


**Fig. 7.7** Overlap between human-SARS-CoV protein interactions identified in four separate studies. The specific study and the methods used to identify these interactions are also shown

been identified as interacting with SARS-CoV-2. The remaining 361 PPIs represent newly discovered human proteins that interact with the virus. Of the seven human proteins that were identified in all four studies, three of them are involved in protein processing in the endoplasmic reticulum, three function in granule and endosome assembly, and two are chaperone proteins (Chen et al. 2023).

An interesting observation was that 24 human proteins were found within the S protein network, suggesting SARS-CoV-2 has other potential ways to enter human cells beyond interactions with ACE2 and TMPRSS2. Two of these proteins, coronin 1C31 and stonin 2, are expressed on the cell membrane, and coronin 1C31 is highly expressed in the lung (Martina et al. 2001; Shimada et al. 2013). Two other proteins identified within the S protein network were epiplakin1 and sperm antigen with calponin homology and coiled-coil domains 1 like (SPECC1L). Both proteins are involved in the formation of cell junctions (Clemente-Moragón et al. 2021) and could represent pathways that permit SARS-CoV-2 to move between cells without interacting with ACE2.

A previously uncharacterized human transcription factor, zinc finger 579 (ZNF579), was found to be associated with ORF3a. This viral accessory protein mediates viral release via lysosomal exocytosis pathway. Interaction between ORF3a and ZNF579 inhibits the repression of heat shock protein family A member 6 (HSP6A) expression (Chen et al. 2023). HSP6A is a member of the HSP70 family of molecular chaperones and functions in viral entry, replication, assembly, and release. By increasing the expression of HSP6A in infected cells, SARS-CoV-2 ensures an adequate supply of molecular chaperones to assist in the production of viral proteins and the ultimate assembly of viral progeny.

## 7.9 Intelligent Drug Repurposing Through Protein-Protein Interaction Data

Proteomics has demonstrated the ability to generate PPIs on a grand scale. What used to be done on a single-protein-at-a-time basis is now performed in highly parallel experiments that produce hundreds of potential PPIs. I think the major question that needs to be addressed is "What do we do with all this data?" Databases of these PPIs have been created and made available for public use and testing. But for proteomics to continue to thrive, more specific uses need to be developed. What this study did was take the PPIs discovered between SARS-CoV-2 and human cells and screened them against a database containing 2900 FDA-approved or investigational drugs to determine if any of them could be repurposed to treat COVID-19. These 2900 candidate drugs were filtered down to 189 that interacted with proteins that were near the SARS-CoV-2 protein within the network map. What this criterion means is that the protein had to bind directly or within close proximity to the viral protein. If the protein was part of the SARS-CoV-2 protein complex but only distantly through its interactions with other human proteins, then blocking it with a drug would not have a strong neutralizing effect on the virus's actions. Of these 189 candidate drugs, 44 had clinical trials for SARS-CoV-2, and 23 had close proximity to the viral host factors and had demonstrated anti-SARS-CoV-2 activity in at least two National Center for Advancing Translational Sciences (NCATS) assays. The drug discovery institute NCATS was established in 2011 to develop high-throughput assays to measure the activity of compounds against biological targets with the aim of identifying potential therapeutic candidates.

The nonselective β-blocker carvedilol and thiazide diuretic hydrochlorothiazide (HCTZ) were identified as potential treatments for COVID-19. Their potential use was supported by their strong network proximity to SARS-CoV-2 host factors, empirical data showing anti-SARS-CoV-2 activity, and reduced SARS-CoV-2 positive test risk obtained from patient electronic health records (EHRs). Data from EHRs are available for these drugs because both carvedilol and HCTZ are FDA-approved drugs that have been prescribed to patients for conditions other than COVID-19. This study simply used the proteomic PPI data to examine known drugs for the purpose of determining if they could be repurposed for another use. The selective β-blocker metoprolol was also identified in the PPI data and has shown efficacy in treating COVID-19 in a small clinical trial (Chen et al. 2004).

Global PPI data not only is useful for identifying drugs that could potentially be repurposed but also helps understand their mechanism of action. Approximately 237 of the host proteins identified in the PPI database had interactions with known targets of the antihypertensive drug carvedilol (Chen et al. 2023). The most populated pathways for these proteins included protein translation, mRNA splicing, glucose metabolism, and neutrophil-mediated immunity. The inclusion of protein translation and mRNA splicing suggests that carvedilol may inhibit SARS-CoV-2 replication through perturbation of these two pathways. Carvedilol is also known to inhibit the expression of vascular cell adhesion molecule-1 (VCAM1), a protein involved in immune responses and inflammation whose levels are elevated in COVID-19 (Rajoria et al. 2024). While the PPI dataset does not provide enough information to determine carvedilol's mechanism of inhibiting VCAM1 expression, at minimum it provides the basis for testing the hypothesis that the drug could be repurposed to treat COVID-19.

Are studies that produce global PPIs between individual proteins and their hosts without flaws? Absolutely not. False-positive results are common since these experiments rely on overexpression of bait proteins that may not mimic their wild-type abundance or conformation. Fortunately, virus proteins are expressed at a high level within their hosts, so AP-MS studies closely reflect their wild-type abundance. One issue though that may not seem obvious is that these large databases of PPIs are most often compared to results obtained from only examining a specific interaction. While this comparison can validate specific results in a large dataset, it is an inefficient use of data. As illustrated in this study, the next phase in proteomics will be the use of existing and development of novel computational tools to generate integrated databases that can compare large datasets about different features of a system. For example, in the previous study datasets of PPIs were compared with datasets of FDA-approved drugs. These two datasets were analyzed to identify approved drugs that could be repurposed, or used off-label, to treat COVID-19. The variety of different data resources that could be combined and analyzed included genomics, proteomics, metabolomics, genome-wide association studies, drug-target associations, protein structure, patient data, etc. Strategic selection of relevant databases and proper software algorithms offer a powerful approach to apply systems biology to solving complex medical problems ranging from pandemics to cancer.

## 7.10 Proteomic Databases of COVID-19 Studies

As much of the previous chapter described the collection of large amounts of proteomic data that is often deposited within publicly available databases, I thought it would be prudent to describe some of the more popular proteomic databases containing COVID-19 information.

### *7.10.1 CoVProt*

Like genomics, proteomic studies generate huge amounts of data. Combined with the number of manuscripts published on COVID-19, it can be difficult to summarize all the data related to COVID-19, especially patient data. The Web-based portal CoVProt was developed as an information repository to help investigators evaluate MS-based proteomics data from other studies (Zhang et al. 2023). The aim of this database is to curate MS-based proteomics data comparing proteins identified in a variety of clinical samples taken from SARS-CoV-2-infected and noninfected patients. CoVProt contains information covering a total of 202 clinical samples acquired from plasma, lung tissue, naso−/oropharyngeal swabs, semen, and bronchoalveolar lavage fluid. There are 37,125 unique proteins that map to 2725 unique proteins listed in the database. To construct CoVProt, relevant datasets were taken from peer-reviewed studies and reanalyzed using a single tool. The data is listed by sample type, and the relative abundance of proteins correlated with patient status is provided. The database is currently constructing tools for data visualization and access to additional clinical parameters of each patient. The goal of CoVProt is to facilitate the identification of COVID-19 biomarkers that can be used diagnostically and therapeutically. The aim is also to use the data to determine the effect of SARS-CoV-2 on various cell functions to better discern how the virus affects the body.

### *7.10.2 COVIDpro*

COVIDpro is a Web-based curated database containing results generated from 41 studies that analyzed 19 different types of clinical samples taken from 3077 patients at 32 hospitals (Gowthaman et al. 2021). The databases contain 14,403 uniquely quantified proteins that were identified in these clinical samples. The clinical samples were acquired from patients with a variety of conditions ranging from non-severe to fatal COVID-19 along with healthy patients and those with non-pulmonary and pulmonary fibrosis to compare protein abundances between samples. The types of samples analyzed include the expected serum, plasma, bronchoalveolar lavage fluid, sputum, and nasopharynx swabs but also include a formalin-fixed paraffin-embedded (FFPE) tissues taken from lung, liver, heart, kidney, spleen, testis, and thyroid tissues.

The protein data can be analyzed using four different analysis modules: (i) data exploration, (ii) analysis by protein, (iii) analysis by pathway, and (iv) integrative analysis. The data exploration module provides information related to a study's number of patients, samples, MS files acquired, and number of unique proteins quantified. In this module, the studies currently with data in this database are listed under a "Hospital" tab on the left-hand side. Selecting hospital sites then lists the projects that were performed using samples acquired at this site. The types of samples and the preparation and MS methods used to quantify the protein are listed in many of the hospital sites. Finally, disease subgroups associated with the samples are provided along with pie charts that show the percentage of quantified proteins from each subgroup. This module also allows the user to select a specific protein and see which studies were able to quantify that protein, providing a measure of the overlap between the studies. The analysis by protein module allows the user to select a specific protein. The abundance of that protein is then displayed in boxplots based on the severity of the COVID-19 symptoms. The analysis by pathway module shows the fold change of proteins in two samples being compared. The proteins are shown in the context of the protein pathway(s) they function within. The p-value related to the identified fold change is also provided. The integrative analysis module allows user to view differential protein expression via a volcano plot map. The data in each plot is also linked to Gene Ontology (GO) and Kyoto Encyclopedia of Genes and Genomes (KEGG) databases.

### *7.10.3 CoViMAPP*

COVID-19 Meta-Analysis of Plasma Proteins (CoViMAPP) is a resource that combines the results of multiple independent analyses (i.e., a meta-analysis) that measured changes in circulating proteins because of COVID-19 infection (Babačić et al. 2023). The data is compiled from 21 independent MS studies of serum and plasma. Among the features of CoViMAPP are a curve showing the standardized mean difference (SMD) of proteins that had a statistically significant alteration in at least two COVID-19 patients compared to healthy controls. The SMDs can be presented based on the number of studies (ranging from 2 to 21) in which specific proteins were observed. The database contains a hyperlink to Genecards—The Human Gene Database, which provides a description of the proteins within CoViMAPP. A tab showing the summary receiver operating characteristics (SROC) that provides the sensitivity and specificity of a selected protein for diagnosing COVID-19 is provided. This tab also provides diagnostic odds ratios (DORs), which represents the ratio of the odds of an individual with COVID-19 testing positive for a specific protein versus testing negative. The DOR is given for each study in which the protein was identified.

### *7.10.4 CoV3D*

Coronaviruses, most recently SARS-CoV-2 which causes COVID-19, are a major target for vaccine and drug design. This CoV3D database provides up-to-date SARS-CoV-2, SARS-CoV, and MERS-CoV structures, focused on antibody recognition, host receptor binding, drug targeting, and diversity (Behzadi and Gajdács 2021). The structures are updated weekly from the Protein Data Bank. The available structures are divided into various categories: spike structures, protease structures, peptide-MHC structures, glycosylated spike models, and spike interfaces (SARS-CoV-2). The spike interface page shows images of the spike receptor binding domain binding to other proteins such as ACE2. The images can be enlarged and rotated allowing the viewer to identify where the binding partners interact. The site also contains structures of antibodies in complex with spike proteins. CoV3D currently contains 1777 structures related to the SARS-CoV-2 spike protein. The peptide-MHC structures include peptides from the spike, nucleocapsid, membrane, and replicase proteins bound to MHC proteins. These peptides represent the targets of anti-SARS-CoV-2 T-cell receptors. Beyond protein structures, the database also contains sequence information listing known polymorphisms in the spike glycoprotein sequence along with tertiary structure models that allow the user to locate specific amino acids residues within specific regions of the protein. The purpose of CoV3D is to provide a centralized database of that the research community can access to study coronavirus protein structures rather than having to search through resources such as the Worldwide Protein Data Bank (Behzadi and Gajdács 2021).

## 7.11 Conclusions

Hopefully the COVID-19 is a "once-in-a-lifetime" event and we never have to go through something of its magnitude again. It was truly one of those events that changed how the global community lives. While much pain and division came from the pandemic, it did bring one major benefit. The pace of progress in many technological areas was astonishing. Diagnostic tests and vaccines were developed, tested, produced, and distributed at a pace never seen before. More relevant to this book, efforts to apply proteomics to study SARS-CoV-2 were greatly accelerated. From not having a single peer-reviewed publication prior to 2020 (since it was not identified until November 2019), there have been over 2100 articles listed on PubMed that include the keywords "proteomics" and "SARS-CoV-2."

The pandemic had a major effect on how proteomics was applied to studying viruses. While much of the focus was identifying PPIs between the virus and its host, the scope of studies started to expand, and researchers initiated creative ways of using this data. For example, PPIs identified using *both* Y2H and AP-MS were published within single manuscripts. Protein interaction studies performed comprehensive comparisons between their results and similar studies. While this step seems minor, it is quite critical for validating proteomic results. Comparisons were not only done between PPI studies, but results were also compared to drug studies to find treatments for SARS-CoV-2 infection. These comparisons show how important tools such as artificial intelligence (AI) will be in the future of proteomics. One of AI's greatest strengths is analyzing large, complex datasets. By applying AI to datasets of diverse information, more reliable hypotheses can be formulated. More reliable hypotheses increase the efficiency and success of validation studies, which will undoubtedly increase the rate at which SARS-CoV-2 (or any virus) knowledge is gained.

## References

Babačić H, Christ W, Araújo JE, Mermelekas G, Sharma N, Tynell J et al (2023) Comprehensive proteomics and meta-analysis of COVID-19 host response. Nat Commun 14:5921–5928

Behzadi P, Gajdács M (2021) Worldwide Protein Data Bank (wwPDB): a virtual treasure for research in biotechnology. Eur J Microbiol Immunol 11:77–86

Beltrami S, Rizzo S, Caccuri F, Rizzo R, Bortolotti D, Schiuma G (2023) SARS-CoV-2 systemic effects: new clues. Microorganisms 11:1209–1215

Brito GC, Andrews DW (2011) Removing bias against membrane proteins in interaction networks. BMC Syst Biol 5:169–193

Bustin SA, Nolan T (2020) RT-qPCR testing of SARS-CoV-2: a primer. Int J Mol Sci 21:3004–3012

Candido KL, Eich CR, de Fariña LO, Kadowaki MK, da Conceição Silva JL, Maller A et al (2022) Spike protein of SARS-CoV-2 variants: a brief review and practical implications. Braz J Microbiol 53:1133–1157

Cao C, Cai Z, Xiao X, Rao J, Chen J, Hu N et al (2021) The architecture of the SARS-CoV-2 RNA genome inside virion. Nat Commun 12:3917–3930

Cardozo KHM, Lebkuchen A, Okai GG, Schuch RA, Viana LG, Olive AN et al (2020) Establishing a mass spectrometry-based system for rapid detection of SARS-CoV-2 in large clinical sample cohorts. Nat Commun 11:6201–6213

Chen JW, Lin FY, Chen YH, Wu TC, Chen YL, Lin SJ (2004) Carvedilol inhibits tumor necrosis factor-alpha-induced endothelial transcription factor activation, adhesion molecule expression, and adhesiveness to human mononuclear cells. Arterioscler Thromb Vasc Biol 24:2075–2081

Chen DY, Turcinovic J, Feng S, Kenney DJ, Chin CV, Choudhary MC et al (2023) Cell culture systems for isolation of SARS-CoV-2 clinical isolates and generation of recombinant virus. iScience 26:106634–106655

Clemente-Moragón A, Martínez-Milla J, Oliver E, Santos A, Flandes J, Fernández I et al (2021) Metoprolol in critically ill patients with COVID-19. J Am Coll Cardiol 78:1001–1011

Coronaviridae Study Group of the International Committee on Taxonomy of Viruses (2020) The species severe acute respiratory syndrome-related coronavirus: classifying 2019-nCoV and naming it SARS-CoV-2. Nat Microbiol 5:536–544

Datta S, Chen DY, Tavares AH, Reyes-Robles T, Ryu KA, Khan N et al (2023) High-resolution photocatalytic mapping of SARS-CoV-2 spike interactions on the cell surface. Cell Chem Biol 30:1313–1322.e7

de Haan CA, Smeets M, Vernooij F, Vennema H, Rottier PJ (1999) Mapping of the coronavirus membrane protein domains involved in interaction with the spike protein. J Virol 73:7441–7452

Diamond MS, Farzan M (2013) The broad-spectrum antiviral functions of IFIT and IFITM proteins. Nat Rev Immunol 13:46–57

Drouin N, Elfrink HL, Boers SA, van Hugten S, Wessels E, de Vries JJC et al (2024) A targeted LC-MRM$^3$ proteomic approach for the diagnosis of SARS-CoV-2 infection in nasopharyngeal swabs. Mol Cell Proteomics 2:100805–100820

Du L, Bouzidi MS, Gala A, Deiter F, Billaud JN, Yeung ST et al (2023) Human galectin-9 potently enhances SARS-CoV-2 replication and inflammation in airway epithelial cells. J Mol Cell Biol 15:mjad030

Earnest JT, Hantak MP, Li K, McCray PB Jr, Perlman S, Gallagher T (2017) The tetraspanin CD9 facilitates MERS-coronavirus entry by scaffolding host cell receptors and proteases. PLoS Pathog 13:e1006546–e1006567

Evans JP, Liu SL (2021) Role of host factors in SARS-CoV-2 entry. J Biol Chem 297:100847–100856

Fajgenbaum DC, June CH (2020) Cytokine storm. N Engl J Med 383:2255–2273

Fénéant L, Levy S, Cocquerel L (2014) CD81 and hepatitis C virus (HCV) infection. Viruses 6:535–572

Filbin MR, Mehta A, Schneider AM, Kays KR, Guess JR, Gentili M et al (2021) Longitudinal proteomic analysis of severe COVID-19 reveals survival-associated signatures, tissue-specific cell death, and cell-cell interactions. Cell Rep Med 2:100287–100310

Gerard L, Lecocq M, Bouzin C, Hoton D, Schmit G, Pereira JP et al (2021) Increased angiotensin-converting enzyme 2 and loss of alveolar type ii cells in COVID-19-related acute respiratory distress syndrome. Am J Respir Crit Care Med 204:1024–1034

Gold L, Walker JJ, Wilcox SK, Williams S (2012) Advances in human proteomics at high scale with the SOMAscan proteomics platform. Nat Biotechnol 29:543–549

Gordon DE, Jang GM, Bouhaddou M, Xu J, Obernier K, White KM et al (2020) A SARS-CoV-2 protein interaction map reveals targets for drug repurposing. Nature 583:459–468

Gowthaman R, Guest JD, Yin R, Adolf-Bryfogle J, Schief WR, Pierce BG (2021) CoV3D: a database of high resolution coronavirus protein structures. Nucleic Acids Res 49:D282–D287

Guo J, Guo S, Lu S, Gong J, Wang L, Ding L et al (2023) The development of proximity labeling technology and its applications in mammals, plants, and microorganisms. Cell Commun Signal 21:269–290

Hikmet F, Méar L, Edvinsson Å, Micke P, Uhlén M, Lindskog C (2020) The protein expression profile of ACE2 in human tissues. Mol Syst Biol 16:e9610–e9615

Hoffmann M, Kleine-Weber H, Schroeder S, Krüger N, Herrler T, Erichsen S et al (2020) SARS-CoV-2 cell entry depends on ACE2 and TMPRSS2 and is blocked by a clinically proven protease inhibitor. Cell 181:271–280.e8

Hoshijima M (2006) Mechanical stress-strain sensors embedded in cardiac cytoskeleton: Z disk, titin, and associated structures. Am J Physiol Heart Circ Physiol 290:H1313–H1325

https://en.wikipedia.org/wiki/The_Greatest_Canadian. Accessed May 2025

https://theweek.com/speedreads/905623/dr-birx-faces-200000-coronavirus-deaths-even-everything-almost-perfectly. Accessed May 2025

https://www.cdc.gov/museum/timeline/covid19.html. Accessed May 2025

https://www.tfri.ca/about/terry-fox. Accessed May 2025

https://www.twinkl.com/search?q=terry+fox&c=226&r=teacher&fl=12&fco=4213. Accessed May 2025

https://www.worldometers.info/coronavirus/coronavirus-death-toll/. Accessed May 2025

Huang Y, Chen J, Chen S, Huang C, Li B, Li J et al (2024) Molecular characterization of SARS-CoV-2 nucleocapsid protein. Front Cell Infect Microbiol 14:1415885–1415896

Ismailia R, Hsao JC, Chung CS, Chen CH, Hsu PW, Liao CL et al (2012) Integrin β1 mediates vaccinia virus entry through activation of PI3K/Akt signaling. J Virol 86:6677–6687

Jackson CB, Farzan M, Chen B, Choe H (2022) Mechanisms of SARS-CoV-2 entry into cells. Nat Rev Mol Cell Biol 23:3–20

Kang JA, Kim YJ, Jeon YJ (2022) The diverse repertoire of ISG15: more intricate than initially thought. Exp Mol Med 54:1779–1792

Karlsson M, Zhang C, Méar L, Zhong W, Digre A, Katona B et al (2023) Quantitative multiorgan proteomics of fatal COVID-19 uncovers tissue-specific effects beyond inflammation. EMBO Mol Med 15:e17459–e17479

Katze MG, He Y, Gale M Jr (2002) Viruses and interferon: a fight for supremacy. Nat Rev Immunol 2:675–687

Lai CC, Wang CY, Ko WC, Hsueh PR (2021) In vitro diagnostics of coronavirus disease 2019: technologies and application. J Microbiol Immunol Infect 54:164–174

Lamers MM, Haagmans BL (2022) SARS-CoV-2 pathogenesis. Nat Rev Microbiol 20:270–284

Leisman DE, Ronner L, Pinotti R, Taylor MD, Sinha P, Calfee CS et al (2020) Cytokine elevation in severe and critical COVID-19: a rapid systematic review, meta-analysis, and comparison with other inflammatory syndromes. Lancet Respir Med 8:1233–1244

Li J, Guo M, Tian X, Wang X, Yang X, Wu P et al (2021) Virus-host interactome and proteomic survey reveal potential virulence factors influencing SARS-CoV-2 pathogenesis. Med 2:99–112.e7

Liu DX, Fung TS, Chong KK, Shukla A, Hilgenfeld R (2014) Accessory proteins of SARS-CoV and other coronaviruses. Antivir Res 109:97–109

Luo H, Charpentier T, Wang X, Qi S, Han B, Wu T et al (2011) Efnb1 and Efnb2 proteins regulate thymocyte development, peripheral T cell differentiation, and antiviral immune responses and are essential for interleukin-6 (IL-6) signaling. J Biol Chem 286:41135–41152

Machala EA, Avdic S, Stern L, Zajonc DM, Benedict CA, Blyth E et al (2019) Restriction of human cytomegalovirus infection by galectin-9. J Virol 93:e01746-18

Malone B, Urakova N, Snijder EJ, Campbell EA (2022) Structures and functions of coronavirus replication-transcription complexes and their relevance for SARS-CoV-2 drug design. Nat Rev Mol Cell Biol 23:21–39

Marquez J, Weldemariam MM, Dong J, Hayashi J, Kane MA, Serrero G (2024) Effect of PTFGRN expression on the proteomic profile of A431 cells and determination of the PTGFRN interactome. ACS Omega 9:14381–14387

Martina JA, Bonangelino CJ, Aguilar RC, Bonifacino JS (2001) Stonin 2: an adaptor-like protein that interacts with components of the endocytic machinery. J Cell Biol 153:1111–1120

Maxwell CB, Sandhu JK, Cao TH, McCann GP, Ng LL, Jones DJL (2023) The edge effect in high-throughput proteomics: a cautionary tale. J Am Soc Mass Spectrom 34:1065–1072

May DG, Martin-Sancho L, Anschau V, Liu S, Chrisopulos RJ, Scott KL et al (2022) A BioID-derived proximity interactome for SARS-CoV-2 proteins. Viruses 14:611–632

Meng Z, Veenstra TD (2011) Targeted mass spectrometry approaches for protein biomarker verification. J Proteome 74:2650–2659

Moustaqil M, Ollivier E, Chiu HP, Van Tol S, Rudolffi-Soto P, Stevens C et al (2021) SARS-CoV-2 proteases PLpro and 3CLpro cleave IRF3 and critical modulators of inflammatory pathways (NLRP12 and TAB1): implications for disease presentation across species. Emerg Microbes Infect 10:178–195

Nie X, Qian L, Sun R, Huang B, Dong X, Xiao Q et al (2021) Multi-organ proteomic landscape of COVID-19 autopsies. Cell 184:775–791

Oikonomou KG, Zachou K, Dalekos GN (2011) Alpha-actinin: a multidisciplinary protein with important role in B-cell driven autoimmunity. Autoimmun Rev 10:389–396

Partridge LJ, Urwin L, Nicklin MJH, James DC, Green LR, Monk PN (2021) ACE2-independent interaction of SARS-CoV-2 spike protein with human epithelial cells is inhibited by unfractionated heparin. Cells 10:1419–1434

Pereira F (2020) Evolutionary dynamics of the SARS-CoV-2 ORF8 accessory gene. Infect Genet Evol 85:104525–104535

Rajoria S, Kavuru SR, Pyda HS, Bihani S, Borishetty D, Biswas D et al (2024) CoVProt: toward a mass spectrometry data portal for COVID-19 proteomics research and development. OMICS 28:24–31

Shimada H, Nambu-Niibori A, Wilson-Morifuji M, Mizuguchi S, Araki N, Sumiyoshi H et al (2013) Epiplakin modifies the motility of the HeLa cells and accumulates at the outer surfaces of 3-D cell clusters. J Dermatol 40:249–258

Sievers BL, Cheng MTK, Csiba K, Meng B, Gupta RK (2024) SARS-CoV-2 and innate immunity: the good, the bad, and the "goldilocks". Cell Mol Immunol 21:171–183

Sijts EJ, Kloetzel PM (2011) The role of the proteasome in the generation of MHC class I ligands and immune responses. Cell Mol Life Sci 68:1491–1502

Simons P, Rinaldi DA, Bondu V, Kell AM, Bradfute S, Lidke DS et al (2021) Integrin activation is an essential component of SARS-CoV-2 infection. Sci Rep 11:20398–20409

Stukalov A, Girault V, Grass V, Karayel O, Bergant V, Urban C et al (2021) Multilevel proteomics reveals host perturbations by SARS-CoV-2 and SARS-CoV. Nature 594:246–252

Sullivan KD, Galbraith MD, Kinning KT, Bartsch KW, Levinsky NC, Araya P et al (2021) The COVIDome explorer researcher portal. Cell Rep 36:109527–109548

Tang YW, Schmitz JE, Persing DH, Stratton CW (2020) Laboratory diagnosis of COVID-19: current issues and challenges. J Clin Microbiol 58:e00512-20

V'kovski P, Kratzel A, Steiner S, Stalder H, Thiel V (2021) Coronavirus biology and replication: implications for SARS-CoV-2. Nat Rev Microbiol 19:155–170

Vasta GR (2009) Roles of galectins in infection. Nat Rev Microbiol 7:424–438

Veenstra BT, Veenstra TD (2024) Proteomic applications in identifying protein-protein interactions. Adv Protein Chem Struct Biol 138:1–48

Veluswamy P, Wacker M, Stavridis D, Reichel T, Schmidt H, Scherner M et al (2021) The SARS-CoV-2/receptor axis in heart and blood vessels: a crisp update on COVID-19 disease with cardiovascular complications. Viruses 13:1346–1379

Wan Y, Shang J, Graham R, Baric RS, Li F (2020) Receptor recognition by the novel coronavirus from Wuhan: an analysis based on decade-long structural studies of SARS coronavirus. J Virol 94:e00127–e00120

Wang K, Chen W, Zhang Z, Deng Y, Lian JQ, Du P et al (2020) CD147-spike protein is a novel route for SARS-CoV-2 infection to host cells. Signal Transduct Target Ther 5:283–292

Wenk D, Zuo C, Kislinger T, Sepiashvili L (2024) Recent developments in mass-spectrometry-based targeted proteomics of clinical cancer biomarkers. Clin Proteomics 21:6–20

Xavier CP, Rastetter RH, Stumpf M, Rosentreter A, Müller R, Reimann J et al (2009) Structural and functional diversity of novel coronin 1C (CRN2) isoforms in muscle. J Mol Biol 393:287–299

Yeung ST, Premeaux TA, Du L, Niki T, Pillai SK, Khanna KM et al (2022) Galectin-9 protects humanized-ACE2 immunocompetent mice from SARS-CoV-2 infection. Front Immunol 13:1011185–1011196

Younes N, Al-Sadeq DW, Al-Jighefee H, Younes S, Al-Jamal O, Daas HI et al (2020) Challenges in laboratory diagnosis of the novel coronavirus SARS-CoV-2. Viruses 12:582–608

Zhang HX, Xu ZS, Lin H, Li M, Xia T, Cui K et al (2018) TRIM27 mediates STAT3 activation at retromer-positive structures to promote colitis and colitis-associated carcinogenesis. Nat Commun 9:3441–3456

Zhang S, Wang L, Cheng G (2022) The battle between host and SARS-CoV-2: innate immunity and viral evasion strategies. Mol Ther 30:1869–1884

Zhang F, Luna A, Tan T, Chen Y, Sander C, Guo T (2023) COVIDpro: database for mining protein dysregulation in patients with COVID-19. J Proteome Res 22:2847–2859

Zhou W, Sui Z, Liu J, He Y, Yuan H, Sun Y et al (2023a) High-sensitivity detection toward SARS-CoV-2 S1 glycoprotein by parallel reaction monitoring mass spectrometry. Anal Chem 95:8752–8757

Zhou Y, Liu Y, Gupta S, Paramo MI, Hou Y, Mao C et al (2023b) A comprehensive SARS-CoV-2-human protein-protein interactome reveals COVID-19 pathobiology and potential host therapeutic targets. Nat Biotechnol 41:128–139

Zhu N, Zhang D, Wang W, Li X, Yang B, Song J et al (2020) A novel coronavirus from patients with pneumonia in China, 2019. N Engl J Med 382:727–733

# Chapter 8
# Proteomic Characterization of HIV Infection

**Abstract** In the summer of 1981, young gay men began falling ill and dying of opportunistic infections. The strange thing was that these infections should have been easily rebuffed by their immune systems (Greene WC, Eur J Immunol 37(Suppl 1):S94–S102, 2007). These men become emaciated and developed dark purple lesions primarily on their arms and faces. These lesions were due to Kaposi's sarcoma, a relatively rare and aggressive form of cancer (Antman K and Chang Y, N Engl J Med 342:1027–1038, 2000). Physicians in cities such as New York, Los Angeles, and San Francisco were dumbfounded by the presence of young men who appeared elderly (Greene WC, Eur J Immunol 37(Suppl 1):S94–S102, 2007; Premeaux TA and Ndhlovu LC, Curr Opin HIV AIDS 18:237–245, 2023). Fear initially spread throughout the gay male community as this group was disproportionately impacted. As their health rapidly deteriorated, doctors were unable to prevent patients from dying from opportunistic infections and malignancies. The condition was recognized in 1981 as acquired immune deficiency syndrome (AIDS) (CDC, MMWR Morb Mortal Wkly Rep 30:305–308, 1981). As it became more publicized, the shock and stigma surrounding individuals who died from AIDS grew. The first major figure to reveal their AIDS diagnosis was Rock Hudson. This revelation was followed by other well-known individuals including Robert Reed (i.e., the father on the Brady Bunch), Arthur Ashe, and Freddie Mercury (webmd.com/hiv-aids/ss/slideshow-celebrities-hiv-aids).

**Keywords** Acquired immune deficiency syndrome • Human immunodeficiency virus • HIV replication • Biomarkers of HIV infection • Host cell proteome • Antiretroviral therapy • Proximity extension assay • Comorbidity biomarkers • Microbiome

## 8.1 Introduction

The causative agent of AIDS was recognized in the early 1980s as the human immunodeficiency virus (HIV) by the laboratories of the French virologist Luc Antoine Montagnier and American physician-scientist Robert C. Gallo (Barré-Sinoussi et al.

 
T. D. Veenstra, *Proteomic Applications in Virology*, Advances in Experimental Medicine and Biology 1511, https://doi.org/10.1007/978-3-032-22340-1_8

1983; Gallo et al. 1984). In 2008 Luc Antoine Montagnier and Françoise Barré--Sinoussi (who worked in Dr. Montagnier's lab) were awarded the Nobel Prize for this discovery. While HIV is primarily spread via sexual contact, there are several other means by which the virus can be transmitted between individuals. These include mother to fetus, sharing of contaminated drug paraphernalia such as needles and syringes, and the receiving of contaminated blood, blood products, organs, or tissues. Fortunately, screening procedures have made the likelihood of receiving HIV from blood highly unlikely. For the virus to be transmitted, fluids containing HIV must enter the bloodstream through a mucous membrane, open wound or sore, or direct injection. Unlike many other viruses, HIV cannot be transmitted via air or water, insects, or body fluids such as saliva, tears, sweat, feces, or urine that is not contaminated with blood from a person with HIV. Studies have shown HIV is a "fragile" virus that does not survive for long outside the body. For example, in 25 °C tap water free HIV or cell-associated HIV loses 90% of its infectivity within 1–2 h and 99.9% by 8 h. Polio virus, in contrast, loses no infectivity after 24 h under the same conditions (Moore 1993).

Since its identification four decades ago, the main group form of HIV-1 has infected an estimated 88 million people and caused an estimated 43 million deaths as of the end of 2023 (www.who.int/data/gho/data/themes/hiv-aids). Developing countries have experienced the greatest HIV/AIDS morbidity and mortality with the highest prevalence rates found among young adults in sub-Saharan Africa (www.unaids.org/). Developing countries have suffered the most since access to therapies (e.g., antiretroviral treatment) is not as available as in developed countries, and no vaccine or curative treatment is currently available. Regardless of whether HIV is considered a pandemic (www.unaids.org/en/resources/presscentre/pressreleaseandstatementarchive/2024/july/20240722_global-aids-update) or epidemic (www.who.int/news--room/fact-sheets/detail/hiv-aids), AIDS will likely pose a significant public health threat for many years to come.

## 8.2 Human Immunodeficiency Virus

Before discussing how proteomics has furthered our understanding of HIV, it is worth describing the structure and life cycle of the virus in greater detail. HIV is a retrovirus, meaning that its genomic material is RNA. Each HIV particle contains two copies of the full-length viral genomic positive strand RNA (+ssRNA) (Moore and Hu 2009). Although the two copies are about 99.9% homologous and each encodes the full complement of genetic information required for viral replication, they are not identical (Duchon and Hu 2024). Either strand contains sufficient information to initiate nascent infection alone; however, the viral replication cycle can only be successfully completed if two copies are packaged into the replicated viral particle. The final packaging of the RNAs in the virus is not simply random (Chen et al. 2016). The two genomic RNA strands exist within the virus as a tightly associated compact structure. Their

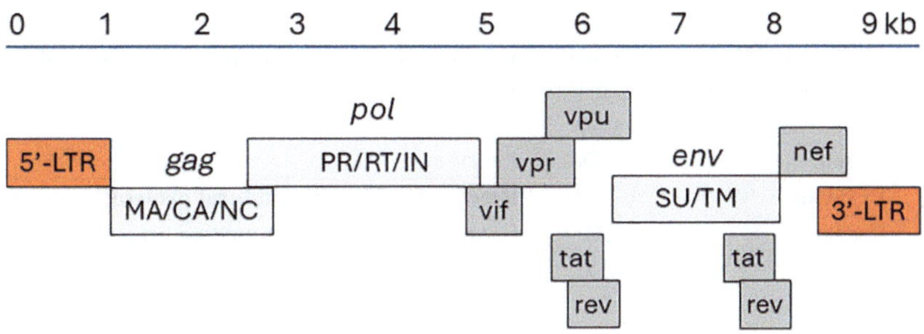


**Fig. 8.1** Structure and organization of the HIV-1 genome. The *gag*, *pol*, and *env* genes encode mainly for structural proteins. LTR, long terminal repeat; *gag*, group-specific antigen; *pol*, polymerase; *env*, envelope. In the case of the regulator genes, the regulator proteins p14 and p19 are encoded by two separate gene regions, *tat* and *rev*. A complete list of the genes, their protein products, and their functions are provided in Table 8.1

dimerization is a highly regulated process and occurs prior to their repackaging into the virus's capsid.

HIV's genome is approximately 9.75 kilobases (kb) in size and contains nine unique open reading frames (ORFs) (see Fig. 8.1) (Watts et al. 2009). These ORFs produce a total of 15 proteins. HIV's RNA genome is transcribed by reverse transcriptase (RT) into proviral DNA, an inactive form of DNA that eventually gets integrated into the host cell's genome (Götte et al. 1999). The viral RNA is subsequently degraded by ribonuclease H (RNase H), which is a component of the virus's (RT) enzyme. The proviral DNA is flanked by long terminal repeat (LTR) segments at the 5′ and 3′ ends. The 5′-LTR codes for a transcriptional promoter region, and the 3′-LTR acts as a transcriptional terminator and polyadenylation site, which serves to increase the DNA's stability (Watts et al. 2009; German Advisory Committee Blood (Arbeitskreis Blut), Subgroup 2016). The first gene encountered reading in the 5′–3′ direction is *gag*, which encodes the matrix (MA), capsid (CA), nucleocapsid (NC), and p6 proteins. The next gene is *pol*, which codes for the enzymes RT, protease (PR), integrase (IN), and RNase H. The *env* gene which follows codes for the envelope glycoproteins gp120 and gp41. The HIV genome also encodes for many regulatory proteins including negative regulatory factor (Nef), viral infectivity factor (Vif), virus protein r (Vpr), and virus protein unique (Vpu), which function in viral replication, budding, and pathogenesis. Two other regulatory proteins, transactivator protein (Tat) and RNA splicing regulator (Rev), are critical for initiating HIV replication.

The *gag*, *pol*, and *env* genes are not directly translated into the individual proteins described above. The *gag* gene produces the polyprotein precursor Pr55Gag, which is proteolytically processed to generate the MA, CA, NC, and p6 proteins (Ono and Freed 1999). The *pol* gene produces the polyprotein precursor, Pr160GagPol, which is processed to generate the PR, RT, IN, and RNase H enzymes (Yu et al. 2020). The polyprotein expressed from the *env* gene encodes a 30 amino acid signal peptide and is proteolytic processed to produce the gp120 and gp41 glycoproteins (Upadhyay et al. 2020). While sometimes abbreviated as surface protein (SU) and transmembrane

**Table 8.1** List of HIV-1 genes, proteins, and their function. The genes *gag*, *pol*, and *env* produce the proproteins Pr55Gag, Pr160GagPol, and PrG160 that are proteolyzed to form the inner structural proteins, viral enzymes, and envelope proteins of the virus, respectively

| Gene | Protein | MW (kDa) | Function |
|---|---|---|---|
| ***gag*** | **Pr55Gag** | | **Precursor of inner structural proteins** |
| | Capsid | p24 | Formation of capsid surrounding genome |
| | Matrix | p17 | Formation of inner membrane layer |
| | Nucleoprotein | p7 | Formation of nucleoprotein/RNA complex |
| | p6 | p6 | Involved in virus particle release |
| ***pol*** | **Pr160GagPol** | | **Precursor of viral enzymes** |
| | Protease | p10 | Proteolytic cleavage of PrGag and Pr160Gag-Pol precursors into structural and viral proteins |
| | Reverse transcriptase | p51 | Transcription of HIV RNA into proviral DNA |
| | RNase H | p15 | Degradation of viral RNA in RNA/DNA replication complex |
| | Integrase | p32 | Integration of proviral DNA into host genome |
| ***env*** | **PrGp160** | | **Precursor of envelope proteins** |
| | Surface glycoprotein (gp120) | p120 | RNA polymerase |
| | Transmembrane protein (gp41) | p41 | Helicase/triphosphatase activity |
| *tat* | Transactivator protein | p14 | Activates transcription of viral genes |
| *rev* | RNA splicing regulator | p19 | Regulates export of non- and partially spliced viral mRNA |
| *nef* | Negative regulating factor | p27 | Enhancement of viral infectivity, downregulation of CD4 and HLA on target cells |
| *vif* | Viral infectivity protein | p23 | Virus production in vivo |
| *vpr* | Virus protein r | p15 | Facilitates viral infectivity, affects cell cycle |
| *vpu* | Virus protein unique | p16 | Virus particle release, CD4 degradation, and intracellular trafficking |
| *vpx* | Virus protein x | p15 | Early steps of virus replication of HIV-2 |
| *tev* | Tat/rev protein | p26 | Tat-Env-Tev protein that regulates activity of Tat and Rev |

protein (TM), these proteins are better known as gp120 and gp41, respectively. A complete list of the genes, the proteins they code for, and the function of these proteins are provided in Table 8.1.

### *8.2.1 The Structure of HIV*

HIV is a roughly spherical, enveloped virus with a diameter of approximately 100 nm (see Fig. 8.2) (German Advisory Committee Blood (Arbeitskreis Blut), Subgroup 2016). Embedded within the envelope are 72 trimers of the envelope proteins gp120

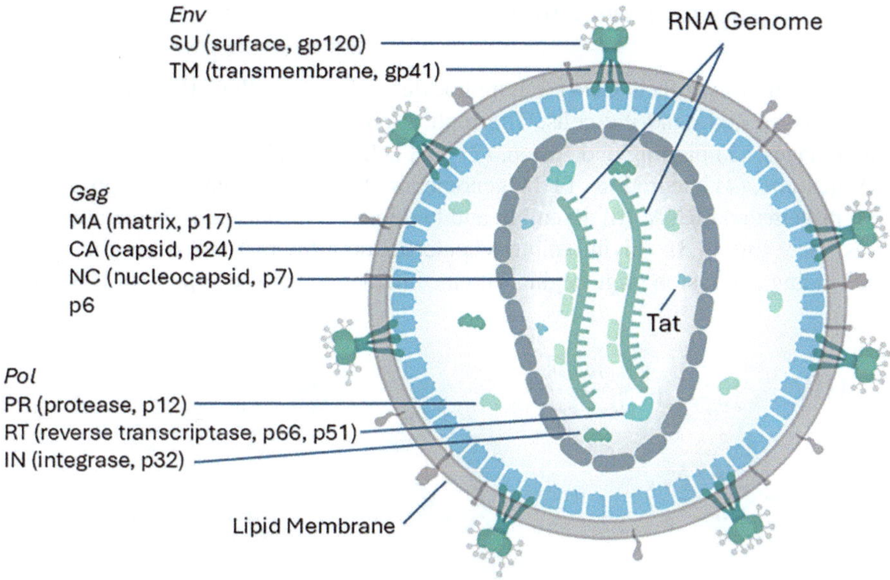


**Fig. 8.2** Structure of the human immunodeficiency virus (HIV). "Creative Commons Diagram of the HI-virion" by Thomas Splettstoesser is licensed under CC BY 4.0. Labels have been changed

and gp41 (Gelderblom 1991). The gp120 trimers are bound to the gp41 trimers, which are anchored into the outer lipid membrane. This lipid membrane with the gp120/gp41 knobs covers the outer capsid membrane, which is composed of MA proteins. The MA proteins are attached to the lipid bilayer through myristate groups attached at their N-terminus (i.e., N-myristoylation) and a region of highly basic amino acids that interact with specific membrane phospholipids (German Advisory Committee Blood (Arbeitskreis Blut), Subgroup 2016; Olety and Ono 2014). Within this outer capsid is another capsid that houses the viral genome along with enzymes essential for HIV's replication (i.e., RT and IN) and other proteins involved in its life cycle (i.e., Vif., Tat, Rev., etc.). The capsid loosely resembles a SpaceX or Blue Origin landing capsule. The capsid's tapered pole is anchored to the outer capsid membrane (German Advisory Committee Blood (Arbeitskreis Blut), Subgroup 2016; Campbell and Hope 2015).

During maturation of HIV, the Pr55Gag and Pr160GagPol polyproteins are cleaved into individual proteins at the end of the budding and release of the virions from their host cells (German Advisory Committee Blood (Arbeitskreis Blut), Subgroup 2016). This processing produces a viable virus that is capable of infecting other host cells and fully replicating. The importance of gp120 to the infectivity of the virus is illustrated by the fact that HIV particles that have lost these surface proteins via mechanical shearing during the budding and release process are no longer infective (McKeating et al. 1991).

### 8.2.2 Replication and Life Cycle of HIV

While it primarily infects CD4+ T lymphocytes (CD4+ T cells), HIV can also infect macrophages and dendritic cells. The replication and life cycle of HIV has seven major stages: (1) binding and fusion, (2) host cell entry, (3) reverse transcription of HIV genome, (4) integration of HIV genome into host cell genome, (5) expression of HIV proteins and RNA, (6) assembly and budding of virions, and (7) maturation of virus (see Fig. 8.3) (hivinfo.nih.gov/understanding-hiv/fact-sheets/hiv-life-cycle). The binding and fusion stage begins with the interaction between gp120 of HIV and a

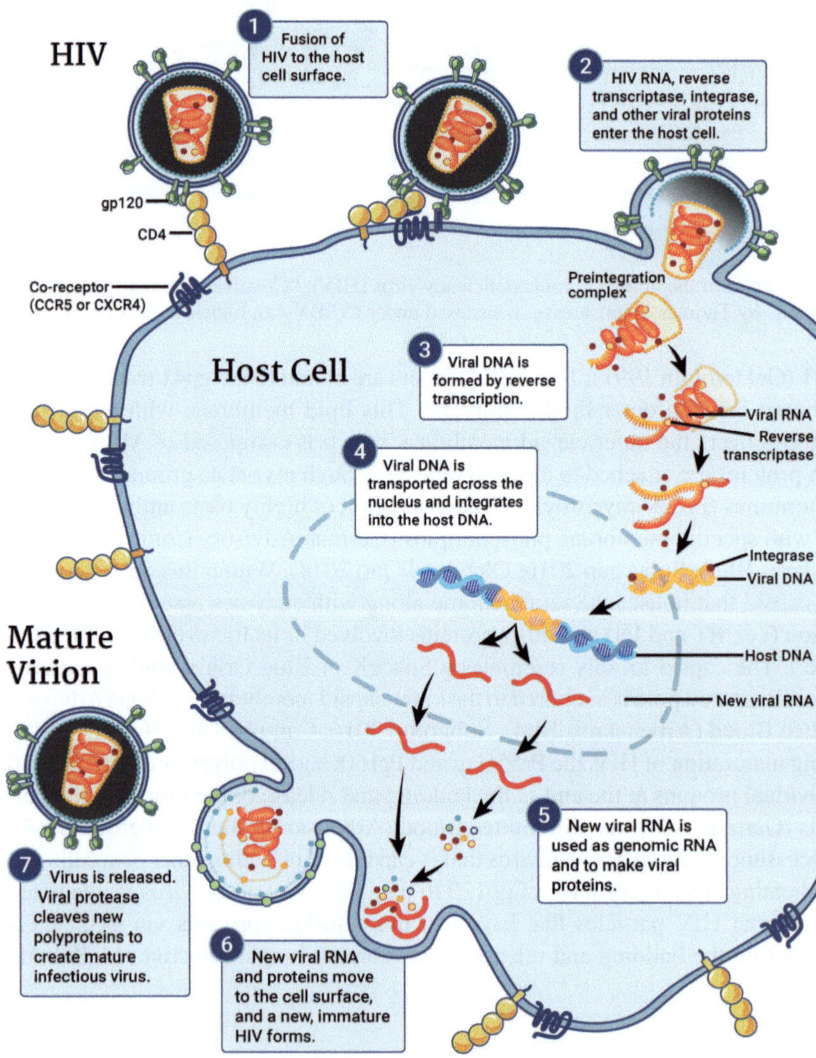


**Fig. 8.3** Human immunodeficiency virus (HIV) life cycle

CD4+ T cell, which invokes a structural change that exposes a chemokine-binding domain that results in an additional interaction between gp120 and a chemokine co-receptor (usually CCR5 or CXCR4) on the surface of the host cell. This stabilized interaction results in the penetration of the host cell membrane by gp41, which undergoes a conformational change into a hairpin shape. This "hook-like" interaction causes the virus and host cell membranes to fuse together, which enables HIV's capsid to enter the cytoplasm. Once the capsid enters, its components (HIV RNA, RT, IN, PR, and RNase H) are released into the cytoplasm. Reverse transcriptase quickly copies the +ssRNA into an antisense complementary DNA (cDNA) molecule. The RNase H and DNA-dependent DNA polymerase activity of RT serves to degrade the viral RNA during cDNA synthesis and produces a sense DNA strand from the antisense cDNA strand, respectively. The cDNA and sense DNA strands come together to form a double-stranded (ds) DNA molecule that enters the host cell's nucleus. At this point, IN incorporates the dsDNA into the cell's genome.

If activated, RNA polymerase II will produce a 9-kb mRNA using the integrated viral DNA as a template (Hu and Hughes 2012). Some of the transcribed HIV RNAs will be repackaged into newly formed HIV particles, and some will be translated into HIV proteins. Regardless of what it is used for, the HIV mRNA moves out of the nucleus. The mRNA being translated moves into the rough endoplasmic reticulum where the cell's ribosomes use it as a template to produce proteins. This 9-kb mRNA is used to produce a variety of HIV components. It produces the Pr55Gag polyprotein, which is cleaved by HIV's protease to produce the MA (p17), capsid (CA; p24), and nucleocapsid (NC; p7) proteins (Smith et al. 1990). It also produces the Pr160GagPol polyprotein that is cleaved into not only producing matrix and capsid proteins but also PR (p10), RT (p51), and IN (p32) (Smith et al. 1990). The 9-kb parent mRNA is also spliced into a 4-kb and 2-kb mRNA. The 4-kb mRNA is used to produce a polyprotein that is cleaved into the gp120 and gp41 envelope proteins. It is also used to translate the Vif, Vpr, and Vpu regulatory proteins. The 2-kb mRNA is used to synthesize the Tat, Rev., and Naf regulatory proteins.

The Pr55Gag and Pr160GagPol polyproteins are not cleaved into their individual proteins until they leave the endoplasmic reticulum and enter the Golgi complex (Balasubramaniam and Freed 2011; Speck et al. 2000). These proteins that will eventually become part of the mature virus are then transported to the plasma membrane where they remain until HIV buds and releases from the cell. Assembly of the nascent HIV particles occurs within specialized microdomains within the plasma membrane. The Pr160GagPol polyprotein mediates the assembly of the new virions (Lerner et al. 2022). It promotes this assembly by creating the necessary protein-protein interactions (PPIs) that concentrate the viral envelope proteins, its genomic RNA, and necessary proteins together. The polyprotein also stabilizes the binding of this complex to the cell's plasma membrane. It is important to note that the polyproteins are not proteolyzed into their final forms at this stage as this will occur after the particles are released and then mature. As the concentration of the HIV components increases at the plasma membrane, they begin to push outward. While the polyprotein can bind and assemble the HIV particles itself, a series of proteins that are part of the host ESCRT (endosomal sorting complexes required for transport) machinery mediate the release of the virions along with the force generated by the viral particles pushing on the membrane (Hurley and Cada 2018). As the immature virus particle pushes itself out of the host cell during the budding process, it acquires its lipid envelope and the embedded gp120/gp41 complexes that remained within the cell membrane as the virus and host cell membranes fused during viral entry. Once the virus has been released from the host cell encapsulated within its envelope, it begins the final process of maturation that will enable it to infect other cells.

Concomitant with the release of the immature virus particles, the HIV PR enzyme cleaves the Gag polyproteins into their mature forms (Mattei et al. 2014; Wiegers et al. 1998). As the polyproteins are cleaved, they are released from the envelope of the virus and form the capsid. Within the capsid are packaged the mature forms of the RT, NC, and IN proteins along with two copies of the RNA genome. This mature virus is now capable of infecting other host cells.

## 8.3 Proteomics and HIV

HIV has been a major focus in virology for over 40 years now. One of the first manuscripts published that has the terms “HIV” and “proteomics” is a study that utilized enzyme-linked immunosorbent assay (ELISAs), two-dimensional polyacrylamide gel electrophoresis (2D-PAGE), immunoblotting, and mass spectrometry (MS) to identify a serological protein that was believed to be diagnostic for individuals with tuberculosis (TB) (Hendrickson et al. 2000). This study began from the discovery of an 88-kDa antigen that elicited a strong antibody response in serum acquired from patients with TB. The monoclonal antibody IT-57 was shown to react with this 88-kDa *mycobacterium tuberculosis* (*Mtb*) antigen although the antigen’s identity was not known. The investigators separated sera samples taken from TB- and HIV-positive individuals using 2D-PAGE. The gels were immunoblotted using IT-57, and the protein spot that showed positive reactivity was correlated with its position on a silver-stained 2D-PAGE gel. The spot was excised from the gel and analyzed using tandem mass spectrometry ($MS^2$). From this analysis two peptides, DELQAQLDK and NYTAPGGGQFTLPGR, were identified. Analysis of the published *M. tuberculosis* H37Rv genomic sequence revealed that both peptides originated from Mtb81. An ELISA developed using recombinantly expressed Mtb81 was able to detect antibodies in 92% of TB patients who were also seropositive for HIV and 57% of patients who were only TB positive. This result suggested that Mtb81 could be a promising antigen for diagnosing TB, especially for patients coinfected with HIV.

Why did I bring up this study that was published over 25 years ago? It serves as the starting point to illustrate the progress that has been made in HIV proteomics. When compared to the following studies, it demonstrates the remarkable progress that has been made considering that Mtb81 was identified through only two peptides.

### 8.3.1 Circulating Biomarkers of HIV Infection

If “the eyes are the window of the soul” in that they reveal a person’s true emotional nature and innermost thoughts, then “the blood may be the window of flesh” in that its contents reveal a person’s true physical condition. The components of blood (i.e., cells, metabolites, ions, proteins, etc.) exist within a normal range in healthy individuals. If their levels are altered, the body quickly responds by bringing its components back to a state of homeostasis. In the context of this book, changes in the blood proteome have been observed in different infections. Understanding these changes helps in the development of diagnostics as well as prophylactic and therapeutic strategies. In a major longitudinal study (see Fig. 8.4), blood plasma samples from 54 participants (34 male, 20 females) from Central and East Africa and South Africa were analyzed to compare their proteomes before, during, and after hyperacute HIV-1 infection (Nazziwa et al. 2024). A total of 157 samples were acquired from three time points: no (a median of 62 days prior to) HIV-1 infection (Visit 0; V0), 10–14 days (median 10 days) after HIV-1 infection (Visit 1; V1), and 15–39 (median 31 days) days after HIV-1 infection (Visit 2; V2). All subsequent models developed from the proteomic data were adjusted for age and cohort (i.e., the IAVI cohort consisting of male patients infected with HIV-1 sub-subtype A1 or the Durban cohort consisting of female patients infected with HIV-1 subtype C). The aim of the study was to understand changes in the plasma proteome that occurs during hyperacute HIV-1 infection—the period from when virus enters the bloodstream (i.e., viremia) to peak viral load (Ndhlovu et al. 2015). Recognizing this period is important because it allows for timely initiation of antiretroviral therapy (ART), which improves health outcomes, reduces risk of transmission, and lowers the risk of developing AIDS. In blood, HIV viremia is detectable about 1 week after infection. Peak viral load (VL) occurs about 1 month after infection at which point viremia gradually decreases and stabilizes over the following 30–35 days.

### 8.3.2 Analysis of Plasma Proteome at Different Stages of HIV Infection

Each plasma sample was analyzed twice using data-independent acquisition--sequential window acquisition of all theoretical mass spectra (DIA-SWATH-MS). In the DIA-SWATH-MS method, which is described in greater detail in Chap. 9, all peptides within a defined mass range are fragmented, and spectral libraries of known peptide fragmentation spectra are used to identify the peptides (Gillet et al. 2012; Rotello and Veenstra 2021). This method provides a deeper proteome coverage than conventional data-dependent MS methods that select and fragment individual peptides based on their signal intensity. In one analysis, neat plasma sample was analyzed, and in the other the 14 most abundant proteins were depleted from the plasma sample prior to DIA-SWATH-MS analysis. Removing the top 14 most abundant

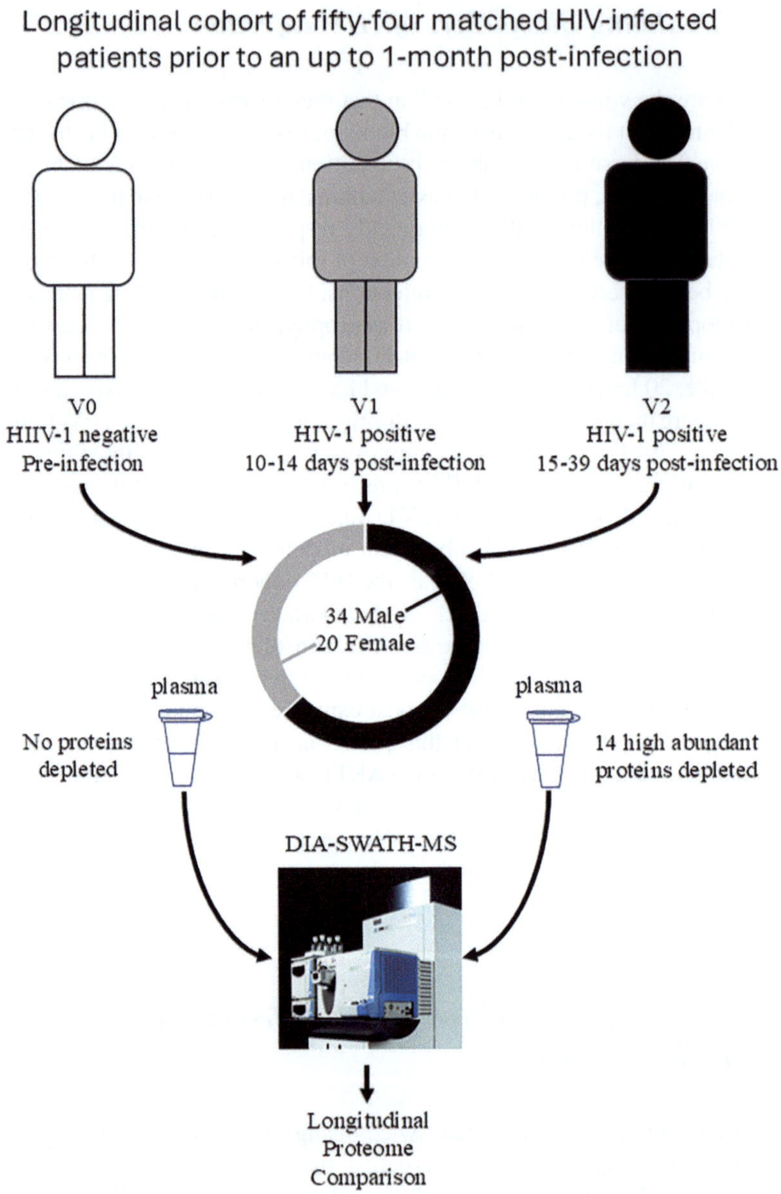


**Fig. 8.4** Flowchart describing sampling and proteomic workflow employed in the longitudinal analysis of 54 individuals pre- (V0) and post-HIV infection (V1 and V2). Fifty-four individuals from two distinct geographical regions provided three matched samples each. Plasma samples from 34 males and 20 females were processed for sequential window acquisition of all theoretical mass spectra (SWATH-MS) analysis with and without prior depletion of 14 high abundance proteins. Cross comparison of abundances was performed between the plasma proteins obtained from the three groups (V0, V1, and V2)

proteins increases the likelihood of lower abundant proteins being detected in the proteomic analysis. These top 14 most abundant proteins constitute approximately 90–95% of the total protein content of plasma and include such proteins as albumin, various immunoglobulins, fibrinogen, and transferrin. Lower abundant proteins are hypothesized to be more useful as clinical biomarkers for observing changes in patient samples during HIV-1 progression. Protein quantitation was performed by combining the ratios from multiple peptides between pairs of samples to provide a single optimal quantitative ratio for each protein (Nazziwa et al. 2024).

A total of 1336 of the detected proteins were used to generate proteomic models that reflect the patients' status before (V0), during (V1), and after (V2) hyperacute HIV-1 infection (hA-HIV). Of these, 1028 had unique UniProt IDs that classified 242 as secreted proteins known to circulate in blood, and 618 were classified as intracellular proteins that leaked or were released from tissues or necrotic cells. The remaining proteins were either secreted elsewhere or associated with immunoglobulins (Consortium TU 2022).

Longitudinal protein expression profiles of 83,643 protein combination values from all three time points (V0, V1, and V2) were conducted across the study participants, resulting in a total of 1336 profiles. The data was analyzed using a variety of algorithms that classified the data based on overall quantitative trends of specific proteins. K-means and hierarchical clustering were used to determine longitudinal patterns within specific patients across the time points. These analyses identified six different expression profiles. Two of the profiles revealed a decline during hA-HIV with a rebound to preinfection levels. These profiles were named "rapid decrease-rapid increase" and "slight decrease-rapid increase." Two models, termed "gradual decrease" and "rapid decrease-slight increase," revealed decreased levels during and after hA-HIV. A single model showed an increase during hA-HIV followed by a decline to preinfection levels ("rapid increase-rapid decrease"), and the final model revealed a sustained increase during and after hA-HIV ("persistent increase").

The functions of the proteins in the "rapid decrease-rapid increase" group, which accounted for 7% of the measured proteins, were primarily cell adhesion and extracellular structure organization (Nazziwa et al. 2024). Proteins in the "gradual decrease" group were primarily involved in activation of immune response, cytoskeleton organization, protein transport, and regulation of apoptosis. Proteins within the "persistent increase" category were associated with epithelial development, complement activation, cellular extravasation (i.e., movement of cells from circulation into tissue), and viral entry into host.

### 8.3.3 The Effect of HIV Infection on Functional Pathways

While differences in the overall profiles were recognized, it would be more useful to be able to identify specific proteins that reflect the stage of HIV infection. To get closer to this result, proteins that were significantly differentially abundant between study visits (i.e., V0 vs V1, V1 vs. V2, and V0 vs. V2) within individuals were identified. A total of identified 97, 168 and 149 differentially expressed proteins between visits V1–V0, V2–V0, and V2–V1, respectively, were identified. Of these, 2 (V1–V0), 18 (V2–V0), and 8 (V2–V1) proteins exhibited substantial changes with a log2 fold change of at least greater than 1 (see Fig. 8.5). Of these proteins, 8 are secreted to the blood or extracellular matrix, while the remaining 20 enter plasma via leakage through the cell membrane.

Analyzing the 97 differentially abundant proteins between V1 and V0, 51 were of higher abundance, and 46 were of lower abundance in V0. The proteins of higher abundance in V0 included 28 and 23 classified as secreted to blood and leakage proteins, respectively. Proteins that were of higher abundance in V1 included those related to immune responses (both innate and adaptive), stress response, transport, protein maturation, and regulation of vesicle-mediated transport. Of the 46 proteins that were of lower abundance in V0, 18 were classified as secreted to blood. Thirteen proteins were linked to regulation of hydrolase/peptidase activity and establishment of endothelial barrier. Other changes observed between V1 and V0 included activation of tissue damage markers that surprisingly were associated with the esophagus mucosa and heart.

When comparing V2 with V0, 77 proteins were upregulated with most of them (75%) associated with leakage proteins and most of the rest linked to functions such as blood coagulation, fibrin clot formation, protein maturation, positive regulation of cell-substrate adhesion, metal ion responses, positive regulation of substrate adhesion-dependent cell spreading, and cell adhesion. Of the 91 proteins of lower abundance in V2, 48 were classified as leakage proteins, and the remaining 53 were associated with regulation of peptidase activity, inflammatory responses, stress responses, wound healing, immune responses, cytokine production of the tumor necrosis factor (TNF) superfamily, and neutrophil chemotaxis. Analysis of tissue-specific transcriptional signatures indicated that muscle skeletal tissues and lungs damage markers were activated while simultaneously suppressed in blood. This result clearly shows that substantial tissue damage and cellular stress responses occur as patients progress from a state of non-infection (V0) to hA-HIV.

Comparison of V1 and V2 showed that 149 proteins were differentially abundant between these two time points with 86 being of higher abundance in V2. Of these proteins, 28 were related to gene expression, protein maturation, and macromolecule biosynthesis. Of the proteins with lower abundance in V2, 53 were associated with inflammatory, stress, and innate immune responses, as well as detoxification, responses to biotic stimulus, complement activation, and neutrophil chemotaxis. Proteins that increased in abundance between V2 and V1 were associated with damage signatures in adipose, cervical, muscle skeletal, and lung tissue. Proteins that decreased in abundance over the same period were associated with tissue damage in the skin and whole blood. Overall, these findings indicate that HIV-1 infection induces a myriad of dynamic changes in protein expression that go beyond immune responses, influencing various biological processes such as cell mobility, metabolic function, and apoptosis.

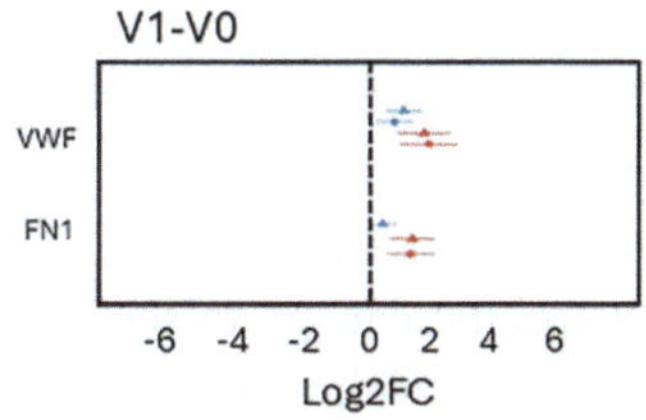


| Protein | Description | Biological Process |
|---|---|---|
| **V1-V0; Increased in V1** | | |
| **VWF** | Von Willebrand Factor | Blood coagulation, Cell adhesion |
| **FN1** | Fibronectin 1 | Acute phase, Angiogenesis, Cell adhesion |

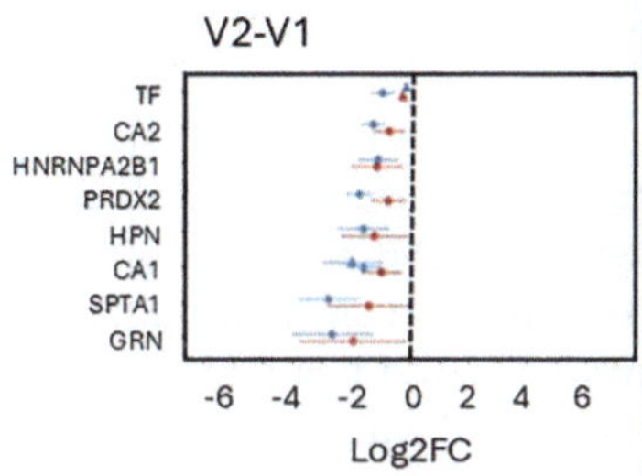


| Protein | Description | Biological Process |
|---|---|---|
| **V2-V1; Decreased in V1** | | |
| **TF** | Transferrin | Immunity, Ion transport |
| **CA2** | Carbonic anhydrase 2 | One-carbon metabolic process, carbon dioxide transport |
| **HNRNPA2B1** | Heterogeneous nuclear ribonucleoprotein A2/B1 | Host-virus interaction, mRNA processing/splicing/ transport |
| **PRDX2** | Peroxiredoxin 2 | T-cell proliferation |
| **CA1** | Carbonic anhydrase 1 | One-carbon metabolic process, carbon dioxide transport |
| **HPN** | Hepsin | Negative regulation of apoptotic process |
| **CA1** | Carbonic anhydrase 1 | One-carbon metabolic process, carbon dioxide transport |
| **SPTA1** | Spectrin α1 | Erythrocytic cell shape |
| **GRN** | Granulin precursor | Astrocyte activation<br>Immune response |

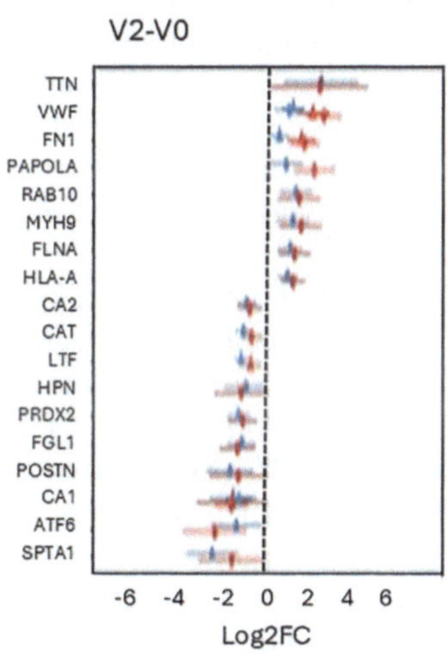


| Protein | Description | Biological Process |
|---|---|---|
| **V2-V0; Increased in V2** | | |
| **TTN** | Titin | Cardiac muscle cell development |
| **VWF** | Von Willebrand Factor | Blood coagulation, Cell adhesion |
| **FN1** | Fibronectin 1 | Acute phase, Angiogenesis, Cell adhesion |
| **PAPOLA** | Poly(A) polymerase α | mRNA processing |
| **RAB10** | Ras-related protein RAB10 | member RAS oncogene family |
| **FLNA** | Filamin A | Cilium biogenesis/degradation, actin cytoskeleton reorganization |
| **HLA-A** | Major histocompatibility complex, class I, A | Adaptive immunity, Host-virus interaction, Innate immunity |
| **MYH9** | Myosin heavy chain 9 | Cell adhesion, Cell shape |
| **V2-V0; Decreased in V2** | | |
| **LTF** | Lactotransferrin | Immunity, Ion transport, Osteogenesis, Transcription regulation |
| **PRDX2** | Peroxiredoxin 2 | T-cell proliferation |
| **FGL1** | Fibrinogen like 1 | Adaptive immunity, regulation of T-cell activation |
| **ATF6** | Activating transcription factor 6 | Transcription regulation |
| **CA2** | Carbonic anhydrase 2 | One-carbon metabolic process, carbon dioxide transport |
| **CAT** | Catalase | Cellular detoxification of hydrogen peroxide |
| **HPN** | Hepsin | Negative regulation of apoptotic process |
| **POSTN** | Periostin | Cell adhesion |
| **CA1** | Carbonic anhydrase 1 | One-carbon metabolic process, carbon dioxide transport |
| **SPTA1** | Spectrin α1 | Erythrocytic cell shape |

**Fig. 8.5** Plots indicating $\log_2$ fold abundance changes of key proteins affected by HIV infection (left panels). Protein abundances were compared between preinfection (V0) samples and those acquired approximately 2 weeks (V1) and 1 month (V2) post-HIV infection. The results include analysis of plasma samples that had been depleted of 14 high abundance proteins and those that had not been depleted. The tables across from each plot list the protein name and its biological function. Adapted from ref. (Vadaq et al. 2023) under the terms and conditions of the Creative Commons Attribution (CC BY) license (https://creativecommons.org/licenses/by/4.0/)

### 8.3.4 Biomarkers Associated with Different Stages of HIV Infection

While the previous analysis revealed functional classes of proteins whose abundance changed over the course of HIV infection, they are not specific enough to serve as clinical markers for diagnosis or prognosis of HIV. It is known that hA-HIV triggers a complex network of cellular and tissue signaling events that are often accompanied by systemic release of cytokines and acute retroviral syndrome (ARS) in many individuals (Cohen et al. 2011; Lama et al. 2025; McMichael et al. 2010). Of the 33 individuals in the IAVI cohort (i.e., male patients infected with HIV-1 sub-subtype A1), 20 were diagnosed with ARS based on 11 symptoms (e.g., headache, fever, diarrhea, skin rash, oral ulcers, etc.) with the requirement they have at least 9 of these symptoms. Linear mixed-effects models were used to determine if any of the differentially abundant proteins correlated with ARS. The result showed that three proteins whose abundance differed between V1 and V0 were associated with ARS. These proteins were the cytoskeleton protein zyxin (ZYX) that modulates inflammatory response in endothelial cells, the pulmonary surfactant anti-inflammatory lung protein secretoglobin family 1A member 1A (SCGB1A1), and the leukocyte immunoglobulin-like receptor subfamily A member 3 (LILRA3), a receptor that is secreted by macrophages and monocytes and functions to regulate immune responses and inflammation during infection. The abundance of both ZYX and SCGB1A1 was about eight times lower in patients with ARS compared to those without the condition, while the abundance of LILRA3 was about two times higher in ARS.

As a pulmonary surfactant protein, the decreased levels of SCGB1A1 are likely associated with increased lung inflammation and decreased epithelial integrity observed in HIV patients (Martinu et al. 2023). It has been shown that binding of cytoplasmic dsRNA by retinoic acid-inducible gene I-like receptors induces ZYX, which binds to mitochondrial antiviral signaling protein (MASP) (Kouwaki et al. 2017). Activation of MASP triggers type I interferon expression, which is the primary immune response during hA-HIV infection (Bender et al. 2015). Decreases in the levels of ZYX would severely diminish the cell's primary response to infection and accelerate HIV progression. Measuring proteins such as ZYX and SCGB1A1 could serve as sentinels of increasing HIV progression, which would indicate the correct time for treatment intervention or signal the development of drug resistance if current medications are being used.

### 8.3.5 Plasma Biomarkers for Monitoring Viral Load

HIV is a chronic condition in which patients can experience temporary increases in viral load. Although temporary increases can occur even when patients are being treated with ARTs, measuring viral load is still important for guiding therapy and monitoring its effectiveness. During times of high viral load, when HIV is actively replicating, there is an increased risk of opportunistic infection (Ekwaru et al. 2013). A low viral load suggests the virus is latent and/or the treatment is effective. While studies have shown that the levels of certain cytokines correlate with viral load in patients, these measurements are only effective in 66% of patients at 12-month set points after infection (Roberts et al. 2010; Turk et al. 2018).

To identify better indicators of viral load, the plasma proteome of 45 of the study's participants that was measured at various time points was reanalyzed comparing the proteins' abundances with viral load. These participants had high viral loads following HIV infection, which subsequently decreased and reached a set point about 50 days after initial infection. Hierarchical clustering created two groups of participants based on their viral load profiles measured during their first year of infection. The clusters were labeled as viral controllers and non-controllers. Viral controllers had viral loads below 10,000 copies/mL, while non-controllers had viral loads greater than 10,000 copies/mL. Neither group was taking ARTs. Linear mixed-effect models were employed to correlate viral load with changes in plasma protein abundance over time.

The protein α-soluble N-ethylmaleimide-sensitive factor attachment protein (NAPA) was more than twice as abundant in viral load controllers compared to non--controllers. This protein is involved in intracellular vesicle trafficking and membrane fusion (Cui et al. 2022). As part of this function, it interacts with HIV's Gag protein to promote viral particle assembly and budding (Klingler et al. 2020). NAPA's abundance was increased at both V1 and V2 time points (when patients were HIV-1 positive) compared to V0 (i.e., HIV-1 negative). The abundance of another protein, the GTP-binding nuclear protein Ran (RAN), was about three times higher in viral load controllers at both V1 and V2. RAN's function involves regulation of nucleocytoplasmic transport, mitotic spindle assembly, and nuclear envelope formation (Ozugergin and Piekny 2021). This protein plays a critical role in HIV replication as it interacts with HIV-1 Rev. protein (Li et al. 2005). This interaction facilitates the export of unspliced viral RNA into the host cell cytoplasm where it can be translated into viral proteins or packaged into new virions. A third protein, inter-α-trypsin inhibitor heavy chain H4 (ITIH4), was approximately four times lower within the plasma of viral load controllers compared to non-controllers. ITIH4 is an abundant acute-phase protease inhibitor that inhibits enzymes linked to the development of angioedema (i.e., swelling) (Troldborg et al. 2024). It functions to stabilize the extracellular matrix and inhibit inflammatory processes that result in tissue damage and disease progression in HIV-1 infection individuals (Hamm et al. 2008).

On the surface, the increase in both NAPA and RAN in viral load controllers seems contradictory as these proteins play roles in HIV genome integration, viral assembly, and budding. What always needs to be remembered when discussing viral replication, however, is that the main responsibility of host cell proteins lies within maintaining the integrity of host cell processes. Their increased abundance suggests that the host cell is increasing the translation of these proteins to maintain normal cellular processes that are being impacted by HIV's presence. If both NAPA and RAN remained at their normal cellular levels, HIV would usurp most of their available activity, and viral load would increase. Keeping the cell's integrity intact would prevent HIV replication from occurring and keep the viral load under control. Likewise, the lower abundance of ITIH4 would seem to suggest that inflammation is poorly controlled and HIV load should be increasing. The low ITIH4 expression and low viral load are reflective of a reduced inflammatory state, which is characteristic of individuals with better HIV load control.

### 8.3.6 Plasma Biomarkers of HIV Disease Progression

The available plasma proteomic data was also analyzed to identify markers of HIV disease progression, which is currently measured by tracking CD4+ T-cell count and viral load (Boyer et al. 2013). In this analysis, only CD4+ T-cell count was used. Disease progression is a critical measure as it indicates how well a therapy is working, if drug resistance has occurred, or if a patient's immune system is faltering. Participants in the study contributed 13 samples for CD4+ T-cell count over a 12-month period following their estimated date of infection (Nazziwa et al. 2024). They were grouped into two categories: fast progressors (CD4+ T-cell count less than 500 cells/mm$^3$ within 12 months) and slow progressors (CD4+ T-cell count greater than 500 cells/mm$^3$ within 12 months). Of the eligible participants, 42 and 12 were classified as fast and slow progressors, respectively. Disease progression was serially compared between each time point (i.e., V1–V0 and V2–V0) and hazard ratios (i.e., a comparison of the rate of an event occurring in one group). When pairwise comparisons of the groups V1–V0 and V2–V0 were made, six and seven proteins were associated with faster disease progression, respectively (see Fig. 8.6). This result is surprising considering that only three proteins were associated with difference in viral load. Comparison between V1 and V0 showed that increased levels of hepsin (HPN), protein kinase C β (PRKCβ), corticotropin-releasing hormone-binding protein (CRHBP), proteasome subunit β type-6 (PSMB6), thioredoxin domain-containing protein 5 (TXNDC5), and apolipoprotein C4 (APOC4) in V1 were associated with an increased risk of HIV-1 disease progression. Decreased levels of glutathione S-transferase μ2 (GSTM2) was associated with a faster disease progression. Comparison between V2 and V0 showed that increased levels of integrin subunit β3 (ITGB3), heat shock 70 kDa protein 8 (HSPA8), D-dopachrome tautomerase-like protein (DDTL), and ubiquitin B (UBB) in V2 were associated with an increased risk of disease progression. Decreased levels of CD84 and latent-transforming growth factor β-binding protein 1 (LTBP1) were associated with decreased risk of disease progression. Each of these proteins either binds to a component of HIV-1 or is involved in a cellular function that effects HIV replication (see Table 8.2).

The direct correlation between increased hepsin abundance and faster HIV progression is consistent with hepsin's ability to suppress the induction of type I interferons (Hsin et al. 2021). As the innate immune system's major defense against viral infection, increased suppression of type I interferons allows HIV to avoid this barrier to its replication. The association between increased PRKCβ abundance and faster HIV progression is most likely linked to nuclear factor-κB (NF-κB) activation

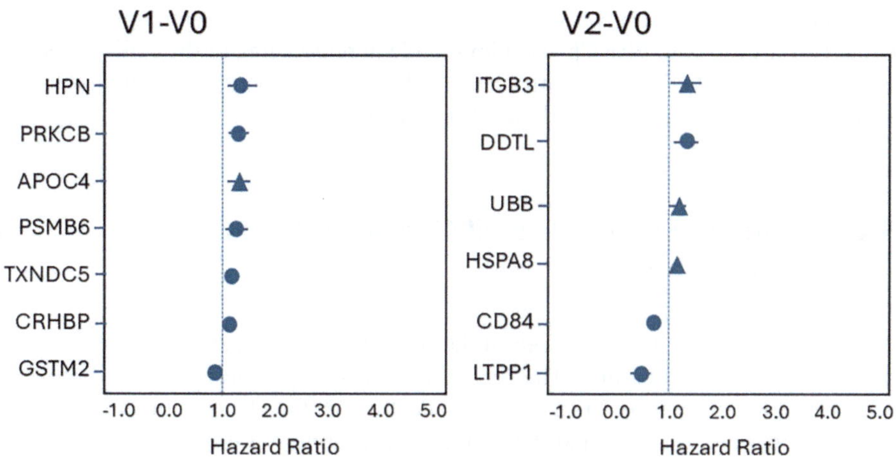


**Fig. 8.6** Hazard ratios (HR) for proteins differentially expressed in groups V1 (median 10 days post-HIV infection) and V2 (median 31 days post-HIV infection), compared to V0 (prior to HIV infection). Adapted from ref. (Nazziwa et al. 2024) under the terms and conditions of the Creative Commons Attribution (CC BY) license (https://creativecommons.org/licenses/by/4.0/)

**Table 8.2** Proteins whose abundance was associated with HIV progression as measured by CD4+ cell count. The function of each protein as it relates to HIV is provided

| Increased in abundance | HIV progression | Function |
|---|---|---|
| Hepsin | Increased | Suppresses antiviral immune responses |
| Protein Kinase Cβ | Increased | Regulates various cell processes that lead to HIV-1 activation |
| Corticotropin-releasing hormone--binding protein | Increased | Elevated in cases where HIV-1 causes dysregulation of HPA axis |
| Proteasome subunit β type-6 | Increased | Interacts with HIV-1 nef protein to inhibit viral degradation |
| Thioredoxin domain-containing protein 5 | Increased | Facilitates HIV-1 Gag protein stability |
| Apolipoprotein C4 | Increased | Enhanced HIV-1 entry |
| Integrin subunit β3 | Increased | Interacts with HIV-1 Tat protein |
| Heat shock 70 kDa protein 8 | Increased | Functions in viral entry, replication, and nuclear export |
| D-dopachrome tautomerase like protein | Increased | Pro-inflammatory cytokine |
| Ubiquitin B | Increased | Targets Tat protein for degradation |
| Decreased in abundance | HIV Progression | Function |
| Glutathione S-transferase μ2 | Increased | Detoxification enzyme |
| CD84 | Decreased | Maintains HIV latency |
| Latent-transforming growth factor β-binding protein 1 | Decreased | Establishes and maintains HIV latency |

(Hokello et al. 2021). NF-κB results in enhanced viral transcription through it binding to the HIV-1 promoter. PRKCβ also plays a role in remodeling of the cell's cytoskeleton in such a way to enhance virus entry. Enhancing viral entry (and along with its viral replication) serves to exacerbate CD4 T-cell reduction.

## 8.4 Identifying Biomarkers of Comorbidities in People Living with HIV

Generating the amount of data seen in the previous study illustrates the strides that have been made in proteomics. The beginning of this chapter described the discovery of a protein for diagnosing TB in HIV patients through the identification of only two peptides. Forward 20 years and the quantitation of almost 1400 proteins from 157 plasma samples identified markers for measuring HIV infection, viral load, and progression. By analyzing samples from patients prior to, during, and after hA-HIV, the dynamics of viral infection could be assessed at both the individual and population levels. Admittedly these studies that provide comprehensive coverage of complex human proteome samples have limitations related to variation among individuals (i.e., genetics, demographics, lifestyles, etc.) and sample analysis since it is impossible to normalize for every protein measurement. Regardless of these limitations, the study reliably and consistently identified several proteins that were associated with diagnosis, viral load, and progression of HIV-1. These types of studies increase our understanding of how HIV affects individuals and provide direction for validating the value of these predictive and prognostic biomarkers.

Writing this book is analogous to a global proteomic study providing more data than can be completely analyzed. Proteomics has grown so exponentially in the past three decades that it is impossible to adequately cover all the various approaches that have been applied to the study of HIV. That being said, my hope is to provide the reader with a view of various technologies that have been used in proteomics and let them imagine how these can be applied to their research interests. As has been mentioned throughout this book, MS dominates proteomics, but I want to take this opportunity to discuss another study that uses a non-MS approach that can be used to not only identify biomarkers but also validate them in a discovery cohort.

The most well-known treatment for individuals living with HIV is antiretroviral therapy (ART) (Kemnic and Gulick 2025). ART involves regularly taking a combination of drugs that prevent HIV from replicating, thereby reducing the viral load to an undetectable level and almost eliminating the risk of transmission. Lowering the HIV viral load enables the body to produce more infection-fighting CD4+ T cells, which enables the immune system to fight off opportunistic infections. While ART is not a cure, it can help people infected with HIV live longer, healthier lives. Unfortunately, even when taking ART, HIV-infected individuals have persistent inflammation that can lead to comorbidities not related to AIDS. A study conducted between 2011 and 2016 showed that 21-year-olds with HIV infection and having a CD4+ T-cell count greater than 500/μL who initiated ART had a life expectancy of 57.4 years compared to 64.2 years for uninfected individuals (Marcus et al. 2020). A study conducted between 2000 and 2003 showed that 21-year-olds with HIV on ART had an average of 11.3 years free of comorbidities, while uninfected individuals could expect to have 26.6 years free of comorbidities (Marcus et al. 2020). The major comorbidities associated with HIV patients on ART include chronic liver disease, kidney disease, diabetes, chronic lung disease, cancer, and cardiovascular disease (Webel et al. 2021). These comorbidities currently represent the leading cause of death among HIV patients living in high income countries.

These comorbidities tend to result from persistent inflammation as people living with HIV on ART demonstrate increased levels of inflammation markers including high-sensitivity C-reactive protein (hsCRP), interleukin-6 (IL-6), and tumor necrosis factor alpha (TNF-α), along with the soluble immune activation markers CD14 and sCD163 (Mabhida et al. 2024; Obare et al. 2024; Williams et al. 2018). Unfortunately, the underlying pathology that mediates chronic immune dysregulation remains unclear, hindering the development of novel treatment strategies to mitigate the risk of these non-AIDs-related comorbidities in people living with HIV taking long-term ART.

### 8.4.1 Plasma Proteomic Analysis Using Proximity Extension Assays

To better understand the mechanisms driving these comorbidities, the concentrations of 1472 "targeted" proteins were measured in plasma acquired from two separate cohorts (i.e., discovery and replication) of virally suppressed people living with HIV and healthy controls (Vadaq et al. 2023). The term "targeted" means that the technology being used directly measures a specific subset of proteins rather than simply casting a broad net and attempting to measure as many proteins as possible. The targeted proteins included those associated with inflammation, oncology, neurology, and cardiovascular processes. The discovery cohort consisted of 205 HIV cases and 120 healthy controls. The replication cohort consisted of 639 HIV cases and 99 healthy controls (see Fig. 8.7).

Part of the purpose of this book is to educate scientists, specifically virologists, on how various proteomic technologies impact the understanding of both the virus and its effects on its host. While proteomics continues to be dominated by MS because of its unbiased detection and broad coverage, other methods have been developed that provide quantitation of specific targeted proteins. While targeted analysis suggests the measurement of only a small number of proteins, proteomic technologies have been developed that can target thousands of proteins in a single experiment. While protein arrays are mostly associated with broad targeted analysis, this study used a method based on proximity extension assay (PEA) technology (Moaddel et al. 2021). This technology, also popularly known as Olink®, combines antibody-based immunoassays and quantitative real-time polymerase chain reaction (qPCR) or next-generation

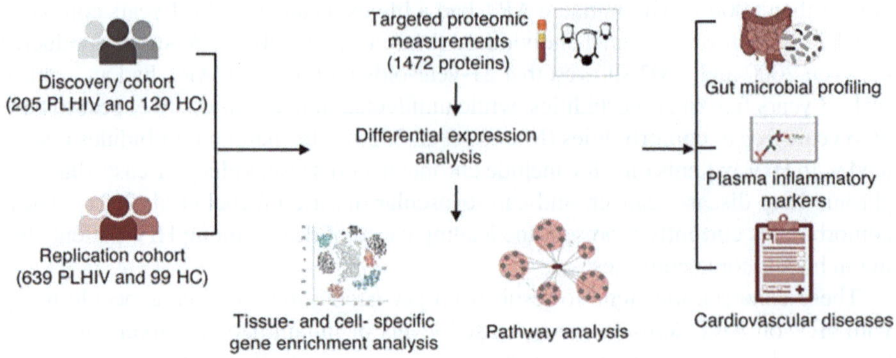


**Fig. 8.7** Study design using proximity extension assay (PEA) to identify biomarkers for comorbidities in patients with HIV. The abundance of specific proteins within the library of PEA antibody constructs was measured in a discovery and replication cohort of patients. Each cohort was composed of samples from HIV-infected individuals and healthy controls. Changes in protein abundances were measured and the results grouped into functional pathways. Adapted from ref. (Vadaq et al. 2023) under the terms and conditions of the Creative Commons Attribution (CC BY) license (https://creativecommons.org/licenses/by/4.0/)

sequencing (NGS) platforms. With a library of more than 5300 validated protein biomarker targets, adoption of this technology has been very strong, leading to over 1400 scientific publications.

The foundation of PEA technology is based on the use of two matched antibodies that simultaneously bind to a specific protein target (see Fig. 8.8) (Lundberg et al. 2011). The matched antibodies are uniquely labeled with complementary DNA oligonucleotides. If the protein target is present, the two antibodies come into close proximity as they bind. The binding brings the DNA oligonucleotides into proximity and they hybridize. This double-stranded DNA serves as a template for a DNA polymerase-dependent extension step that produces an amplicon (i.e., a piece of amplified DNA or RNA) that is unique to each antibody pair (and therefore the specific protein they bind to). The unique amplicons (often referred to as "barcodes") are immediately quantified using microfluidic qPCR. Proteins that are more highly abundant will bind more antibody and produce more amplicons. Advanced Olink® methods that use NGS for amplicon detection enable the abundance of up to 5400 protein targets to be measured. In this method the intensity of the signal generated during the incorporation of fluorescent DNA is used as a measure of the amount of protein in the sample being analyzed.

### *8.4.2 Identification of Comorbidity Biomarkers*

The plasma proteome data obtained using PEA technology was analyzed using principal component analysis (PCA). Little overlap was observed between the samples acquired from the patients living with HIV and healthy controls in both the discovery and replication groups (Vadaq et al. 2023). This result proves that each group has a unique plasma proteomic profile and it was likely that differentially abundant proteins

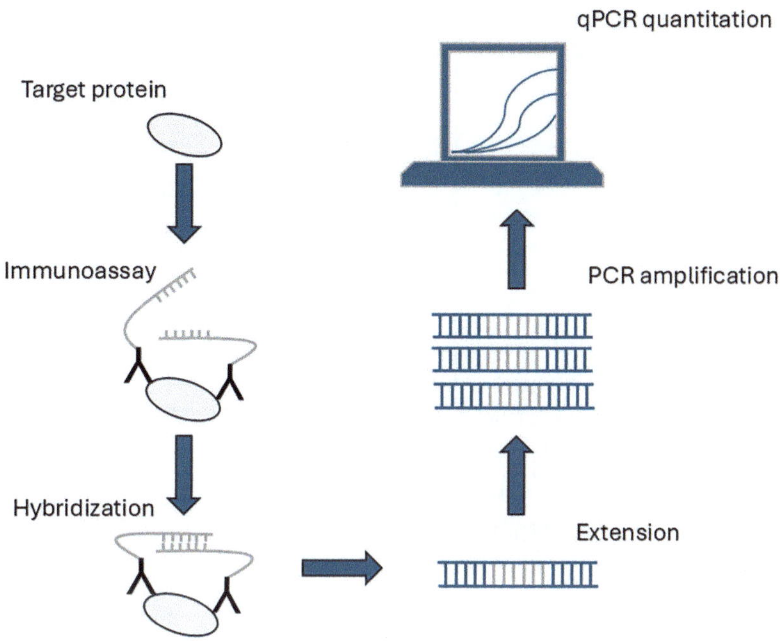


**Fig. 8.8** Proximity extension assay (Olink®) with quantitative polymerase chain reaction (qPCR) detection. The key to quantitating a specific protein using proximity extension assay (PEA) technology uses two matched antibodies that bind to a specific protein target. The two antibodies are labeled with complementary DNA oligonucleotides that hybridize when they come into proximity. The double-stranded DNA is extended using a polymerase and subjected to qPCR to quantitate the relative amount of DNA that originally hybridized. Since the amount of DNA generated during qPCR directly correlates to the amount of DNA that hybridizes, it also directly correlates with the amount of protein the antibodies bind to

could be identified that could help in recognizing the pathophysiology behind comorbidities observed in people living with HIV.

Almost all these proteins were more highly abundant in people living with HIV compared to healthy controls (see Fig. 8.9). Only 10 of the 276 proteins showed a lower abundance in people living with HIV. Several proteins were at least 1.5-fold differentially abundant between cohorts. These proteins were involved in lipid uptake, transport, and metabolism. Cytosolic fatty acid-binding proteins (FABPs) 1, 2, and 5 along with carboxylesterase 3 (CES3) and retinol-binding protein 2 (RBP2) had the highest fold change among differentially abundant proteins. FABP2 (also known as intestinal fatty acid binding protein) and another differentially abundant protein, regenerating family member 3α (REG3A), are both highly expressed in the intestinal epithelium and have been recognized as gut damage markers in people living with HIV. While FABP2's role in lipid absorption and transport in the intestine in lipid metabolism is well established, REG3A has also been implicated in glucose and lipid homeostasis. When given endogenous REG3A, nutritional and genetic mouse models of obesity and type 2 diabetes have reduced blood lipid levels (Gonzalez et al. 2023). This effect is believed to be a result of REG3A increasing insulin sensitivity.

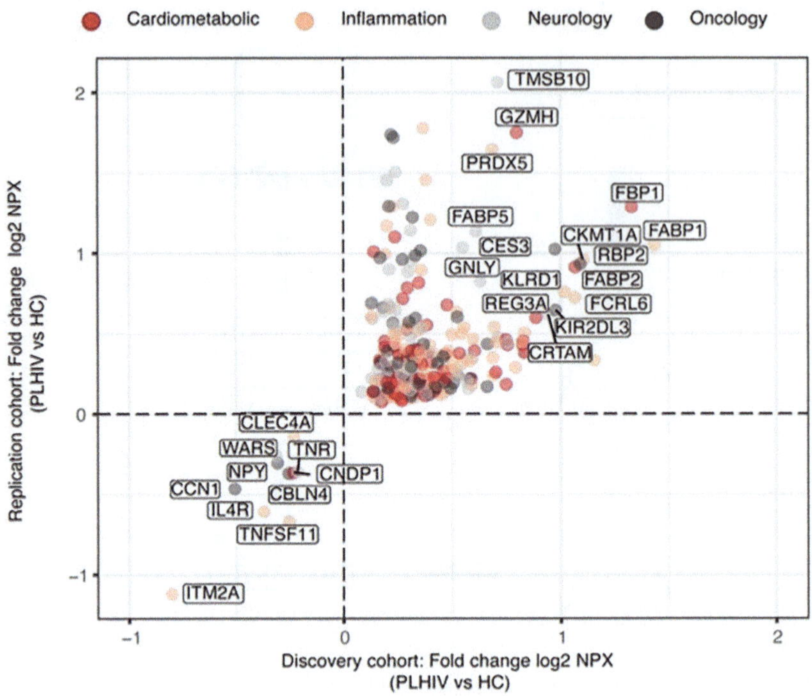


**Fig. 8.9** Four-quadrant plot of the fold change of differentially expressed proteins in discovery and replication cohorts of plasma acquired from people living with HIV (PLHIV) and healthy controls (HC). Only proteins with a $\log_2$ fold abundance change of at least 1.5 were included in the analysis. Adapted from ref. (Vadaq et al. 2023) under the terms and conditions of the Creative Commons Attribution (CC BY) license (https://creativecommons.org/licenses/by/4.0/)

Several proteins associated with immune function were among those that showed a significant increase in abundance in plasma acquired from people living with HIV and healthy controls. These included proteins associated with natural killer cell (GZMH, KLRD1, FCRL6, GNLY, KIR2DL3) and T-cell function (CRTAM, GLNY, KIR2DL3). Other proteins that had higher abundance in people living with HIV included the gluconeogenesis regulatory enzyme fructose-1,6-bisphosphatase 1 (FBP1), the mitochondrial creatine kinases 1A and 1B (CKMT1A and CKMT1B), and the antioxidant enzyme peroxiredoxin 5 (PRDX5). Thymosin β10 (TMSB10), a protein involved in actin regulation and cytoskeleton organization (Zhang et al. 2017), was also found to be upregulated in people living with HIV. While not a classic tumor suppressor or proto-oncogene, this protein has been shown to have both tumor-suppressive and cancer-promoting functions depending on the specific cancer type (Zhang et al. 2017; Lee et al. 2005). Of the ten proteins that were downregulated in patients with HIV, five were associated with behavioral or psychiatric disorder (NPY, CNDP1, CBLN4, WARS, and TNR), four with inflammatory responses (CLEC4A, IL4R, RANKL, and ITM2A), and one with gut mucosal healing (CCN1). Receptor activator of nuclear factor-κB ligand (RANKL) is a well-known cytokine that plays a critical role in osteoclast differentiation and activation (Mueller and Hess 2012).

### 8.4.3 *Tissue and Cell Origin of Comorbidity Biomarkers*

Assignment of the tissue and cellular origin of all 276 proteins that were upregulated in people living with HIV confirmed the dysregulation of proteins associated with both immune function and lipid metabolism. Proteins assigned to intestine and lymphoid tissue were significantly enriched. At the cellular level, there was a significant number of proteins originating from proximal enterocytes and intestinal goblet cells along with dendritic, T, and plasma cells. A pathway enrichment analysis of this group of differentially abundant proteins revealed three highly populated clusters: (i) antigen presentation and innate immune system, (ii) cytokine and chemokine receptor interaction, and (iii) lipid metabolism. While enrichment of pathways associated with immune function may have been expected, the extent to which proteins involved in lipid metabolism are dysregulated may come as a surprise.

### 8.4.4 *Association of Putative Biomarkers with Cardiovascular Disease*

To evaluate the association between people living with HIV being treated with ART and cardiovascular disease (CVD), the relative abundance of approximately 1500 proteins associated with CVD in the discovery and replication cohorts was analyzed. Twenty-seven proteins were of higher abundance in plasma samples taken from people living with HIV compared to healthy controls. These proteins included the well-known cardiac injury biomarker N-terminal prohormone of brain natriuretic peptide (NT pro-BNP) (Seino and Otsuka 2015). During a 5-year follow-up of those living with HIV, 10% experienced a cardiovascular event. Ninety-nine proteins were found to be associated with the probability of an individual suffering a cardiovascular event. Of these, 29 were upregulated in people living with HIV. The proteins that had the strongest association with CVD were involved in lipid metabolism (FABP4, FABP5, and arylsulfatase A), immune response (PLAUR, RELT, IL1RN, GDF15, and MZB1), and extracellular matrix organization and/or degradation (COL18A1, COL6A3, MIA, and TIMP4). Also identified in this group was neurofilament light polypeptide (NEFL) and ectodysplasin-A2 (EDA2R), which are associated with axon development and apoptosis, respectively.

### 8.4.5 *What Was Found*

Today people living with HIV lead healthier and longer lives than those in the early days of this pandemic owing to treatments such as ARTs. Unfortunately, ARTs are not a cure, and HIV patients people still suffer comorbidities at much higher rates than noninfected individuals. While identifying diagnostic and prognostic biomarkers of

HIV is critical, so is identifying biomarkers associated with specific comorbidities. A surprising result of this PEA study was that only a few proteins that showed the highest fold change in abundance were associated with immune response. Rather it was proteins associated with lipid absorption and transport that are highly expressed in the intestines that showed the greatest difference in abundance. These proteins were of higher abundance in HIV patients indicating an increased energy demand. But what is driving this increased demand? The answer to that may be related to other groups of proteins that showed a significant difference in abundance between the two cohorts. Many of these proteins are associated with immunity. Immune activation places greater energy and biosynthesis demand on the body, which could be supplied by higher lipid absorption within the intestines.

The focus of this part of the PEA study was to determine if proteomic profiling of plasma from patients living with HIV that were receiving ART could reveal biomarkers associated with inflammation, cardiometabolic, oncologic, or neurologic comorbidities. These comorbidities were selected because proteins within these categories are targeted as part of the Olink® Explore 1536 platform.

To identify the risk of HIV-infected patients developing CVD, the overlap between proteins found to be upregulated in patients living with HIV compared to healthy controls was compared to proteins known to be associated with CVD at baseline and those associated with the risk of developing CVD. Six proteins (GDF15, NEFL, PLAUR, RELT, COL6A3, and EDA2R) were all upregulated in plasma from people living with HIV on ART compared to healthy controls and are associated with the presence and risk of developing CVD (see Fig. 8.10). GDF15, RELT, and PLAUR are associated with at least three CVD-related traits. While NEFL is a well-known marker of neuronal injury, it has also been shown to be elevated in individuals following an ischemic cerebrovascular event (Pedersen et al. 2019). It has also been used to predict stroke incidents in patients with diabetes mellitus. The Framingham Heart Study has described NEFL's prognostic value for coronary heart disease and heart failure. While reports are scarce, RELT (a TNF receptor) has been associated with myocardial infarction in elderly male patients (Carro and Kaski 2011). Neither COL6A3 (the most abundantly expressed collagen in adipose tissue) nor EDA2R (a member of the TNF receptor family) had previously been associated with CVD. Taken together, the results of this study suggest there is value in using these six proteins to diagnose and monitor possible CVD in people living with HIV receiving ART.

### 8.4.6 Correlation Between Putative HIV Biomarkers and the Microbiome

One of the next great strides in personalized medicine will be correlating data from various omic studies (i.e., genomics, transcriptomics, proteomics, metabolomics, etc.) While not directly part of the analysis, this study evaluated associations between the plasma proteome and gut microbiome of people living with HIV. It is well established that gut mucosal homeostasis is maintained via direct interactions between innate

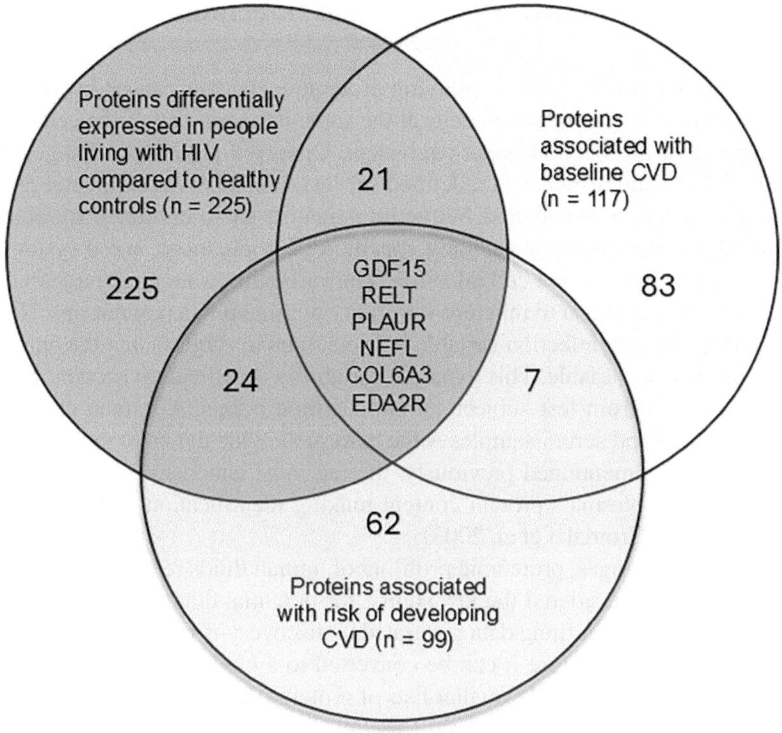


**Fig. 8.10** Venn diagram showing the number of overlapping proteins identified in three groups: (i) proteins that are differentially expressed in plasma from people living with HIV compared to healthy controls, (ii) proteins associated with cardiovascular disease (CVD) at baseline, and (iii) proteins associated with risk of developing CVD. Adapted from ref. (Vadaq et al. 2023) under the terms and conditions of the Creative Commons Attribution (CC BY) license (https://creativecommons.org/licenses/by/4.0/)

immune cells and the gut microbiome (Zheng et al. 2020). This study found that the abundances of several immune-related proteins, in particular, IL-32, FABP32, and CCL11, correlated with increased levels of bacterial species *Prevotella* sp. *CAG 1092*, *Prevotella* sp. *CAG 279*, and *Prevotella* sp. *CAG 520 (*Vadaq et al. 2023*)*. *Prevotella* is associated with a plant-based diet and contributes to fiber digestion and the production of short-chain fatty acids (De Filippis et al. 2016). Increase levels of *Prevotella* have also been associated with gut inflammation, and one of the main species of this genus, *Prevotella copri*, is believed to contribute to chronic inflammation in patients with HIV (Dillon et al. 2016). What strengthens the association between these proteomic components and the microbiome is correlation only exists between these proteins associated with intestine and lymphoid tissue. Other intestine- and lymphoid tissue-specific proteins showed association with higher levels of microbial translocation and inflammation markers substantiating the interaction among bacterial dysbiosis and translocation, gut barrier integrity, and systemic immune responses in people living the HIV that were receiving ART.

## 8.5 Challenges in Investigating Viral Infections

There are many challenges when studying proteomic changes caused by viral infections. The virus does not infect all cells at the same moment, and of the cells they do infect, the extent of infection is not equivalent. Observed proteomic changes do not represent the specific response of a defined subset of cells but rather the net response of the entire system being studied. While investigators try to minimize variability by using in vitro cell cultures or isolating specific cell populations, these systems still cannot eliminate inconsistent cell infection. The variability is more extreme in human samples, which are subject to inherent variability within human populations. Not only is the response to viral infection variable between human subjects, but the duration of the response is also variable. This dynamic variability is difficult to account for when samples are taken from test subjects at similar time points. A unique challenge to analyzing plasma and serum samples is the extremely wide dynamic range of protein concentrations. As mentioned previously, the ten most abundant proteins constitute about 90–95% of plasma's protein content making identification of low abundance proteins difficult (Tirumalai et al. 2003).

With all its challenges, proteomic profiling of human fluids remains a focused area of research as it is considered the key source for potential diagnostic and prognostic biomarkers. The key to turning data generated in discovery-driven studies into clinical tests is validation. But before it can be converted to a clinical test, discovery-driven data must be filtered to provide smaller lists of proteins that have the greatest probably of surviving the rigorous tests required for clinical validation. No company wants to take an unreasonable monetary risk to develop a test that may not achieve the standards required for marketable status. Bioinformatics has made extraordinary progress in developing tools, refining algorithms, and integrating databases so that fewer false positives are generated. The increasing development of artificial intelligence will strengthen bioinformatic's impact on proteomic data quality.

## 8.6 The Need for Canonical Proteins in Proteomics

It may not seem obvious, but the technology used to acquire global proteomic data has become one of the most robust features of proteomics. Mass spectrometers and the algorithms for conducting database searching have become commonplace in the past couple of decades. Twenty years ago, only a few laboratories around the world could generate high-accuracy MS measurements on proteome samples; however, this capability is now routinely available. The widespread adoption of global proteomics technology has increased the accuracy of the measurements and made the sources of systematic errors predictable. What could make the data even more reliable, however, are reliable normalization factors (Kas 2004). One approach for normalization involves a spiked-in approach where a standard (e.g., protein or other molecule) is added to each sample at a known concentration (Ogata et al. 2022). While this approach has its benefits, it also adds steps to the analytical process, and when using MS, the signal of the standard can vary due to ionization effects of other components that co-elute into the mass spectrometer along with it (this concept will make more sense to the reader with expertise in MS). Unfortunately, a spiked-in standard cannot offset the temporal variability of proteins within human samples. For example, serum albumin range between 3.5 and 5.5 g/dL between individuals, and its concentration can fluctuate hourly or daily within a single individual (Owens et al. 1983). Averaging the results for a specific protein from many samples can result in a large standard deviation, which could exceed a relatively small change resulting from a pathological process. This scenario could potentially cause many important pathological effects to be overlooked.

### *8.6.1 Normalization Using Canonical Proteins*

There is no acceptable number of patient samples that will alleviate this normalization problem. In fact, increasing sample size can increase variation, making small differences that may be clinically relevant more difficult to recognize. If the sample size is small, however, it is difficult to make any statistically relevant conclusions. It has been proposed that highly specific proteins (termed "canonical") could serve as markers and allow comparison between different datasets acquired using different patient cohorts and technologies (Guillon et al. 2024). For a protein to be considered "canonical," its behavior during infection would have to be well established. A canonical marker would be akin to creatinine clearance. Creatinine is a waste product of muscle metabolism, and its clearance is a marker for estimating glomerular filtration rate (GFR) and kidney function. Since it is generally produced at a relatively constant rate, it is used to normalize the measurement of other urinary markers of kidney function since creatinine's concentration controls for variation in urine flow rate. Discovering "canonical" markers is no less difficult than finding other biomarkers, but studies using HIV-1-infected macrophages have taken this issue in a positive direction and provides a roadmap on how canonical markers can be recognized (Grabowska et al. 2021).

There have been several proteomic studies on the effect of HIV-1 infection on macrophages that focused on measuring difference in protein abundances between infected and noninfected cells (Luo et al. 2003; Kadiu et al. 2012; Donnelly and Ciborowski 2016). Many times, hundreds of proteins were found to be differentially abundant resulting in an enormous amount of comparative data. While their overlap between samples was not absolute, there were some proteins that showed consistent change between the studies and could potentially function as canonical markers. These potential markers were metalloproteinase 9 (MMP-9) and signal transducer and activator of transcription 1 (STAT-1) (Ciborowski et al. 2004; Rivera et al. 2016). Decreased MMP-9 production by HIV-1-infected macrophages has been reported in clinical analyses (Ciborowski et al. 2004; Ghorpade et al. 2001; Winkler et al. 2012).

On the contrary, elevated levels of STAT-1 are recognized as a trait of innate immune activation and active HIV-1 replication (Appelberg et al. 2017; Giunta et al. 2006). An increase in its expression is consistent with its function of transcribing IFN-stimulated genes that push the cell into an antiviral state. STAT1 is also elevated in all HIV-associated dementia studies (i.e., neuroAIDS) including those examining postmortem human samples and samples from nonhuman primates infected with simian immunodeficiency virus (Winkler et al. 2012; Maze et al. 2019). Regardless of the experimental system, canonical proteins that can serve as reference points to normalize the levels of other proteins will increase the confidence in large-scale proteomic analyses. There are numerous examples in basic research where canonical proteins are used to normalize experimental results. These include using GAPDH or β-actin in Western blotting, the addition of known concentrations of isotope-labeled standards to measure protein or metabolite concentrations in biofluids, and amino acid analysis to determine the mass fraction of tau proteins in cerebrospinal fluid (Pottiez et al. 2017).

## 8.7 Conclusions

While HIV remains a pandemic, improvements in its diagnosis and treatment have helped control its spread and manage the health of those living with the virus. A safe and effective HIV vaccine would be the optimal solution for controlling this pandemic. Researchers have been actively trying to develop an HIV vaccine for almost 40 years, yet none have proven effective. HIV is notorious for its ability to rapidly mutate and evade an adaptive immune response, so any vaccine is quickly rendered obsolete. The virus does not induce a broad reactive antibody response, and even if it did it generates latent viral pools early in the infection process that can activate in the future.

Without an effective vaccine, people get infected with the HIV virus and remain infected for the rest of their lives. Forty years ago, this infection led to AIDS, which led to the patient's death. Fortunately, today HIV is no longer a death sentence. While infected individuals will get ill, prompt diagnosis and ART therapy prior to HIV damaging their immune system dramatically increase the probability of them having a normal life expectancy. Antiretroviral therapy can also inhibit disease progression and death in individuals diagnosed later and already suffering from HIV-related immunodeficiency.

Unfortunately, ART is not a cure. It limits HIV replication only when patients adhere to the medication. Once they cease ART, viral replication starts again. Since HIV inserts its viral genes into the genome of long-lived memory T cells, the information to replicate itself becomes invisible to the immune system. This proviral DNA is a key component of the latent HIV reservoir that can reactivate and generate an eruption of viral infection. Its ability to form a latent reservoir makes HIV infection lifelong. Currently less than ten people have been cured of HIV (Mallapaty 2024). The cure required eliminating all traces of the virus from their systems by replacing all their immune cells with donor cells that are HIV resistant.

It is unlikely that proteomics can solve the difficulties associated with HIV vaccine development, since its major contribution is recognition of highly antigenic epitopes within the virus. If the virus rapidly mutates, antigenic epitopes must be reidentified and new vaccines developed. Proteomic scientists have focused their efforts toward deciphering the mechanism of HIV infection, replication, and transmission while studying how HIV affects the human proteome. While not discussed in this chapter, scientists have used differential affinity purification methods to identify PPIs involving HIV and host proteins at various stages of the HIV life cycle (Jäger et al. 2001; Dyer et al. 2004; Arthos et al. 2008). Knowing how HIV invades, replicates in, and releases from host cells in greater detail provides more potential drug targets for inhibiting infection and/or transmission. This chapter focused more studies measuring what HIV infection does to the human body. While the emergence of HIV out of latency, also known as seroconversion, can be detected through flu-like symptoms, swollen lymph nodes, oral thrush, etc., the most telling sign of seroconversion is a drop in CD4+ T-cell count (de Wolf et al. 1997). Beyond helping identify druggable targets to prevent infection, proteomics can identify early biomarkers of seroconversion. These biomarkers would enable immediate treatment to minimize viral replication. Since people living with HIV are at risk for developing various comorbidities, the biomarkers could also indicate the comorbidity the patient is at most risk of developing. These panels of biomarkers could direct both early treatment of HIV seroconversion and its associated comorbidity to minimize patient suffering. While proteomics has not yet identified clinically validated biomarkers, its progress almost ensures that in the near future changes in the physiology of a person living with HIV will be recognized well before symptoms negatively impact the patient's lifestyle.

## References


Antman K, Chang Y (2000) Kaposi's sarcoma. N Engl J Med 342:1027–1038

Appelberg KS, Wallet MA, Taylor JP, Cash MN, Sleasman JW, Goodenow MM (2017) HIV-1 infection primes macrophages through STAT signaling to promote enhanced inflammation and viral replication. AIDS Res Hum Retrovir 33:690–702

Arthos J, Cicala C, Martinelli E, Macleod K, Van Ryk D, Wei D et al (2008) HIV-1 envelope protein binds to and signals through integrin alpha4beta7, the gut mucosal homing receptor for peripheral T cells. Nat Immunol 9:301–309

Balasubramaniam M, Freed EO (2011) New insights into HIV assembly and trafficking. Physiology (Bethesda) 26:236–251

Barré-Sinoussi F, Chermann JC, Rey F, Nugeyre MT, Chamaret S, Gruest J et al (1983) Isolation of a T-lymphotropic retrovirus from a patient at risk for acquired immune deficiency syndrome (AIDS). Science 220:868–871

Barré-Sinoussi F, Abdool Karim SS, Albert J, Bekker LG, Beyrer C, Cahn P et al (2018) Expert consensus statement on the science of HIV in the context of criminal law. J Int AIDS Soc 21:e25161–e25172

Bender S, Reuter A, Eberle F, Einhorn E, Binder M, Bartenschlager R (2015) Activation of type I and III interferon response by mitochondrial and peroxisomal MAVS and inhibition by hepatitis C virus. PLoS Pathog 11:e1005264

Boyer S, March L, Kouanfack C, Laborde-Balen G, Marino P, Aghokeng AF et al (2013) Monitoring of HIV viral load, CD4 cell count, and clinical assessment versus clinical monitoring alone for antiretroviral therapy in low-resource settings (Stratall ANRS 12110/ESTHER): a cost-effectiveness analysis. Lancet Infect Dis 13:577–586

Campbell EM, Hope TJ (2015) HIV-1 capsid: the multifaceted key player in HIV-1 infection. Nat Rev Microbiol 13:471–483

Carro A, Kaski JC (2011) Myocardial infarction in the elderly. Aging Dis 2:116–137

CDC (1981) Kaposi's sarcoma and pneumocystis pneumonia among homosexual men—New York City and California. MMWR Morb Mortal Wkly Rep 30:305–308

Chen J, Rahman SA, Nikolaitchik OA, Grunwald D, Sardo L, Burdick RC et al (2016) HIV-1 RNA genome dimerizes on the plasma membrane in the presence of gag protein. Proc Natl Acad Sci USA 113:E201–E208

Ciborowski P, Enose Y, Mack A, Fladseth M, Gendelman HE (2004) Diminished matrix metalloproteinase 9 secretion in human immunodeficiency virus-infected mononuclear phagocytes: modulation of innate immunity and implications for neurological disease. J Neuroimmunol 157:11–16

Cohen MS, Shaw GM, McMichael AJ, Haynes BF (2011) Acute HIV-1 infection. N Engl J Med 364:1943–1954

Consortium TU (2022) UniProt: the universal protein knowledgebase in 2023. Nucleic Acids Res 51:D523–D531

Cui L, Li H, Xi Y, Hu Q, Liu H, Fan J et al (2022) Vesicle trafficking and vesicle fusion: mechanisms, biological functions, and their implications for potential disease therapy. Mol Biomed 3:29–61

De Filippis F, Pellegrini N, Vannini L, Jeffery IB, La Storia A, Laghi L et al (2016) High-level adherence to a Mediterranean diet beneficially impacts the gut microbiota and associated metabolome. Gut 65:1812–1821

de Wolf F, Spijkerman I, Schellekens PT, Langendam M, Kuiken C, Bakker M et al (1997) AIDS prognosis based on HIV-1 RNA, CD4+ T-cell count and function: markers with reciprocal predictive value over time after seroconversion. AIDS 11:1799–1806

Dillon SM, Lee EJ, Kotter CV, Austin GL, Gianella S, Siewe B et al (2016) Gut dendritic cell activation links an altered colonic microbiome to mucosal and systemic T-cell activation in untreated HIV-1 infection. Mucosal Immunol 9:24–37

Donnelly MR, Ciborowski P (2016) Proteomics, biomarkers, and HIV-1: a current perspective. Proteomics Clin Appl 10:110–125

Duchon A, Hu W-S (2024) HIV-1 RNA genome packaging: it's G-rated. MBio 15:e0086123–e0086139

Dyer MD, Murali TM, Sobral BW (2004) The landscape of human proteins interacting with viruses and other pathogens. PLoS Pathog 4:e32–e51

Ekwaru JP, Campbell J, Malamba S, Moore DM, Were W, Mermin J (2013) The effect of opportunistic illness on HIV RNA viral load and CD4+ T cell count among HIV-positive adults taking antiretroviral therapy. J Int AIDS Soc 16:17355–17360

Gallo RC, Salahuddin SZ, Popovic M, Shearer GM, Kaplan M, Haynes BF et al (1984) Frequent detection and isolation of cytopathic retroviruses (HTLV-III) from patients with AIDS and at risk for AIDS. Science 224:500–503

Gelderblom HR (1991) Assembly and morphology of HIV: potential effect of structure on viral function. AIDS 5:617–637

German Advisory Committee Blood (Arbeitskreis Blut), Subgroup (2016) 'Assessment of pathogens transmissible by blood'. Human Immunodeficiency Virus (HIV). Transfus Med Hemother 43:203–222

Ghorpade A, Persidskaia R, Suryadevara R, Che M, Liu XJ, Persidsky Y et al (2001) Mononuclear phagocyte differentiation, activation, and viral infection regulate matrix metalloproteinase

expression: implications for human immunodeficiency virus type 1-associated dementia. J Virol 75:6572–6583
Gillet LC, Navarro P, Tate S, Röst H, Selevsek N, Reiter L et al (2012) Targeted data extraction of the MS/MS spectra generated by data-independent acquisition: a new concept for consistent and accurate proteome analysis. Mol Cell Proteomics 11:O111.016717
Giunta B, Obregon D, Hou H, Zeng J, Sun N, Nikolic V et al (2006) EGCG mitigates neurotoxicity mediated by HIV-1 proteins gp120 and tat in the presence of IFN-gamma: role of JAK/STAT1 signaling and implications for HIV-associated dementia. Brain Res 1123:216–225
Gonzalez P, Dos Santos A, Darnaud M, Moniaux N, Rapoud D, Lacoste C et al (2023) Antimicrobial protein REG3A regulates glucose homeostasis and insulin resistance in obese diabetic mice. Commun Biol 6:269–285
Götte M, Li X, Wainberg MA (1999) HIV-1 reverse transcription: a brief overview focused on structure-function relationships among molecules involved in initiation of the reaction. Arch Biochem Biophys 365:199–210
Grabowska K, Harwood E, Ciborowski P (2021) HIV and proteomics: what we have learned from high throughput studies. Proteomics Clin Appl 15:e2000040
Greene WC (2007) A history of AIDS: looking back to see ahead. Eur J Immunol 37(Suppl 1):S94–S102
Guillon C, Pichereaux C, Lazar I, Chaoui K, Mouton-Barbosa E, Liauzun M et al (2024) Mass spectrometry-based workflow for the identification and quantification of alternative and canonical proteins in pancreatic cancer cells. Cells 13:1966–1981
Hamm A, Veeck J, Bektas N, Wild PJ, Hartmann A, Heindrichs U et al (2008) Frequent expression loss of Inter-alpha-trypsin inhibitor heavy chain (ITIH) genes in multiple human solid tumors: a systematic expression analysis. BMC Cancer 8:25–39
Hendrickson RC, Douglass JF, Reynolds LD, McNeill PD, Carter D, Reed SG et al (2000) Mass spectrometric identification of mtb81, a novel serological marker for tuberculosis. J Clin Microbiol 38:2354–2361
hivinfo.nih.gov/understanding-hiv/fact-sheets/hiv-life-cycle. Accessed July 2025
Hokello J, Lakhikumar Sharma A, Tyagi M (2021) AP-1 and NF-κB synergize to transcriptionally activate latent HIV upon T-cell receptor activation. FEBS Lett 595:577–594
Hsin F, Hsu YC, Tsai YF, Lin SW, Liu HM (2021) The transmembrane serine protease hepsin suppresses type I interferon induction by cleaving STING. Sci Signal 14:eabb4752
Hu WS, Hughes SH (2012) HIV-1 reverse transcription. Cold Spring Harb Perspect Med 2:a006882
Hurley JH, Cada AK (2018) Inside job: how the ESCRTs release HIV-1 from infected cells. Biochem Soc Trans 46:1029–1036
Jäger S, Gulbahce N, Cimermancic P, Kane J, He N, Chou S et al (2001) Purification and characterization of HIV-human protein complexes. Methods 53:13–19
Kadiu I, Narayanasamy P, Dash PK, Zhang W, Gendelman HE (2012) Biochemical and biologic characterization of exosomes and microvesicles as facilitators of HIV-1 infection in macrophages. J Immunol 189:744–754
Kas K (2004) On the technicalities of discovering and applying protein biomarkers for cancer prevention. Eur J Cancer Prev 13:437–446
Kemnic TR, Gulick PG (2025) HIV antiretroviral therapy. [updated 2022 Sep 20]. In: StatPearls [internet]. StatPearls Publishing, Treasure Island (FL). Available from: https://www.ncbi.nlm.nih.gov/books/NBK513308/
Klingler J, Anton H, Réal E, Zeiger M, Moog C, Mély Y et al (2020) How HIV-1 Gag manipulates its host cell proteins: a focus on interactors of the nucleocapsid domain. Viruses 12:888–929
Kouwaki T, Okamoto M, Tsukamoto H, Fukushima Y, Matsumoto M, Seya T et al (2017) Zyxin stabilizes RIG-I and MAVS interactions and promotes type I interferon response. Sci Rep 7:11905–11917
Lama JR, Bender Ignacio RA, Duerr A (2025) Acute retroviral syndrome. Curr Opin HIV AIDS 20:186–192

Lee SH, Son MJ, Oh SH, Rho SB, Park K, Kim YJ et al (2005) Thymosin {beta}(10) inhibits angiogenesis and tumor growth by interfering with Ras function. Cancer Res 65:137–148
Lerner G, Weaver N, Anokhin B, Spearman P (2022) Advances in HIV-1 assembly. Viruses 14:478–511
Li L, Li HS, Pauza CD, Bukrinsky M, Zhao RY (2005) Roles of HIV-1 auxiliary proteins in viral pathogenesis and host-pathogen interactions. Cell Res 15:923–934
Lundberg M, Eriksson A, Tran B, Assarsson E, Fredriksson S (2011) Homogeneous antibody-based proximity extension assays provide sensitive and specific detection of low-abundant proteins in human blood. Nucleic Acids Res 39:e102–e109
Luo X, Carlson KA, Wojna V, Mayo R, Biskup TM, Stoner J et al (2003) Macrophage proteomic fingerprinting predicts HIV-1-associated cognitive impairment. Neurology 60:1931–1937
Mabhida SE, Mchiza ZJ, Mokgalaboni K, Hanser S, Choshi J, Mokoena H et al (2024) High-sensitivity C-reactive protein among people living with HIV on highly active antiretroviral therapy: a systemic review and meta-analysis. BMC Infect Dis 24:160–176
Mallapaty S (2024) Seventh patient 'cured' of HIV: why scientists are excited. Nature 632:235–236
Marcus JL, Leyden WA, Alexeeff SE, Anderson AN, Hechter RC, Hu H et al (2020) Comparison of overall and comorbidity-free life expectancy between insured adults with and without HIV infection, 2000-2016. JAMA Netw Open 3:e207954
Martinu T, Todd JL, Gelman AE, Guerra S, Palmer SM (2023) Club cell secretory protein in lung disease: emerging concepts and potential therapeutics. Annu Rev Med 74:427–441
Mattei S, Anders M, Konvalinka J, Kräusslich HG, Briggs JA, Müller B (2014) Induced maturation of human immunodeficiency virus. J Virol 88:13722–13731
Maze EA, Ham C, Kelly J, Ussher L, Almond N, Towers GJ et al (2019) (2019) variable baseline *Papio cynocephalus* endogenous retrovirus (PcEV) expression is upregulated in acutely SIV-infected macaques and correlated to STAT1 expression in the spleen. Front Immunol 10:901–915
McKeating JA, McKnight A, Moore JP (1991) Differential loss of envelope glycoprotein gp120 from virions of human immunodeficiency virus type 1 isolates: effects on infectivity and neutralization. J Virol 65:852–860
McMichael AJ, Borrow P, Tomaras GD, Goonetilleke N, Haynes BF (2010) The immune response during acute HIV-1 infection: clues for vaccine development. Nat Rev Immunol 10:11–23
Moaddel R, Ubaida-Mohien C, Tanaka T, Lyashkov A, Basisty N, Schilling B et al (2021) Proteomics in aging research: a roadmap to clinical, translational research. Aging Cell 20:e13325–e13338
Moore BE (1993) Survival of human immunodeficiency virus (HIV), HIV-infected lymphocytes, and poliovirus in water. Appl Environ Microbiol 59:1437–1443
Moore MD, Hu WS (2009) HIV-1 RNA dimerization: it takes two to tango. AIDS Rev 11:91–102
Mueller CG, Hess E (2012) Emerging functions of RANKL in lymphoid tissues. Front Immunol 3:261–267
Nazziwa J, Freyhult E, Hong MG, Johansson E, Årman F, Hare J et al (2024) Dynamics of the blood plasma proteome during hyperacute HIV-1 infection. Nat Commun 15:10593–10606
Ndhlovu ZM, Kamya P, Mewalal N, Kløverpris HN, Nkosi T, Pretorius K et al (2015) Magnitude and kinetics of CD8+ T cell activation during hyperacute HIV infection impact viral set point. Immunity 43:591–604
Obare LM, Temu T, Mallal SA, Wanjalla CN (2024) Inflammation in HIV and its impact on atherosclerotic cardiovascular disease. Circ Res 134:1515–1545
Ogata S, Masuda T, Ito S, Ohtsuki S (2022) Targeted proteomics for cancer biomarker verification and validation. Cancer Biomark 33:427–436
Olety B, Ono A (2014) Roles played by acidic lipids in HIV-1 Gag membrane binding. Virus Res 193:108–115
Ono A, Freed EO (1999) Binding of human immunodeficiency virus type 1 Gag to membrane: role of the matrix amino terminus. J Virol 73:4136–4144

Owens SM, Mayersohn M, Woodworth JR (1983) Phencyclidine blood protein binding: influence of protein, pH and species. J Pharmacol Exp Ther 226:656–660

Ozugergin I, Piekny A (2021) Complementary functions for the Ran gradient during division. Small GTPases 12:177–187

Pedersen A, Stanne TM, Nilsson S, Klasson S, Rosengren L, Holmegaard L et al (2019) Circulating neurofilament light in ischemic stroke: temporal profile and outcome prediction. J Neurol 266:2796–2806

Pottiez G, Yang L, Stewart T, Song N, Aro P, Galasko DR et al (2017) Mass-spectrometry-based method to quantify in parallel tau and amyloid β 1-42 in CSF for the diagnosis of Alzheimer's disease. J Proteome Res 16:1228–1238

Premeaux TA, Ndhlovu LC (2023) Decrypting biological hallmarks of aging in people with HIV. Curr Opin HIV AIDS 18:237–245

Rivera LE, Kraiselburd E, Meléndez LM (2016) Cystatin B and HIV regulate the STAT-1 signaling circuit in HIV-infected and INF-β-treated human macrophages. J Neurovirol 22:666–673

Roberts L, Passmore JA, Williamson C, Little F, Bebell LM, Mlisana K et al (2010) Plasma cytokine levels during acute HIV-1 infection predict HIV disease progression. AIDS 24:819–831

Rotello RJ, Veenstra TD (2021) Mass spectrometry techniques: principles and practices for quantitative proteomics. Curr Protein Pept Sci 22:121–133

Seino Y, Otsuka T (2015) Expanding the role of cardiac biomarkers–natriuretic peptides and troponins–further in pre-Stage A. Hypertens Res 38:11–12

Smith AJ, Cho MI, Hammarskjöld ML, Rekosh D (1990) Human immunodeficiency virus type 1 Pr55gag and Pr160gag-pol expressed from a simian virus 40 late replacement vector are efficiently processed and assembled into virus-like particles. J Virol 64:2743–2750

Speck RR, Flexner C, Tian CJ, Yu XF (2000) Comparison of human immunodeficiency virus type 1 Pr55(Gag) and Pr160(Gag-pol) processing intermediates that accumulate in primary and transformed cells treated with peptidic and nonpeptidic protease inhibitors. Antimicrob Agents Chemother 44:1397–1403

Tirumalai RS, Chan KC, Prieto DA, Issaq HJ, Conrads TP, Veenstra TD (2003) Characterization of the low molecular weight human serum proteome. Mol Cell Proteomics 2:1096–1103

Troldborg A, Godnic-Polai Z, Cervenak L, Hansen AG, Farkas H, Thiel S (2024) Inter-α-trypsin inhibitor heavy chain 4 (ITIH4) as a compensatory protease inhibitor in hereditary angioedema. J Allergy Clin Immunol 154:468–479.e6

Turk G, Ghiglione Y, Hormanstorfer M, Laufer N, Coloccini R, Salido J et al (2018) Biomarkers of progression after HIV acute/early infection: nothing compares to $CD4^+$ T-cell count? Viruses 10:34–47

Upadhyay C, Feyznezhad R, Cao L, Chan KW, Liu K, Yang W et al (2020) Signal peptide of HIV-1 envelope modulates glycosylation impacting exposure of V1V2 and other epitopes. PLoS Pathog 16:e1009185–e1009215

Vadaq N, Zhang Y, Vos WA, Groenendijk AL, Blaauw MJ, van Eekeren LE et al (2023) High-throughput proteomic analysis reveals systemic dysregulation in virally suppressed people living with HIV. JCI Insight 8:e166166–e166182

Watts JM, Dang KK, Gorelick RJ, Leonard CW, Bess JW Jr, Swanstrom R et al (2009) Architecture and secondary structure of an entire HIV-1 RNA genome. Nature 460:711–716

Webel AR, Schexnayder J, Cioe PA, Zuñiga JA (2021) A review of chronic comorbidities in adults living with HIV: state of the science. J Assoc Nurses AIDS Care 32:322–346

webmd.com/hiv-aids/ss/slideshow-celebrities-hiv-aids. Accessed June 2025

Wiegers K, Rutter G, Kottler H, Tessmer U, Hohenberg H, Kräusslich HG (1998) Sequential steps in human immunodeficiency virus particle maturation revealed by alterations of individual Gag polyprotein cleavage sites. J Virol 72:2846–2854

Williams JC, Zhang X, Karki M, Chi YY, Wallet SM, Rudy BJ et al (2018) Soluble CD14, CD163, and CD27 biomarkers distinguish ART-suppressed youth living with HIV from healthy controls. J Leukoc Biol 103:671–680

Winkler JM, Chaudhuri AD, Fox HS (2012) Translating the brain transcriptome in neuroAIDS: from non-human primates to humans. J Neuroimmune Pharmacol 7:372–379

www.unaids.org/. Accessed June 2025

www.unaids.org/en/resources/presscentre/pressreleaseandstatementarchive/2024/july/20240722_global-aids-update. Accessed June 2025

www.who.int/data/gho/data/themes/hiv-aids. Accessed June 2025

www.who.int/news-room/fact-sheets/detail/hiv-aids. Accessed June 2025

Yu FH, Huang KJ, Wang CT (2020) HIV-1 mutant assembly, processing and infectivity expresses pol independent of gag. Viruses 12:54–70

Zhang X, Ren D, Guo L, Wang L, Wu S, Lin C et al (2017) Thymosin beta 10 is a key regulator of tumorigenesis and metastasis and a novel serum marker in breast cancer. Breast Cancer Res 19:15–29

Zheng D, Liwinski T, Elinav E (2020) Interaction between microbiota and immunity in health and disease. Cell Res 30:492–506

# Chapter 9
# Proteomic Analysis of Influenza

**Abstract** The past several years have increased the public's knowledge of various terms related to virology. Besides learning about messenger RNA (mRNA) vaccines, N95 masks, and coronaviruses (CoVs), we became familiar with the term pandemic. Pandemics are not novel. There have been approximately 250 pandemics recorded throughout history since 1200 B.C. with approximately 20 of these resulting in the deaths of more than one million people (https://study.com/learn/lesson/pandemics-in-history.html; Sampath S et al., Cureus 13:e18136, 2021). The earliest recorded pandemic for which there is detailed information is the Justinian plague (Drancourt M and Raoult D, Clin Microbiol Infect 22:911–915, 2016). This pandemic began in 540 A.D., lasted for two centuries, and was a major factor that hastened the fall of the Roman Empire. So what exactly is a pandemic? At its most basic a pandemic is a disease that is simultaneously occurring worldwide, affecting many individuals. This definition does not include anything about the severity of the disease or the population's immunity. While most people would say that pandemics and epidemics are extremely rare, there have been six just in the last two decades (Bhadoria P et al., J Family Med Prim Care 10:2745–2750, 2021). These include the severe acute respiratory syndrome coronavirus 1 (SARS-CoV-1), swine flu, Middle Eastern respiratory syndrome CoV infection, Ebola virus, Zika virus, and the most recent SARS-CoV-2 pandemic (see Table 9.1).

**Keywords** Pandemic • Influenza • Influenza life cycle • Pan-viral therapy • Host-directed therapy targets • Phosphoproteomics • Adjuvants • Radiofrequency adjuvant • Adjuvant mechanism of action

## 9.1 Introduction

Influenza A viruses have been the cause of two major pandemics. The first was the 1918 Spanish flu, and the other was the more recent swine flu pandemic of 2009. The term "swine flu" is a misnomer as the virus was actually a human virus that resulted from a triple reassortment between avian, swine, and human viruses



T. D. Veenstra, *Proteomic Applications in Virology*, Advances in Experimental Medicine and Biology 1511, https://doi.org/10.1007/978-3-032-22340-1_9

**Table 9.1** List of the six pandemics/epidemics that have occurred in the past 20 years

| Pandemic | Years |
|---|---|
| Severe acute respiratory syndrome corona virus infection | 2002–2004 |
| Influenza A H1N1 2009 (swine flu) pandemic | 2009–2010 |
| Middle East respiratory syndrome (MERS) CoV infection | 2012–present |
| Western African Ebola virus epidemic | 2013–2016 |
| Zika virus epidemic | 2015–2016 |
| SARS-CoV-2 infective disease (COVID-19) pandemic | 2019–2023 |

(Trifonov et al. 2009). The official name of the virus was the influenza A(H1N1) pdm09 virus. Beyond conventional symptoms associated with the flu, infection with this virus often led to acute respiratory distress syndrome (ARDS) or pneumonia. It is believed that the influenza A(H1N1)pdm09 virus was in circulation for several months before being identified as a novel strain (Butler 2009). While the virus originated in Mexico, it eventually spread across the globe and affected more than 214 countries. The number of cases reported to the WHO was between 700 million and 1.4 billion, resulting in more than 18,000 deaths (Dawood et al. 2012). The pandemic was successfully controlled after 19 months due to availability of an effective vaccine and antiviral treatment that reduced mortality and morbidity. While humans have adapted to it, the A(H1N1)pdm09 strain did not completely disappear and continues to circulate as a seasonal flu virus, causing illness, hospitalization, and deaths every year (Baldo et al. 2016) (Table 9.1).

Besides pandemics, a more prevalent danger is the potential resurgence of viral outbreaks and epidemics. An outbreak is an episode in which more cases of a disease are diagnosed in a specific geographic area or population over a given period than is normally expected. The increase in the numbers of measles cases observed in 2025 in areas such as Texas and Alberta, Canada, would be defined as an outbreak. An epidemic is often thought of as a widespread outbreak as it affects a greater number of individuals than an outbreak but significantly less than a pandemic (Dias 2024). An epidemic also affects a larger geographic area, potentially encompassing multiple communities, regions, or even nations. Another classification of disease outbreak is endemic, which refers to situations in which a disease is persistently present in a specific geographical region. Examples of endemics include infectious diseases malaria, schistosomiasis, and cholera, which are common in areas with poor sanitation (Rupali 2019). Other factors including natural barriers that produce isolation, the presence of insect vectors, climate stability, and habitat destruction also foster endemics (Zhang et al. 2022). Outbreaks, epidemics, and endemics generally do not spread so that they cause global catastrophes associated with pandemics; nonetheless, they are serious health issues that require attention.

Influenza is considered a seasonal epidemic. With the recent attention given to the COVID-19 pandemic, the yearly devastation caused by influenza has seemed to become an afterthought. Every year tens of millions of people in the USA get sick from influenza resulting in hundreds of thousands of hospitalizations. Preliminary CDC estimates for the 2024–2025 flu season predicted at least 27,000 deaths including 280 pediatric deaths (Reinhart et al. 2025). The cost to US businesses is an estimated $11.2 billion annually.

## 9.2 The Influenza Viruses

There are four recognized influenza virus types: A, B, C, and D (Dangi and Jain 2012). The seasonal influenza epidemics that occur almost every winter in the USA (known as flu season) are caused by influenza A and B. While all four influenza types can be traced to shared *Orthomyxoviridae* evolutionary ancestor, they have diverged significantly over time (Wille and Holmes 2020). Types A and B are more closely related to each other, and types C and D are more closely related to each other than to types A and B (Bouvier and Palese 2008). They have genetically diverged to such an extent that exchange of viral RNA between these viruses does not occur. This divergence has given each type of virus different behaviors and risks associated with humans and other animals. For example, type A can infect humans and a wide variety of animals. Types B and C also infect humans but have only been found in a limited number of animal species (Mostafa et al. 2018). Unfortunately, the wide variation in type A infection allows for antigenic shift that can result in pandemics. Although type B is primarily a human pathogen and can cause seasonal epidemics, it does not result in pandemics. Both influenza C and D are much less common than types A and B. Type C causes mild respiratory infections, while type D does not cause disease in human and is mainly a cattle virus (Liu et al. 2020). A comparison of the structures of influenza viruses type A, B, C, and D is shown in Fig. 9.1.

Influenza A viruses are further characterized by the *subtype* of their surface glycoproteins hemagglutinin (HA) and neuraminidase (NA). The subtype of these two proteins is incorporated into the standard nomenclature of influenza A viruses, which includes virus type, species from which it was isolated (if nonhuman), location at which it was isolated, isolate number, and isolate year (Bouvier and Palese 2008). Using this convention, isolate 2007 of a human influenza A virus obtained in Panama in 1999 with HA subtype 3 and NA subtype 2 would be named A/Panama/2007/1999(H3N2). While 16 HA and 9 NA isotypes have been identified in circulating influenza A, only H1, H2, and H3 and N1 and N2 subtypes have resulted in human epidemics (Clancy 2008). Although other subtypes such as H5, H7, and H9 have infected humans, mostly via direct animal contact, they are not efficiently transmitted among humans and therefore have been unable to generate sustained epidemics.

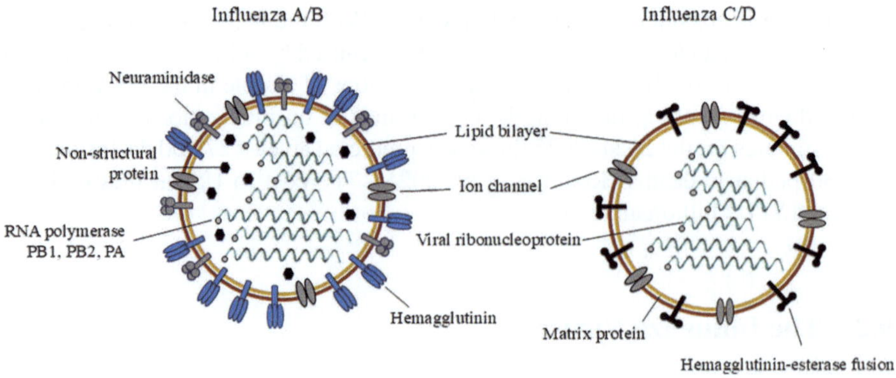


**Fig. 9.1** Comparative structures of influenza virus types A, B, C, and D. The structures of influenza types A and B are closely related as are those of influenza C and D. While influenza types A and B contain hemagglutinin and neuraminidase proteins to facilitate entry and exit from the host cell, types C and D contain the hemagglutinin-esterase fusion protein. While influenza types A and B possess eight negative sense single-stranded RNA segments, types C and D only contain seven

### *9.2.1 The Structure of Influenza Viruses*

Since types A and B are the most common influenza viruses, this section will focus on either of these two. Influenza A and B virions look extremely similar at a microscopic level. Both are spherical or filamentous in shape. The influenza A virion is studded with spikes of HA and NA glycoproteins that project from its lipid membrane (Palese and Shaw 2007). The third major influenza A envelope protein is the matrix (M2) protein, which forms an ion channel that spans the lipid envelope. While the ratio of HA to NA proteins is on the order of 4:1, the ratio of HA to M2 proteins is between 10 and 100 to 1 (Sautto et al. 2018). Beneath this HA-, NA-, and M2-containing lipid envelope is a coat of matrix (M1) proteins. The M1 proteins serve two major functions (Zhu et al. 2025). The first function is to encapsulate the nucleoprotein core of the virus, and the second is to bind the virus's RNA genome. These functions require the M1 proteins to simultaneously bind to the membrane and RNA of the virus.

Within the M1 matrix are the nuclear export protein (NEP), which is also called nonstructural protein 2 (NS2), and the ribonucleoprotein (RNP) complex (Bouvier and Palese 2008). The RNP complex consists of viral RNA segments coated with nucleoprotein (NP) and RNA-dependent RNA polymerase. The RNA polymerase is a heterotrimeric complex composed of PB1, PB2, and PA polymerase subunits of which PB1 and PB2 are basic and PA is acidic.

Influenza B has a similar structure to influenza A with some notable differences. While influenza B possesses the envelope proteins HA and NA, it possesses NB and B Matrix protein 2 (BM2) rather than M2 (Bouvier and Palese 2008; Hatta et al. 2004). The NB protein is an integral membrane protein that acts as an ion channel (Betakova et al. 1996). The BM2 protein is a proton channel that plays an essential role in viral replication (Hatta et al. 2004). Its role in viral replication is believed to be related to uncoating of the virus once it is encapsulated in the endosome of infected host cells. Influenza C virions are compositionally similar to influenza A and B in that they possess a glycoprotein-studded lipid envelope overlying a protein matrix and the RNP complex (Bouvier and Palese 2008). Instead of containing both HA and NA proteins, however, influenza C viruses possess the hemagglutinin--esterase-fusion (HEF) protein, which carries out functions analogous to the HA and NA proteins of influenza A and B (Wang and Veit 2016). The lipid envelope of influenza C contains CM2, a minor protein that serves as an analog to M2 protein of influenza A and B. This protein is involved in the release of viral RNPs during virus uncoating in an infected cell and the repackaging of RNPs during virus assembly prior to exocytosis (Pekosz and Lamb 1998).

### *9.2.2 Genomic Structure of Influenza*

All influenza viruses consist of negative-sense single-stranded segmented RNAs (-ssRNAs). The genomes of both influenza A and B are comprised of eight -ssRNAs, which are numbered in order of decreasing length (Bouvier and Palese 2008). In influenza A and B viruses, RNA segments 1, 3, 4, and 5 encode the PB2, PA, HA, and NP proteins, respectively (Muramoto et al. 2013). While all influenza viruses encode the polymerase subunit PB1 on segment 2, this segment also codes for the accessory protein PB1-F2 in a +1 alternate reading frame in some strains of influenza A (Bouvier and Palese 2008; Pinto et al. 2021). PB1-F2 is a small, 87-amino acid pro-apoptotic protein. Influenza A and B also differ in the expression of segment 6. While influenza A encodes only the NA protein from segment 6, in influenza B this segment encodes both the NA and NB proteins. The NB protein is expressed in a −1 alternate reading frame. The M1 matrix protein is encoded by segment 7 of both influenza A and B. The M2 ion channel is also expressed in influenza A from segment 7 via RNA splicing, while influenza B uses a +2 alternate reading frame to encode its BM2 membrane protein from this segment (Lamb et al. 1981). Influenza A and B both express the NS1 and, by mRNA splicing, the NEP/NS2 proteins from gene segment 8. NS1 protein acts as an interferon antagonist, while NEP/NS2 regulates the transcription and translation of the viral genomes (Lamb et al. 1980; Briedis and Lamb 1982). While influenza C has a similar genomic structure, it contains one fewer segment than influenzas A and B. Influenza C only requires seven -ssRNA segments since the HEF protein replaces the two HA and NA proteins of A and B (Wang and Veit 2016). The genomic segment of influenza A/Puerto Rico/8/1934 (H1N1) is provided in Table 9.2 along with the proteins they encode and their functions. The biology of this strain of influenza A strain is well known, and it has been used to generate seasonal H5N1 strains because of its high yield in embryonic chicken eggs.

**Table 9.2** Genomic segments of influenzaA/Puerto Rico/8/1934 (H1N1) virus and their encoded proteins

| Segment | Segment length (nucleotides) | Protein encoded | Protein function |
|---|---|---|---|
| 1 | 2341 | PB2 | Polymerase subunit; mRNA cap recognition |
| 2 | 2341 | PB<br>PB1-F21 | Polymerase subunit; RNA elongation, endonuclease<br>Pro-apoptotic activity |
| | | PB1-F21 | Pro-apoptotic activity |
| 3 | 2233 | PA | Polymerase subunit; protease activity |
| 4 | 1778 | HA | Surface glycoprotein; receptor binding and fusion activities for viral entry into host cell |
| 5 | 1565 | NP | RNA-binding protein; nuclear import regulation |
| 6 | 1413 | NA | Surface glycoprotein; sialidase activity, virus release |
| 7 | 1027 | M1<br>M2 | Matrix protein; viral RNP interaction, encapsulate core<br>Ion channel; virus uncoating and assembly |
| 8 | 890 | NS1<br>NEP/NS2 | Interferon antagonist protein; regulation of host gene expression<br>Nuclear export of RNA |

## 9.3 The Influenza Virus Life Cycle

Influenza virus infection is a highly contagious virus that can be transmitted through respiratory droplets generated from coughs and sneezes and even simple talking. If these infected drops are inhaled or land in the mouths or nose of nearby people, they can also become infected. The virus can also be transmitted through touching contaminated surfaces and then touching your own face. Fortunately, the life cycle of influenza virus is very well understood. This knowledge has been used to generate effective preventatives (i.e., vaccines) and treatments to lessen the effect of the virus on humans. Even these advances, however, have not eliminated the constant threat that influenza poses on human health.

### *9.3.1 Virus Attachment and Entry*

While a basic depiction of the life cycle of influenza was provided in Chap. 2, the following description provides greater detail of the functions of the various influenza proteins (see Fig. 9.2). The first stage in the entry of an influenza virus into a host cell is the binding of the virus to N-acetylneuraminic (sialic) acid on the host cell surface via recognition by the HA spikes on the viral surface (Eichberg et al. 2022; Sieczkarski and Whittaker 2005). Once influenza attaches to sialic acid, it is endocytosed into the host cell. What needs to be appreciated at this point, however,

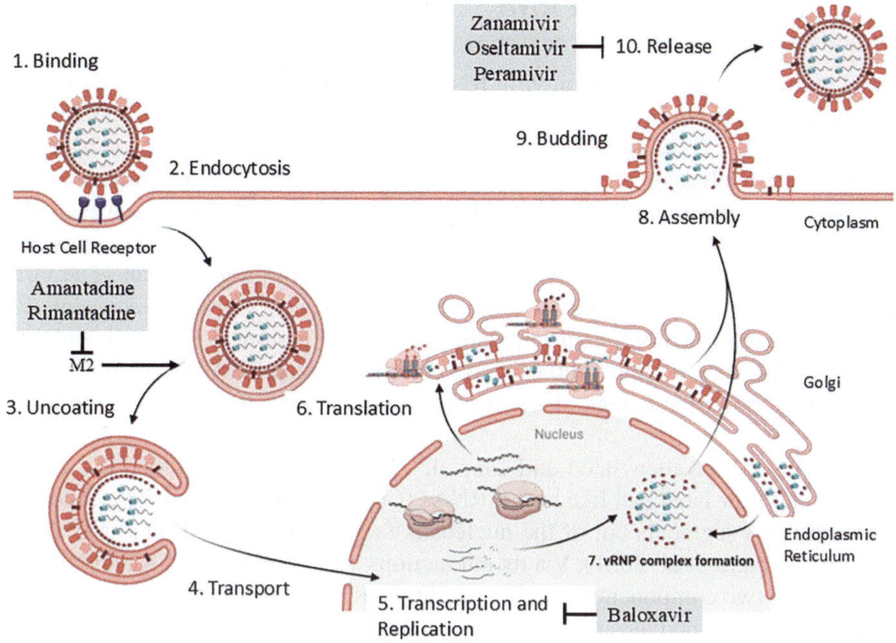


**Fig. 9.2** The life cycle of influenza virus. Stage 1: The virus binds via hemagglutinin on its surface to sialic acid residues on the host cell. Stage 2: The virus is then taken into the host cell via endocytosis and is encapsulated into an endosome. Stage 3: The endosome becomes acidified through the entry of protons via the M2 protein of the virus causing the release of acidification of the endosome and viral interior, and the vRNPs are released in the cytosol. Stage 4: These vRNAs are transported into the nucleus, where viral RNA is transcribed and replicated in mRNAs (Stage 5). Stage 6: After being transported into the host cell cytosol, the viral mRNAs are translated into the various viral proteins required for replication. Stage 7: Nucleoproteins (NPs) and the subunits of the viral polymerase PA, PB1, and PB2 are transported back in the nucleus to form new vRNP complexes. Stage 8: The replicated vRNPs, along with the expressed HA, NA, and M1 proteins, migrate to the cell plasma membrane for assembly into viral progeny. Stage 9: The virions bud from the cell surface and are released by the neuraminidase-catalyzed cleavage of sialic acid bonds holding the virions to the host cell (Stage 10). These virions are now free to infect nearby cells. (Adapted from ref. (Eichberg et al. 2022) under the terms and conditions of the Creative Commons Attribution (CC BY) license (https://creativecommons.org/licenses/by/4.0/))

is that even though the virus has entered the host cell, it is still encased within a membrane, unable to utilize the cell's machinery to replicate. The key to releasing the components necessary for its replication into the cytoplasm of the host cell is the acidity of the endosome (Sieczkarski and Whittaker 2005). This acidity triggers a conformational change in HA, which exposes a fusion peptide that connects the viral envelope to the endosomal membrane. This contact creates a pore through which viral RNPs are released into the host cell cytoplasm. Abundant protons in the endosome are pumped into the virus particle via the M2 ion channel. This internal acidification disrupts protein-protein interactions that connect viral RNPs with the virus's matrix and allow them to be released into the cytoplasm.

### 9.3.2 Synthesis of Viral RNA

Once in the host cell cytoplasm, viral RNPs are trafficked to the nucleus by means of nuclear localization signals (NLSs) on the viral proteins (Sieczkarski and Whittaker 2005). It is critical that the viral genome and its associated proteins enter the host cell nucleus as this is the site of all influenza viral RNA syntheses. This repertoire of RNA includes the capped, polyadenylated mRNA that serves as the template for host-cell translation of viral proteins and the viral RNA segments that will be replicated and eventually become the genomes of progeny virus. The viral RNA-dependent RNA polymerase synthesizes two positive-sense RNAs using the negative-sense viral RNA as a template. These positive-sense RNAs are mRNA templates for viral protein synthesis and complementary RNA intermediates that are used to transcribe copies of negative-sense, genomic viral RNA via the RNA polymerase.

After being polyadenylated and capped, viral mRNAs are exported out of the nucleus and translated just like host mRNAs (Cros and Palese 2003). The export of the viral RNA segments out of the nucleus is facilitated by the M1 and NEP/NS2 proteins (Paragas et al. 2021). Via its interactions with both viral RNA and NP, M1 brings these two components together within the RNP complex. M1 also associates with NEP, which mediates the export of the M1-RNP complex into the cytoplasm via the nuclear pore complex.

### 9.3.3 Synthesis of Viral Proteins

The mRNA produced from the influenza genome is used to translate the envelope proteins HA, NA, and M2 (Bouvier and Palese 2008). After being translated on the rough endoplasmic reticulum, they are folded and trafficked to the Golgi apparatus where they undergo posttranslational modification. These proteins all have sorting signals that mediate their localization to the host cell membrane for virion assembly. As part of the bundling of the viral progeny, the M1 protein brings the RNP-NEP complex into contact with the HA, NA, and M2 proteins for packaging at the host cell membrane (Palese and Shaw 2007).

### 9.3.4 Packaging of RNA and Assembly of Virus

Until it contains a full genome of eight segments, the influenza A or B viruses are not fully infectious. While it was thought to be a random process, recent studies have shown that packaging of the viral RNA segments is a selective process that incorporates signals on all viral RNA segments to ensure that most virions receive a full genomic complement prior to budding and release from the host cell (Noda 2021).

### 9.3.5 Virus Budding and Release

Prior to budding and release, HA, NA, and M2 proteins are transported to the cytoplasmic side of the host cell's plasma membrane (Veit and Thaa 2011; Dou et al. 2018). Their transport to this area is directed by sequences within their primary structure. M1 matrix proteins create a dense layer on the apical side of the lipid bilayer and form a bond between the RNPs of the virus and the viral envelope proteins (Bouvier and Palese 2008). Even when assembly of the virion components is complete and the viral bud is formed, the virus cannot immediately be released from the host cell. As described in the section on viral entry, the HA spikes of influenza bind to sialic acid on the cell surface keeping the virus tethered to the host cell. Fortunately for the virus, NA cleaves terminal sialic acid residues from cell-bound glycoproteins and gangliosides destroying the link between HA and the host cell (Els et al. 1989). This cleavage allows the virions to be released from the host cell.

Understanding the influenza's life cycle enables the design of drugs that block key processes in the viruses' replication (see Fig. 9.2). For example, amantadine and rimantadine block M2 channels on the virus's envelope preventing protons from entering the virus particle. Without this influx of protons, the endosome interior does not acidify, and the virus is not released. Baloxavir marboxil acts within the host cell nucleus by blocking the polymerase acidic (PA) protein subunit of the virus's RNA polymerase complex preventing it from transferring 5′ caps to the viral mRNA primers. Without these caps, viral RNA cannot be transcribed, and replication of the virus is inhibited. The NA inhibitory drugs zanamivir, oseltamivir, and peramivir prevent the release of progeny virus by mimicking sialic acid and blocking NA from clipping sialic acid residues from cell-bound glycoproteins and gangliosides that bind these viruses to the host cell. Preventing their release inhibits the spread of the virus to other cells.

## 9.4 Proteomic Analysis to Identify Pan-Viral Therapies to Treat Influenza A Infections

Influenza A continues to be a tremendous worldwide health and economic burden. Since they are the predominant subtypes that circulate yearly, the pandemic H1N1-like (pH1N1) and H3N2 subtypes are the main targets of the annual flu vaccines. While influenza infection is sustained through human-to-human transmission, there is a constant threat of distinct influenza A strains causing global pandemics via antigenic drift, antigenic shift, and zoonotic transmission (i.e., between animals and humans). While rare, a study has shown that a single mutation in H5N1 could enable zoonotic transmission of this subtype between mammals generating a potential risk of a human pandemic (Lin et al. 2024).

Currently most vaccines and antiviral drugs target influenza proteins, especially HA, NA, and M2. The aim of these treatments is to inhibit influenza entry, assembly, or release to prevent its replication and spread. Antiviral drugs include umifenovir, which binds to HA and inhibits the conformational change necessary for its fusion with the host cell (Kang et al. 2023). Other drugs such as oseltamivir, zanamivir, and peramivir act as NA inhibitors, preventing this protein from cleaving the HA-sialic linkage, preventing influenza virions from being released from the host cell (Ma et al. 2016). Since influenza strains possess highly conserved regions, it can be argued that targeting of viral proteins is an effective strategy. These highly conserved regions also enable the potential development of a universal vaccine or druggable target that can treat all influenza strains.

Over time, distinct influenza strains have emerged with altered transmissibility and pathogenicity that threaten the human population. Yearly development and administration of vaccines continue to be limited in their effectiveness. Although influenza is not a reportable disease, the CDC still estimates that almost 30,000 Americans die from this disease. The increased prevalence of influenza mutations that render the virus resistant to the three classes of commonly used antiviral therapies also limited their effectiveness (Hussain et al. 2017). In addition, antiviral treatments are only effective when administered soon after symptoms appear. The time has arrived to pursue additional targets to combat this continually burdensome virus.

With antigenic drift, and especially antigenic shift, there exists an ever-present threat that a novel influenza virus will evolve that results in a human pandemic. This threat requires a better antiviral strategy that targets human proteins that are necessary for influenza replication and transmission. By targeting host factors necessary for viral replication, these host-directed therapies (HDTs) avoid the possibility of viruses developing drug resistance (Lee et al. 2025). Not only do HDTs have the potential for preventing spread of all strains of influenza, but their efficacy could extend to other virus types since many use the same host pathways. For example, influenza co-circulates with SARS-CoV-2, respiratory syncytial virus (RSV), rhinovirus, and adenovirus (Kampenusa et al. 2024). The combination of influenza and SARS-CoV-2 especially remains a global health emergency. Just imagine the number of lives that could be saved if a pan-specific drug could be developed that targeted human proteins and prevented infection by all these viruses.

### *9.4.1 Proteomic Identification of Host-Directed Therapy Targets*

Like any drug discovery project, the first step is to identify key host proteins that are essential for viral replication, propagation, and transmission. A recent study used affinity purification mass spectrometry (AP-MS) to identify protein-protein interactions (PPIs) between three strains of influenza and SARS-CoV-2 (Haas et al. 2023). Thirteen proteins encoded by pH1N1, H3N2, and H5N1 were cloned into tagged expression vectors, which were subsequently transfected into adenocarcinomic human alveolar basal epithelial (A549), normal human bronchial epithelial (NHBE),

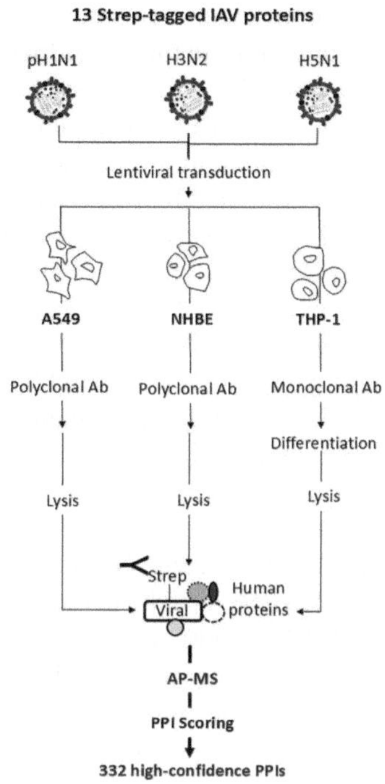


**Fig. 9.3** Design of affinity purification-mass spectrometry experiment to identify host-directed therapy to treat influenza A infection. Thirteen streptavidin-tagged influenza A virus (IAV) proteins from pH1N1, H3N2, and H5N1 were individually transduced by lentivirus to generate stable A549, NHBE, and THP-1 cell lines. Expression of the 13 IAV proteins was induced for 24 h. The cells were lysed, and tagged IAV proteins were extracted using biotin. The affinity-purified IAV proteins and co-purified human proteins were identified using MS. (Adapted from ref. (Haas et al. 2023) under the terms and conditions of the Creative Commons Attribution (CC BY) license (https://creativecommons.org/licenses/by/4.0/))

and differentiated human leukemia monocytic (THP-1) cells (see Fig. 9.3). While all 13 influenza virus proteins are stably expressed in at least 1 cell type, 8 are stably expressed in all 3. Affinity purification of the tagged proteins was performed using Strep-Tactin Sepharose beads. After a series of washing steps to minimize nonspecific binding proteins (while retaining high affinity interactions) the bead-bound proteins were reduced and digested with trypsin. The peptides were recovered from the beads and acidified by adding 1% formic acid. After desalting and drying, the peptides were dissolved in 2% acetonitrile containing 0.1% formic acid. Each peptide sample was analyzed using LC/MS$^2$ employing a capillary column packed with a C18 reversed-phase chromatography resin and an ion-trap mass spectrometer. The raw MS files from all samples were grouped by cell line and searched against the human proteome data available within SwissProt (UniProt Consortium 2025).

The data analysis identified 332 PPIs between all influenza strains and human cell types, which mapped to 214 human proteins (Haas et al. 2023). These 332 high-confidence PPIs included interactions between 108 human and 9 influenza proteins in A549 cells; 88 human and 8 influenza proteins in NHBE cells; and 56 human and 8 influenza proteins in THP-1 cells. Within the individual viral strains, 77 PPIs were observed among 8 pH1N1 proteins, 77 PPIs among 10 H3N2 proteins, and 142 among 11 H5N1 proteins. Twenty-nine PPIs were found in at least two of the three cell lines, with six of these found within all three. The lack of overlap between PPIs within all three cell lines can be explained by two reasons. Firstly, not all viral proteins consistent expressed in all cell types. Secondly, many PPIs identified in other cell lines were below the scoring threshold for classifying them as "high confidence." Among all these interactions, 44 PPIs between viral and human proteins that had not been previously observed were discovered.

The PPIs found within multiple cell types included those between influenza NS1 and cleavage and polyadenylation factor complex members. This interaction is not surprising since NS1 is known to interfere with host cell mRNA export by posttranscriptionally dampening host mRNA expression (Zhang et al. 2018). The NP of influenza was found to interact with spliceosome components and mitochondrial ribosome subunits in NHBE and THP-1 cells. NP has previously been identified as interacting with splicing factor UAP56, which is essential for the formation of viral RNP complexes (Hu et al. 2017). Interactions between NP and heterogeneous nuclear ribonucleoproteins that negatively and positively regulate viral growth via promoting host cell apoptosis or increasing viral RNA synthesis have also been previously observed (Kaur et al. 2022; Castello et al. 2024).

Since this stage of the study was designed to identify any potential HDTs, all PPIs were considered legitimate regardless of how many cell types that were found in. Human interactions shared by all three influenza strains included proteins involved in spliceosomes, the CDC5L complex, the mitochondrial ribosome, the 60 S ribosome, nuclear transport, macroautophagy, and proton transport. The biological processes associated with these PPIs are consistent with known essential entry, replicative, assembly, and budding roles of interacting virus proteins. Strain-specific PPIs were also identified and may represent unique strain-specific co-opting of host protein complexes. For example, NS1 interactions were predominantly identified with pH1N1 and H3N2. Specifically, NS1 of these two strains was shown to interact with four PI3K signaling components, consistent with studies that demonstrate PI3K activation by NS1 infection to modulate host apoptotic response (Haas et al. 2023).

The confidence of the overall PPI database built in this study is bolstered by many of these interactions being previously identified. For example, the vacuolar ATPase catalytic subunit component ATP6V1A, which is important for entry of influenza into the host cell, was found to interact with M2 in all three cell lines used in this study. Not only had this interaction been previously identified, but it was further identified in this study using a reverse immunoprecipitation assay (Zhao et al. 2024; Guinea and Carrasco 1995).

In addition to known PPIs, novel interactions were identified between NEP and host cell proteins involved in cytoskeletal regulation and RNA binding. One specific structural scaffold protein, AHNAK, was of particular interest. AHNAK, also known as desmoykin, orchestrates a range of cellular functions by serving as a hub that brings multiple proteins together (Zhang et al. 2023a). An interaction between AHNAK and H5N1 NEP in NHBE cells was identified and validated in a reciprocal pull-down assay and their colocalization in the cytoplasm. Three novel interactions involving HA with ER proteins that regulate protein folding and processing were also identified (ERLEC1, SEL1L, P4HB) as were eight human proteins that interact with N40 (Haas et al. 2023). Of these, six are involved in protein quality control machinery, demonstrating a potentially novel role of N40 in modulating both human and viral protein expression.

### *9.4.2 Identification of Host Kinase Activity Affected by Influenza Infection*

There are many different aspects to a proteome besides protein abundance and PPIs. To screen as many components as possible that affect pH1N1, H3N2, and H5N1 infection, studies were conducted to identify global changes in both protein abundance and phosphorylation state (Haas et al. 2023). These studies were done over various time points (3, 6, 12, and 18 h postinfection) to enable time-dependent changes in phosphorylation states to be detected (see Fig. 9.4a). Hundreds of significant protein abundance changes and site-specific phosphorylation events were identified. Proteins demonstrating consistently increasing phosphorylation status were involved in RNA splicing and processing, cellular and nuclear membranes, regulation of gene silencing, and innate immune response. These findings are consistent with influenza A's ability to manipulate host RNA machinery to splice and translate viral RNA, viral entry and exit, inhibition of export host mRNA from the nucleus to silence host gene expression, and cellular detection and response to virus.

Examining the changes in the abundance and phosphorylation state of each protein showed weak correlation, leading to the theory that these phosphorylation changes must originate from alterations in kinase activity. By comparing the phosphorylation data to a proteomic catalog of kinase-substrate partners, 13 kinases whose activity was altered upon viral infection were identified (Türei et al. 2021). Five mitogen-activated protein kinase members (MAP 2 K3, MAP 2 K6, MAPKAPK2, MAPKAPK3, and MAPKAPK5) showed increased activity upon H3N2 and H5N1 infection but little change in activity upon pH1N1 infection (see Fig. 9.4b). The same trend was observed for ribosomal protein S6 kinase family members RPS6KB1 and RPS6KB2. Phosphatidylinositol 3-kinase-related kinase (PRKG1) activity was increased by pH1N1 and H3N2 infection but not H5N1. The lack of absolute congruence in these results demonstrates that different strains of viruses possess variable pathogenicities resulting in diverse host responses (Haas et al. 2023).

Comparison of the PPI and phosphoproteomics data revealed 45 human proteins that both interact with and whose phosphorylation state is altered by an influenza A protein. For example, the protein-folding protein calnexin was found to interact

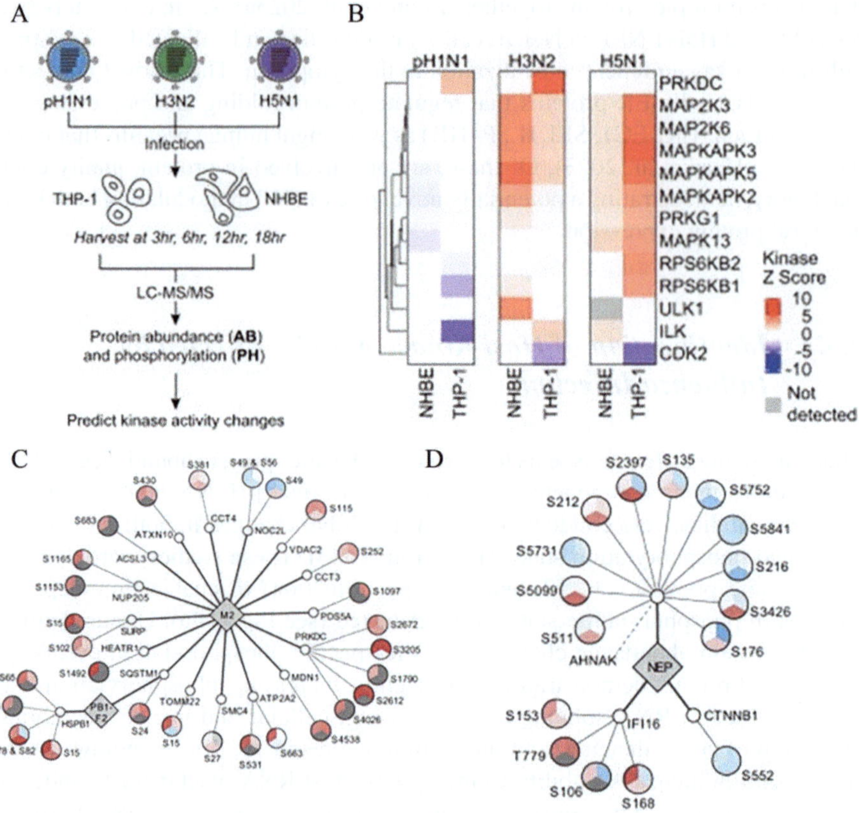


**Fig. 9.4** (**a**) Experimental design workflow for global proteomic profiling to measure both protein abundance (AB) and phosphorylation (PH) changes pH1N1-, H3N2-, or H5N1-infected NHBE and THP-1 cells. (**b**) Heatmap of predicted kinase activity observed in influenza A virus (IAV)-infected NHBE and THP-1 cells. While kinase activity was upregulated in both cell lines infected with H5N1 and H3N2, little change in activity was observed in pH1N1-infected cells. (**c**) Of the proteins that interacted with influenza matrix protein (M2), the DNA damage-sensing kinase (PRKDC) exhibited an increase in phosphorylation at five distinct residues. (**d**) The phosphorylation state of 11 unique residues with desmoykin (AHNAK) was affected by IAV infection. Desmoykin was also shown to interact with the nuclear export protein (NEP) of influenza virus. (Adapted from ref. (Haas et al. 2023) under the terms and conditions of the Creative Commons Attribution (CC BY) license (https://creativecommons.org/licenses/by/4.0/))

with HA of H3N2 and exhibited a downregulation of its phosphorylation state by all three viral strains. The phosphorylation state of SRSF2, a protein involved in pre-mRNA splicing (Kędzierska and Piekiełko-Witkowska 2017) that was found to interact with PA-X, was upregulated by all three IAV strains. Other PPIs showed strain-specific regulation of phosphorylation state. For example, the phosphorylation state of cleavage and polyadenylation specific factor 4 (CPSF4), which helps to polyadenylate the 3′ end of mRNA and interacts with NS1 to block the export of host pre-mRNA from the nucleus, was downregulated in pH1N1- and

H3N2-infected cells but upregulated in H5N1-infected cells (Dubois et al. 2019). Previous studies showing that H5N1 infection can block mRNA export by CPSF4-NS1 complexes to a greater extent than pH1N1 leads credence to this proteomic finding (Kainov et al. 2011; Hale et al. 2012).

While most PPIs correlated with one or two phosphorylation sites whose abundance changed because of viral infection, some contain at least four altered sites. One of these included PRKDC, a DNA-damage sensing kinase that interacts with M2 (Camfield et al. 2024). This protein contained five upregulated phosphorylation sites postinfection with all three influenza viruses (see Fig. 9.4c). Desmoykin (AHNAK), which interacted with NEP, contained 11 influenza-modified phosphorylation sites (see Fig. 9.4d). The exceptional degree of phosphorylation changes seen in these two proteins could highlight their functional importance in viral infection.

### *9.4.3 Recognition of Host-Directed Therapies Against Influenza*

As the goal of this study was to find HDTs, ultimate success would be finding drugs that target human proteins and disrupt PPIs necessary for viral infection, replication, propagation, or transmission. Compounds that targeted a subset of the PPIs and kinases identified in the global proteomic analyses were identified via previously published reports. These compounds were screened for their ability to inhibit viral infection along with pimodivir, which served as a positive control since it is known to possess antiviral activity against all three strains of influenza A tested in this study (Takashita 2021). Approximately 20 host proteins were targetable by at least 1 compound, including 8 PPIs (targeted by total 16 compounds) and 12 kinases (targeted by total 15 compounds), with 2 host proteins identified as both PPIs and kinases (total 29 unique compounds). Sixteen compounds with antiviral activity against at least one influenza strain were identified, with seven compounds demonstrating antiviral activity against all three strains (Haas et al. 2023).

Four compounds showed antiviral activity against at least two influenza A strains and targeted PPIs identified in this study. These compounds were bafilomycin A1, daunorubicin, PACMA 31, and NU7441. Bafilomycin A1, a macrolide antibiotic, targeted the interaction between M2 of influenza A and the V-ATPase subunit ATP6V1A of host cells (Londino et al. 2015). This compound blocks the maturation of autophagosomes and acidification of organelles and inhibits autophagy. Its ability to inhibit autophagy and disrupt organelle pH balance is associated with its ability to trigger cell apoptosis. Research has shown this drug's ability to reduce the levels of various viruses, including SARS-CoV-2, in animal models (Zhang et al. 2023b). Daunorubicin inhibited the interaction between M2 and host cell ABCC1 (an ATP-binding cassette subfamily member). This compound had never been shown to inhibit influenza A infection; however, one study has suggested its efficacy in suppressing hepatitis B viral production (Imai et al. 2018). Daunorubicin is believed to function as a DNA intercalating agent, topoisomerase II inhibitor, and a generator of free radicals (Nitiss 2009). Potent pan-influenza A antiviral activity was observed in infected cells treated with the protein disulfide isomerase inhibitor PACMA 31. While PACMA 31 is known to disrupt viral processes in infected cells by preventing the proper folding of host proteins (Li et al. 2025), its disruption of the interaction between HA and the protein disulfide isomerase P4HB suggests an important role for this host protein in the proper folding of viral HA. NU7441 targets DNA-dependent protein kinase (DNA-PK) inhibiting this enzyme's ability to repair non-homologous breaks within DNA strands (Pascale et al. 2016). This study demonstrated that NU7441 disrupted the interaction between M2 and the DNA-PK PRKDC. While pH1N1 and H3N2 infection was suppressed by NU7441, H5N1 infection was not. This strain specificity was consistent with the phosphoproteomic data, in which PRKDC activity was absent in H5N1-infected cells. The connection between M2, an ion-channel protein involved in virus uncoating and assembly, and PRKDC is not obvious. PRKDC has been associated with the triggering of an interferon response to viral infection that may be blunted by its interaction with M2. Disruption of this interaction using NU7441 may enable this interferon signal to induce an immune response to inhibit viral propagation.

In addition to PRKDC, five additional kinase inhibitors showed anti-influenza A activity. Dinaciclib showed potent antiviral activity against all three strains. This cyclin-dependent kinase CDK2 inhibitor has previously been reported to exhibit antiviral activity for the influenza A strain H7N9 (Perwitasari et al. 2015). The autophagy-activating kinase ULK1 inhibitor MRT68921 inhibited H3N2 and H5N1 activity and decreased pH1N1 infection. By inhibiting ULK1 activity, this novel influenza A antiviral likely downregulates autophagy. Since autophagy is critical for influenza A infection, inhibiting this activity serves to suppress viral infection (Wang et al. 2018). Additionally, lestaurtinib and three inhibitors of four members of the mitogen-activated protein kinase (MAPK) pathway showed antiviral activity. The three compounds lestaurtinib, MAPK-13-IN-1, and PF-3644022, which target members of the mitogen-activated protein kinase (MAPK) pathway, showed antiviral activity. Lestaurtinib, which targets both MAP 2 K3 and MAP 2 K6, decreased H3N2 infection but demonstrated clear antiviral activity against pH1N1 and H5N1. Although it demonstrated activity against all three influenza A strains, MAPK-13-IN-1 treatment did show some strain-specific differences in potency. MAPK-13-IN-1 targets MAPK13 (p38δ) and is often used to study cellular processes such as inflammation, cytoskeletal remodeling, and transcription factor activation. The MAPKAPK2-targeting compound PF-3644022 is known to prevent inflammation by inhibiting TNFα activity (Fiore et al. 2016). This compound demonstrated antiviral activity against H5N1. While these compounds may not individually have the attributes required for an effective host-directed therapeutic, they do suggest that MAPK signaling pathway is an important therapeutic target for preventing viral infection.

### 9.4.4 *Identifying a "Universal" Host-Directed Therapeutic*

Three SARS-CoV−2 antiviral compounds also demonstrated antiviral activity against at least two strains of IAV. Gilteritinib showed antiviral activity against all three influenza A strains. Gilteritinib targets AXL, a kinase that functions upstream of p38, MAP 2K3, and MAP 2K6 (Boytz et al. 2023). While lestaurtinib (which inhibits MAP 2K3 and MAP 2K6) and MAPK13-IN-1 (which inhibits MAPK13 or p38δ) showed antiviral activity against multiple influenza A strains, the two SARS-CoV-2 antiviral p38 inhibitors that were tested (ralimetinib and ARRY-797) had no effect on influenza A infection. These results show that kinase signaling activity is critical for influenza A and SARS-CoV-2 infection and their kinase-altering activities converge upstream of the MAPK pathway. The PIK3-catalytic subunit A and D (PIK3CD and PIK3CD)-targeting compound pictilisib also showed broad antiviral activity against influenza A and SARS-CoV-2; however, strain-specific differences in potency were observed. PIK3CA-targeting compounds have been shown to inhibit infection by two different influenza A strains supporting this kinase as a human-directed therapeutic target to combat influenza A infection (Deinhardt-Emmer et al. 2021). PIK3CA was found to form a PPI with NS1 of pH1N1 in the proteomic screening analysis, but the mechanism by which it regulates infection by H3N2 and H5N1 infection is unclear. MK-2206 showed strong antiviral activity against H5N1 and moderate antiviral activity against pH1N1. This compound disrupts AKT signaling through AKT1, AKT2, and AKT3, and its variant results in the different viral strains may indicate the AKT signaling pathway plays a greater role in H5N1 infectivity (Knowles et al. 2011). In addition to gilteritinib, pictilisib, and MK-2206, dinaciclib and MAPK13-IN-1 (which were discussed as anti-influenza A compounds) have been reported in other studies to possess anti-SARS-CoV-2 activity (Bouhaddou et al. 2020). Taken collectively, this study identifies five compounds that may have antiviral activity across influenza A and SARS-CoV-2 and serve as useful HDTs.

One limitation of proteomics technology is that it does not measure every protein, PPI, phosphorylation site, etc., that exists in a sample. To make matters worse, it is impossible to know which important biological information may have been missed. That is why many studies collating data from multiple different experiments are presently the best option to identify protein targets that warrant further investigation. In this study, the scaffold protein AHNAK was found to interact with NEP of H5N1, pH1N1, and H3N2 showing that this PPI is not strain specific. Infection of cells with H5N1, pH1N1, and H3N2 also revealed 11 sites within AHNA whose phosphorylation state is differentially regulated by these viruses. These results combined with previous results show that patients having loss-of-function variants of the *AHNAK* gene experience severe influenza disease and knockdown of AHNAK decreases both influenza A and SARS-CoV-2 infection point toward its potential as a human drug target (Haas et al. 2023).

### 9.4.5 Connecting Therapy to Mechanism

Comfort in any scientific finding comes with being able to plausibly connect an identification with a plausible mechanism of action. While the role of AHNAK in influenza A and SARS-CoV-2 infection is unknown, it is believed to be related to viral RNA export and/or virus budding (Thiruvaiyaru et al. 2025). Calcium-dependent cell-cell contact formation has been shown to trigger protein kinase B-dependent phosphorylation of AHNAK that results in its translocation out of the nucleus (Sussman et al. 2001). The AHNAK complex then migrates to the plasma membrane and interacts with an S100A10-annexin 2 complex to coordinate changes in the cytoskeleton and membrane (Benaud et al. 2004). In *Salmonella*, this coordinated function is important as AHNAK migration to the membrane in host cells is required for pathogen infection (Jolly et al. 2014). AHNAK's association with cytoskeletal and membrane remodeling may not be critical for influenza A infection but rather serve another cytoskeletal and membrane-associated purpose. The export of influenza A viral ribonucleoprotein (vRNP) complexes from the nucleus to the cytoplasm is facilitated by NEP (Paragas et al. 2021). This protein also facilitates formation and budding of the virus at the plasma membrane. It is plausible that vRNP export and virus formation and budding may be dependent on AHNAK and NEP coordinating the cytoskeletal and membrane remodeling. In support of this model, five other human cytoskeletal proteins were identified as NEP interactors in this proteomic study (Haas et al. 2023). While AHNAK has not previously been found to interact with any SARS-CoV-2 proteins, this study showed that SARS-CoV-2 caused a peak in AHNAK's RNA-binding kinetics early in the infection process. Identification of AHNAK having a role in SARS-CoV-2 infection and influenza A propagation through a role in vRNA production, trafficking, or assembly makes it a prime target for a host-directed therapy.

### 9.4.6 The Job Is Not Yet Finished

Ever since its inception, it was realized that global proteomics provided evidence to build further hypotheses around. These hypotheses would drive the design of further experiments that would substantiate, refute, or result in the modification of the original hypotheses. This pattern would continue until a preponderance of evidence would make validation of a result(s) feasible. One of the challenges with this approach had to do with scalability. While the number of laboratories dedicated to proteomics rapidly increased over the past three decades, few had the capacity or expertise to build upon their findings. For example, while proteomic labs were proficient in generating large datasets of PPIs or phosphoproteins, their capabilities in validating specific results via Western blotting, co-immunoprecipitation, etc., were limited and required collaboration with another lab. In most cases the proteomic findings would be published with the hope that some other group would have interest in further pursuing the results. This model was inefficient and duplicative and prevented many useful results from being validated.

Over time, this model has started to change. Recognizing the need for additional bioanalytical technologies and specialties, proteomic laboratories started to expand in breadth or participated in multi-institutional collaborations to ensure discovery-driven results could be validated. Beyond the requisite MS and bioinformatic technologies, these collaborations included labs that specialized in basic biochemical methods, as well as verification and validation testing. The aforementioned study is a fantastic example of this new multi-institutional collaborative model. The study generated data using AP-MS, phosphoproteomics profiling, database searching, pharmacological screening, and other techniques (e.g., exome sequencing, functional genetics, etc.) to identify druggable human targets that could inhibit influenza infection and propagation. While a HDT was not validated, this study provided the basis for pursuing potential therapies that could be broadly used against multiple viral strains by targeting the host rather than the specific virus. Maybe even more important than these results, the study may have set the standard for what level of organization and collaboration is going to be necessary for proteomics to reach the next level in its pursuit of revolutionizing global health and personalized medicine.

## 9.5 The Role of Adjuvants in Vaccination

When coadministered with a vaccine antigen, adjuvants enhance the immunogenicity of the vaccine (Pulendran et al. 2011). Adjuvants were first used by Alexander Glenny in 1926 who showed that mixing an antigen with aluminum salts induced more antibodies in an inoculated guinea pig (Glenny et al. 1926). Two decades later, water-in-oil emulsions were developed by Freund and his colleagues (Biehl and Vilter 1982). While these Freund adjuvants and bacterial lipopolysaccharides (LPS) adjuvants enhanced immunogenicity, only aluminum adjuvants were licensed for human used through the 1990s (Ko and Kang 2018). Four other nonaluminum-based adjuvants (AS04, AS03, AS01, and CpG ODN 1018) were licensed for human use over the following two decades (Iwasaki and Omer 2020). Since this time several other classes of compounds have shown promise as adjuvants including mineral salts, microbial products, emulsions, saponins, synthetic small-molecule agonists, polymers, nanoparticles, and liposomes (O'Hagan and Valiante 2003).

While the ability of adjuvants to enhance the immunogenicity and persistence of a vaccine is undisputed, their mechanisms of action are not fully understood. Adjuvants are believed to enhance the adaptive immunity of vaccines by activating innate immune cells (Pulendran et al. 2011). This activation begins with adjuvants activating antigen-presenting cells (APCs), which promotes the generation of antigen presentation signals (signal 1) and co-stimulatory signals (signal 2) by activating APCs (see Fig. 9.5). After taking up the vaccine and processing it into small antigenic fragments, an antigen is presented on major histocompatibility complexes on the APC surface. The co-stimulatory molecules arising from the APC include

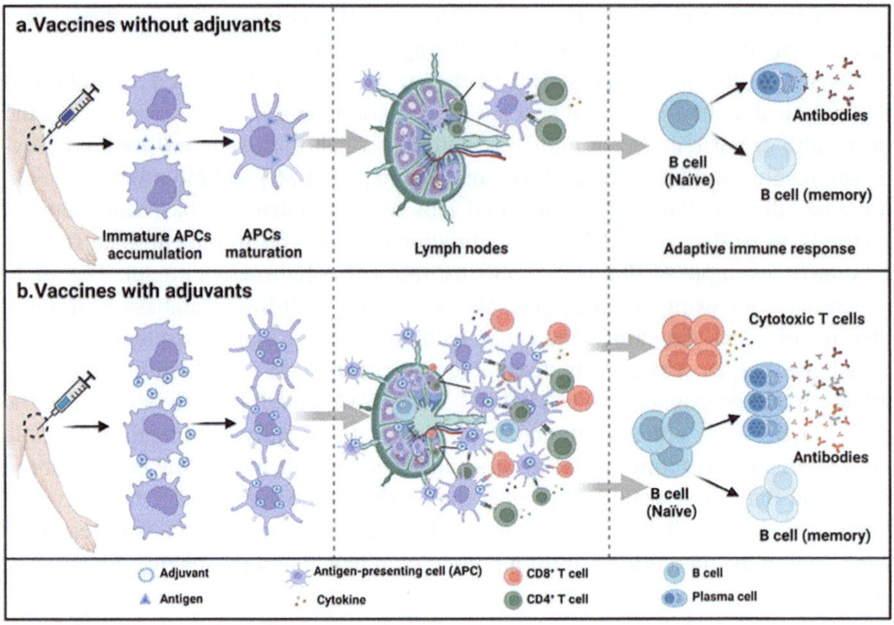


**Fig. 9.5** Adjuvant mechanism of action of enhancement of vaccine immunogenicity. (**a**) Vaccines without adjuvants induce modest production of T helper-polarizing cytokines, antibodies, and activated T cells. (**b**) In contrast, vaccines with adjuvants promote the maturation of more APCs, increase the interaction between APCs and T cells, promote the production of greater numbers and more types of T helper-polarizing cytokines, multifunctional T cells, and antibodies, leading to broad and durable immunity, as well as dose and antigen savings. (Reproduced from ref. (Pradere et al. 2014) under CC BY 4.0)

those that are presented on its surface (e.g., CD40, CD80, CD86, etc.) or secreted (e.g., IL-6, IL-10, IL-12, TNF-α, etc.). These co-stimulatory signals enhance the adaptive immune response through their activation of naïve T cells (Janeway 1989; Pollard and Bijker 2021). Pathogen-associated molecular patterns (PAMPs), damage-associated molecular patterns (DAMPs), and small-molecule agonist of toll-like receptors (TLRs) act as immunostimulants to enhance antigen presentation and co-stimulatory signals production by APCs (Pradere et al. 2014).

Adjuvants are often classified as immunostimulants or delivery systems based on their mechanism of action. Immunostimulant adjuvants' main mechanism of action is to enhance the production of danger signal molecules that bind to specific receptors on APCs resulting in their differentiation and activation (Zhao et al. 2023). This activation causes APCs to cease their phagocytic activity and enhance their antigen presentation capabilities, which results in a concomitant expression of high levels of co-stimulatory signals and cytokines. This enhanced expression of co-stimulatory signals and cytokines not only leads to greater adaptive immune responses but can activate different responses depending on the DAMPs and PAMPs which are released and interact with specific pattern recognition receptors.

Delivery system adjuvants are carrier materials that are loaded with antigen (Bachmann and Jennings 2010). Their purpose is to facilitate antigen presentation by increasing the uptake and presentation of antigens by APCs. As described earlier, antigen presentation involves recognition and internalization of antigens, followed by their loading and presentation by MHCs on the surface of the APC. Delivery systems increase the presentation of antigens on the APC's surface via multiple factors including prolonging the bioavailability of antigens and targeting APCs. This prolonged availability is accomplished via three methods. Injectable hydrogel adjuvant delivery systems provide a sustained release of antigens that allow APCs to continuously present antigens over a longer time (Mohammadzadeh et al. 2025). This elongated time enables B cells to undergo multiple rounds of affinity maturation, which leads to both a larger quantity of antigen-specific antibodies and antibodies with greater affinity for their target. Some delivery systems form antigen/immune niches at the injection site, which results in higher immune activation (Meany et al. 2025). These immune niches are a result of the vaccine forming a localized microenvironment at the site of injection that actively recruits, concentrates, and activates key immune cells such as dendritic cells and macrophages. The recruitment of these APCs increases the uptake of antigens and generates localized inflammatory signals that activate downstream immune responses. The use of vaccine delivery systems that form immune niches has a greater immunogenicity than those that only result in sustained antigen release. Some delivery systems have even incorporated specific cytokines, such as granulocyte-macrophage colony-stimulating factor (GM-CSF), to increase immune cell migration to the site of injection. Some adjuvants used in delivery systems also enhance innate immune cell infiltration into the immune niche. The final method by which adjuvant delivery systems prolong antigen bioavailability and immunogenicity is the protection of the vaccine from enzymatic degradation (Zhao et al. 2023). The most obvious application is the protection of naturally unstable DNA- and RNA-based vaccines. The most popular example is the lipid nanoparticles (LNPs) used to deliver the mRNA COVID-19 vaccines. The LNPs were arguably the major technological development that enabled the rapid, large-scale production of these vaccines, saving countless lives. The net result of any of these methods that increase the vaccines availability is an increase in the number of MHC-antigen signals available for activating cells of the adaptive immune system.

### *9.5.1 Tissue Proteomics to Understand Adjuvant Mechanism*

Recent studies have shown that nonphysical techniques such as lasers and radiofrequency adjuvants (RFAs) can enhance intradermal vaccine-induced immune responses (Li et al. 2022a; Chen 2023). Radiofrequency adjuvants apply radiofrequency (RF) energy to the skin over a short period of time (i.e., 1–2 min). While the exact mechanisms are not known, it is believed that application of RF energy creates thermal stress in the area (Li et al. 2022b). This treatment results in the release of endogenous danger signals, such as heat shock proteins, which enhance immune cell activity. Since it operates within the RF range of the electromagnetic spectrum, RFA is safe to use and does not cause visible or histological damage to the skin (Li et al. 2021).

Compared to the squalene-based oil-in-water nano-emulsion AddaVax, RFA has been shown to provide comparable potency when administered in combination with the pandemic 2009 H1N1 vaccine (94). RFA also significantly enhanced immune responses to influenza nucleoprotein, intradermal ovalbumin, and recombinant hemagglutinin antigen vaccines. It was also able to increase the overall immune response to the low immunogenic H5N1 vaccine (Li et al. 2021; Cao et al. 2018). While chemical adjuvants such as AddaVax, Alum, and MPL cause persistent and intense local inflammation, RFAs only cause low-level, transient inflammation (Cao et al. 2018).

## *9.5.2 Mechanism of Action of Adjuvants*

Improvement of any adjuvant requires understanding their mechanism of action. While various omic technologies have been applied to understand the mechanism of vaccine-induced immune responses, the study of adjuvant models has been almost exclusively limited to transcriptomics. A study of alum, CpG, and MF59 adjuvants showed that these additives induced well over 100 genes (Olafsdottir et al. 2015). Unfortunately, omic studies over the past 25 years have consistently demonstrated a lack of correlation between mRNA and protein levels (Gygi et al. 1999; Payne 2015). To truly understand an adjuvant's mechanism of action requires direct measurement of changes at the protein level.

Changes in the proteome of tissues that were treated with RFA were recently analyzed using sequential windowed acquisition of all theoretical fragment ion mass spectra (SWATH-MS) (see Fig. 9.6). SWATH-MS is a peptide identification method based on the use of LC/MS$^2$ (Gillet et al. 2012; Ludwig et al. 2018). Like a typical LC/MS$^2$ experiment, SWATH-MS begins with the digestion of a proteome sample into peptides using a proteolytic enzyme such as trypsin. The peptides are separated using liquid chromatography coupled to a tandem mass spectrometer. Unlike a typical LC/MS$^2$ experiment in which the mass spectrometer is operated in a data-dependent acquisition (DDA) mode, a SWATH-MS experiment operates the mass spectrometer in a data-*independent* acquisition (DIA) mode. In the conventional DDA mode, peptides are sequentially and individually isolated and

**Fig. 9.6** (continued) separation at a given time point. In this example, the mass range from *m/z* 925 to 950, three co-eluting peptide species are detected (green, red, and blue) in this range, and two are detected in the *m/z* range 1175–1200. (**d**) The MS$^2$ scans obtained using an isolation window of *m/z* 925–950 is a combination of the MS$^2$ spectrum obtained from all three peptides. (Reproduced from ref. (Ludwig et al. 2018) under the terms and conditions of the Creative Commons Attribution (CC BY) license (https://creativecommons.org/licenses/by/4.0/))

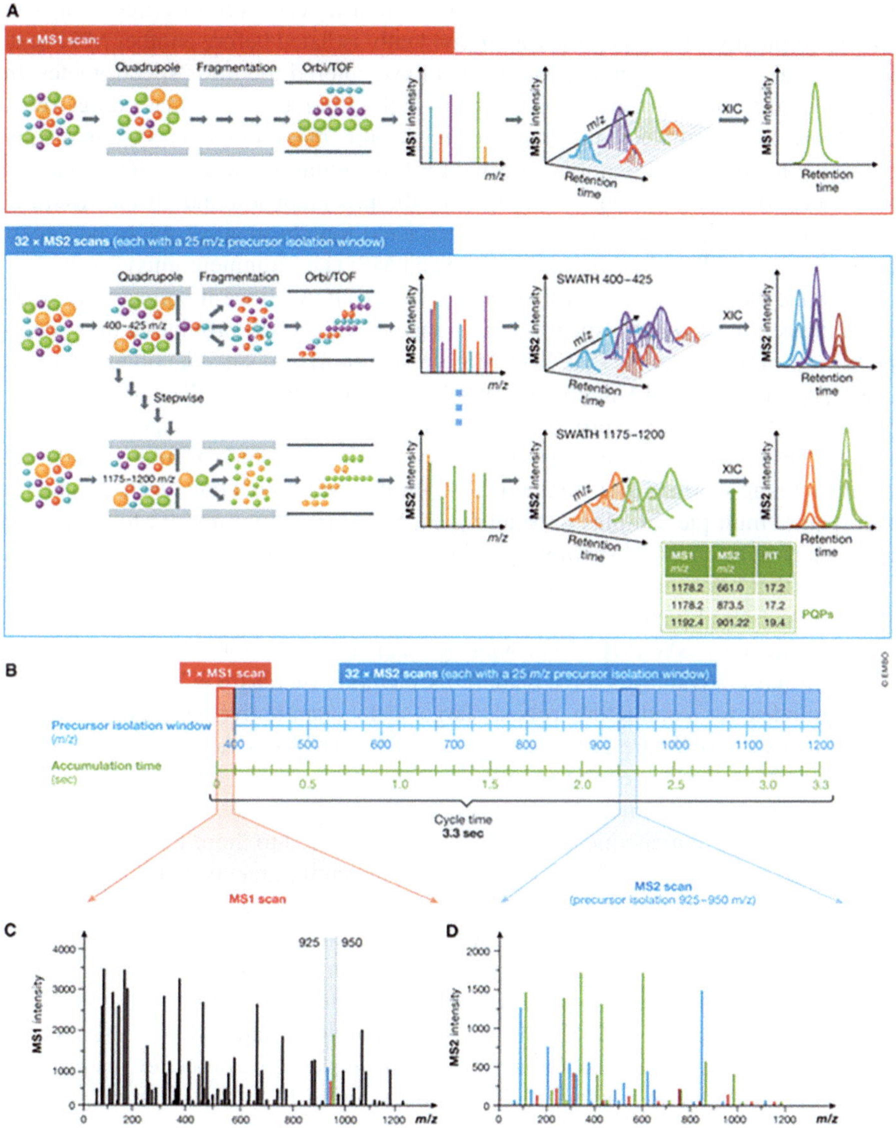


**Fig. 9.6** (**a**) Sequential window acquisition of all theoretical mass spectra (SWATH-MS) measurements are performed using fast scanning hybrid mass spectrometers, typically combining a quadrupole as the first mass analyzer and a high-resolution Orbitrap or time of flight (TOF) as second mass analyzer. In a typical SWATH-MS analysis, a single precursor ion ($MS^1$) spectrum is recorded, followed by the acquisition of several tandem MS ($MS^2$) spectra within a wide precursor isolation window (e.g., 25 *m/z* in this example). By cycling through a series of precursor isolation windows, a comprehensive dataset including information on the precursor ions and their fragments is recorded. This data allows extracted ion chromatograms (XIC) to be generated from both $MS^2$ and $MS^1$ data. (**b**) In the example shown, SWATH-MS data is acquired using *m/z* ranges of 25 (**c**). The $MS^1$ scan detects all peptides that elute into the mass spectrometer via a liquid chromatography

fragmentated for identification based on their intensity observed in a precursor scan (MS1). Typically, 10–25 peptides are individually isolated before another MS1 scan is performed to look for new peptides that have entered the mass spectrometer. In the DIA mode, all the ionized peptides that fall within a specific mass range (e.g., 25–50 *m/z*) are isolated and fragmented. As the mass range is systematically changed, the number of peptides selected for fragmentation increases. The data collected includes information on the mass of the precursor ion and all the resulting fragment ions. This information allows extracted ion chromatograms to be generated at both the MS1 (i.e., precursor ions) and $MS^2$ (i.e., fragment ions) levels.

SWATH-MS provides several advantages over DDA. Since it collects $MS^2$ data of all the peptides within a specific mass range, SWATH-MS provides a more complete view of the proteome. SWATH-MS is also non-biased since it collects data for every peptide unlike DIA methods that select peptides based on their intensity. The SWATH-MS run-to-run reproducibility is much higher since every peptide is analyzed. The intensity criteria for being selected in a DIA study results in different peptides being isolated and fragmented when the sample is reanalyzed. In studies that require multiple samples to be analyzed, SWATH-MS is superior to DIA since it acquires data on all observable peptides. One of the major limitations of DIA methods (that has been mentioned several times throughout this book) is the inconsistent selection of peptides across different samples. Since peptide selection is not based on intensity, SWATH-MS provides greater sensitivity for low abundance proteins. All these advantages have made DIA methods such as SWATH-MS the preferred modes for conducting robust, large-scale quantitative proteomic studies. While DDA methods are still widely used, SWATH-MS is rapidly becoming the preferred choice for proteomic analysis.

To compare proteomic changes induced by RFA, tissue samples from mice treated with different chemical adjuvants were compared to those treated with RFA (see Fig. 9.7) (Li et al. 2022a). For chemical adjuvant treatments, hair was removed from the lateral dorsal skin of the mice, and they were intradermally injected with solutions containing AddaVax, alum, MPL, or a combination MPL/alum adjuvant. For RFA treatment, an RF device equipped with an array of 144 microelectrodes covering a $2 \times 2$ cm$^2$ area was used. For RF treatment, ultrasound coupling medium was applied to the skin, and RF pulses were applied to this area for 1–2 min. A control group of mice was included that followed the exact same procedure as the RFA except the device was not activated. Skin was harvested 18 h after treatment with the different adjuvants. Proteins extracted from the tissues were digested with trypsin and analyzed using SWATH-MS with sequential window sizes of *m/z*.

The number of proteins identified in each group ranged from 1521 to 1953 (see Fig. 9.8). Further analysis revealed that over 25% of the proteins identified in the AddaVax treated tissue showed a greater than 30% change in abundance compared to the control tissue. The percentage of proteins that showed the same relative change in alum-, MPL-, and MPL/alum-treated mice ranged between 10.7% and 13.4%. Only 6.6% of the proteins identified in the RFA-treated tissue showed a 30% abundance change. These results show that RFA had the least impact on the tissue proteome.

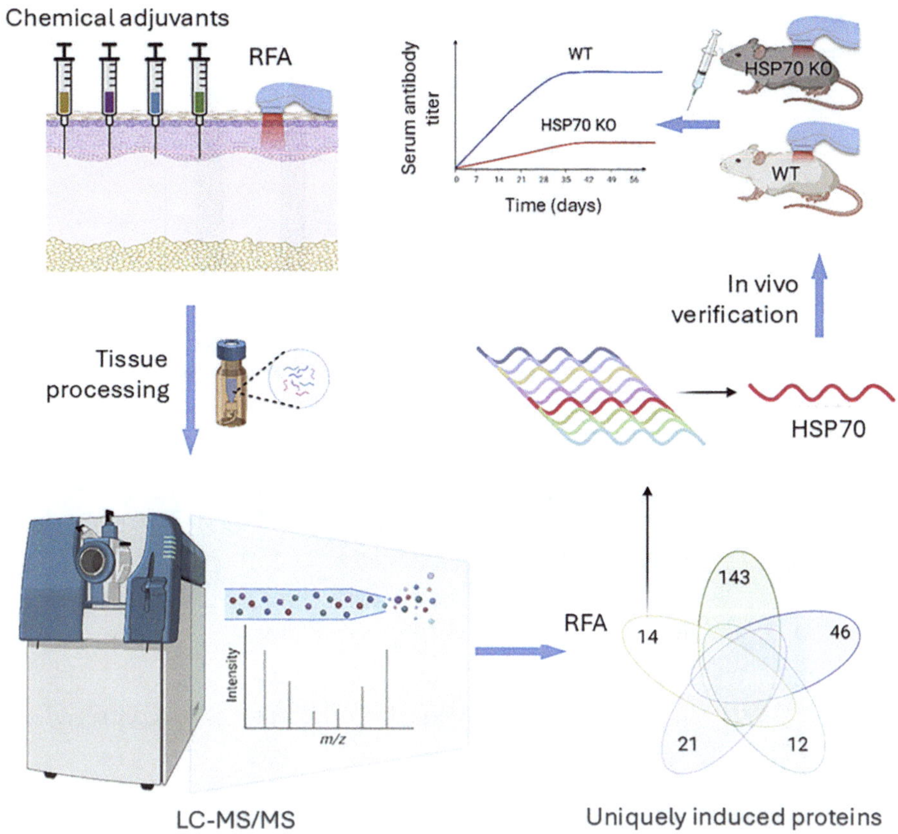


**Fig. 9.7** Proteomic comparison of mice treated with various adjuvants. In the described study, mice were treated with radiofrequency adjuvant and four different chemical adjuvants (AddaVax, alum, MPL, and MPL/alum). The proteome was extracted from skin samples taken from the treatment sites and analyzed using SWATH-MS to identify proteins unique to each treatment. In a separate analysis, wild-type and HSP70 knockout (KO) mice were injected with ovalbumin along with RFA treatment and their anti-ovalbumin antibody titer measured 3 weeks after initial inoculation and 3 weeks after ovalbumin booster. (Reproduced from ref. (Li et al. 2022a) under the terms and conditions of the CC BY-NC-ND license (http://creativecommons.org/licenses/by-nc-nd/4.0/))

The differentially expressed proteins in each treatment were analyzed using Gene Ontology. Most of the proteins across all samples could be classified into four major biological processes: cellular processes, metabolic processes, biological regulation, and response to stimuli. While different adjuvants are expected to activate different proteins, all (both chemical and RFA) adjuvants stimulated *inter-alpha-trypsin inhibitor heavy change H2* (*Itih2*) and *alpha-2-antiplasmin* (also known as *Serpinf2*) expression. Itih2 belongs to a family of plasma protease inhibitors, while Serpinf2 is a serine protease inhibitor that specifically inactivates plasmin. While there was little overlap in the proteins whose abundance was affected by both RFA and the chemical adjuvants, the expression of seven proteins were downregulated by

A

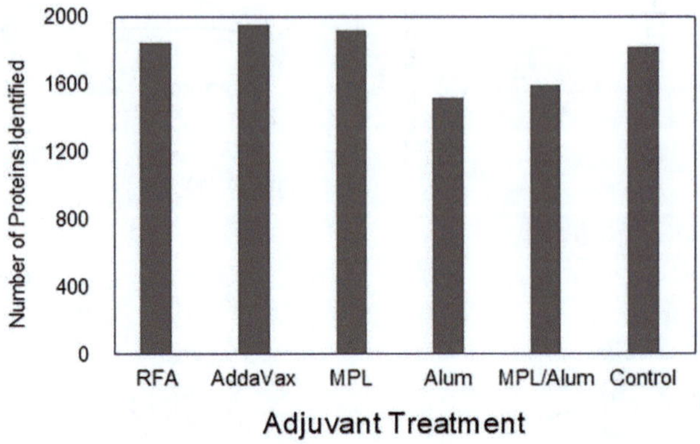


B

| Adjuvant Treatment | Proteins Increased in Abundance | Proteins Decreased in Abundance | Total Number | Total Percentage |
|---|---|---|---|---|
| RFA | 52 | 70 | 122 | 6.6 |
| Alum | 60 | 103 | 163 | 10.7 |
| MPL/Alum | 96 | 118 | 214 | 13.4 |
| MPL | 204 | 36 | 240 | 12.5 |
| AddaVax | 323 | 170 | 493 | 25.2 |

**Fig. 9.8** (**a**) Bar graph comparing the total number of protein identified in tissue extracted from mice subjected to each adjuvant treatment. (**b**) Table showing the number of proteins whose abundance levels changed by at least 30% when compared to the proteome of Sham-treated mice

all four chemical adjuvants. Included within these seven were actin alpha 1 (Acta1), palladin (Palld), and pyridoxal kinase (Pdxk). Acta1 is an actin isoform that is a member of the tissue contractile apparatus (Feng and Marston 2009). Palld controls cell shape, adhesion, and contraction through its interaction with actin microfilaments (Mykkänen et al. 2001). Pdxk is a kinase that phosphorylates vitamin B6 into its active form, pyridoxal 5′-phosphate (PLP). PLP is an essential cofactor for over 140 enzymatic reactions including amino acid metabolism and neurotransmitter synthesis (Parra et al. 2018). These findings show that there are at least some similarities in the mechanism of action of all the adjuvants tested in this study.

There were 12 proteins that demonstrated an increase in abundance when induced by all 4 chemical adjuvants (Li et al. 2022a). Among the 12 proteins, at least 5 proteins have important roles in inflammation and immunity. These include chitinase-like protein 3, which acts like a lectin to activate macrophage, promote eosinophil chemotaxis, influence B-cell antibody production, and elicit both Th1 and Th2 immune responses (Zhao et al. 2020). Cathelicidin is an antimicrobial peptide that when released from lysosomes of macrophages and polymorphonuclear leukocytes (PMNs) directly kills microbes (Kościuczuk et al. 2012). Lipocalin 2 is secreted by immune cells and binds to and sequesters siderophores (i.e., iron-containing molecules) that are secreted by bacteria and are required for their survival (Wilson et al. 2016). H-2 class I histocompatibility antigen Q10 alpha chain serves to present antigens to the immune system (Sullivan et al. 2016). Mitogen-activated protein kinase 14 (Mapk14) is activated during stress and via cytokine signaling (Lo et al. 2014). It plays a major role in proliferation and differentiation of various immune cells including macrophages, neutrophils, microglia, and T cells.

### 9.5.3 Proteins Unique to Radiofrequency Adjuvant Treatment

The focus of the study was to evaluate the mechanism of RFA treatment on host cells. To this effect, 14 proteins were found to be uniquely induced by RFA treatment (see Table 9.3). Discovery-based proteomic studies of this nature routinely identify several protein targets to which hypotheses can be generated. The focus of this study turned to heat shock protein 70-2 (HSP70.2), whose increase in abundance due to RFA treatment was confirmed using Western blotting. RFA treatment, however, did not significantly affect the expression of other heat shock proteins.

The role of HSP70 in RFA effects was tested by immunizing wild-type (WT) and HSP70 knockout (KO) mice with ovalbumin (OVA) in the presence of RFA or Sham

**Table 9.3** Proteins whose expression was affected by radiofrequency adjuvant treatment but not treatment with chemical adjuvants

| Protein | Gene | Function |
|---|---|---|
| Heat shock protein 70-2 | *Hspa1b* | Molecular chaperone |
| NFU1 iron-sulfur cluster scaffold | *Nfu1* | Iron-sulfur scaffold protein |
| Leucine-rich repeat flightless--interacting protein 2 | *Lrrfip2* | Modulates inflammatory response and Wnt signaling |
| Reticulocalbin 1 | *Rcn1* | Regulates endoplasmic reticulum $Ca^{2+}$ levels and stress |
| Set | *Set* | Regulates gene transcription, cell division, and cell mobility |
| Versican | *Vcan* | Extracellular matrix remodeling |
| Tenascin-X | *Tnxb* | Organizes and maintains connective tissue |
| Dead box protein 6 | *Ddx6* | RNA helicase |
| Polysaccharide biosynthesis domain containing 1 | *Pbdc1* | Polysaccharide synthesis |
| Thioredoxin 2 | *Txn2* | Oxidative damage protection |
| EF-hand domain-containing protein D2 | *Efhd2* | Modulate B-cell activity and T-cell signaling |
| Nucleobindin 2 | *Nucb2* | Appetite and glucose homeostasis |
| Serine/arginine-rich splicing factor | *Srsf1* | mRNA stability, export, and translation |
| Filaggrin-2 | *Flg2* | Maintain function of skin barrier |

treatment (see Fig. 9.7). The mice were inoculated, and then a booster shot was given 3 weeks later. Anti-OVA antibody titers were measured 3 weeks after the initial inoculation and then 3 weeks after the booster shot. A significantly higher anti-OVA titer was measured in WT mice that were treated with RFA compared to Sham treatment. In KO mice, however, no significant difference in anti-OVA levels was observed in RFA- and Sham-treated animals after either the initial or booster inoculation. The only viable conclusion was that HSP70.2 plays a crucial role in modulating RFA's effect on OVA immunization.

Since individuals are inoculated with trace amounts of OVA when they receive some egg-based vaccines (Abrams et al. 2024), the more important question is whether HSP70.2 plays a role in the ability of RFA to boost the immunogenicity of the influenza vaccine. Both WT and KO mice were inoculated with the pdm009 vaccine and subjected to either RFA or Sham treatment. A significant increase in pdm009 vaccine-induced hemagglutinin inhibition (HI) and anti-recombinant hemagglutinin antigen (rHA) IgG titer was observed in WT mice subjected to RFA treatment but not in HSP70.2 KO mice Meany et al. 2025. Mice were then given a lethal viral challenge (i.e., ten times the LD50 for the pdm009 virus). A majority of WT and HSP70.2 KO mice inoculated with a lethal pdm009 dose in the Sham group either died or lost at least 25% of their body weight within 9 days (at which point they were humanely euthanized). In contrast, vaccination with pdm009 along with RFA treatment conferred significantly greater protection against body weight loss in WT mice compared to HSP70.2 KO mice. Pdm009-vaccinated and RFA-treated WT mice only lost about 8% of their body weight after 8 days, which was regained by day 10. HSP70.2 KO mice that were vaccinated with pdm009 and treated with RFA lost 16% of their body weight by day 7 and required another 7 days to recover this lost weight. The survival rate of WT mice treated with RFA was significantly increased, but RFA had no effect of survival in HSP70.2 KO mice. While all WT mice in Sham groups died within 10 days, approximately two-thirds of WT mice survived inoculation with pdm009. In HSP70.2 KO mice that were inoculated with a lethal dose of pdm009, 30% of those treated with RFA survived compared to only 16% in the Sham group. Overall, the results demonstrated the effectiveness of RFA treatment along with the crucial role of HSP70.2 in increasing vaccine immunogenicity.

### *9.5.4 Impact on Understanding the Mechanism of Action of Adjuvants*

While the focus of this study was to demonstrate the effectiveness of RFA in increasing vaccine immunogenicity, its greater contribution lies in the use of proteomics to identify proteins that are regulated by adjuvants. Unfortunately, little progress has been made over the past couple of decades in deciphering the mechanism of action of commonly used adjuvants. While the downstream effect of adjuvants such as alum, MF59, and MPL on immune cells is known (Israeli et al. 2009), the genetic and protein pathways that modulate their effects on cells are not well understood (Zhao et al. 2023; Pulendran et al. 2021). This study was one of the first to compare tissues from groups treated with various types and classes of adjuvants to identify proteins that play regulate their effect. There were 12 proteins that were induced by all chemical adjuvants tested—a result that has been corroborated at the mRNA level. So how can comparative tissue proteomics be utilized in the future? As greater amounts of data are collected, a core group of proteins that regulate adjuvant mechanisms continues to be recognized. This data enables two potentially valuable avenues for optimizing adjuvant effectiveness. Firstly, proper pairing of vaccines and adjuvants can maximize the vaccines immunogenicity. Secondly, and admittedly a loftier goal, would be engineering improved next-generation adjuvants. Using molecular design, adjuvants could be engineered to enhance vaccine effectiveness. Enhancing adjuvant efficacy would require specifically activating components of the immune system. Some possible strategies would include engineering molecules that target innate immune receptors, such as toll-like receptors (TLRs) and retinoic acid-inducible gene 1 (RIG-1), are intrinsically immunostimulatory, or form more effective antigen delivery systems. While aluminum salts and oil-in-water emulsions are the most used adjuvants molecules (Facciolà et al. 2022), designer adjuvants could integrate state-of-the-art materials such as nanoparticles, quantum dots, metal-organic frameworks, metamaterials, etc. As tissue proteomics continues to mature, it can be used to decipher the mechanisms of action of designer adjuvants to optimize their efficacy.

## 9.6 Conclusions

While it may have been overshadowed in recent years by the SARS-CoV-2 pandemic, influenza remains a constant threat to human life. While influenza activity generally peaks between December to February in the USA, the timing of its impact can vary year to year (see Fig. 9.9). The months of December, January, and February have collectively accounted for 33 of the peak influenza months over the past 42 years; however, peak months have come as early as October and as late as April. Most people infected with influenza die from pneumonia, which was the 12th leading cause of death in the USA between the years 2018 and 2023 (https://wonder.cdc.gov/controller/datarequest/D158;jsessionid=31F4CB2777713D3A244AC82A1995). A disconcerting fact is that since 2021–2022, fewer Americans are getting vaccination against influenza. In the 2024–2025 season, approximately 28 million fewer doses of influenza vaccine were distributed across the USA than just 3 years previously (https://www.cdc.gov/fluvaxview/dashboard/vaccine-doses-distributed.html).

The aims of this book are twofold. To the proteomic scientist, it aims to provide a simplified description of the structure and mechanism of action of specific viruses. To the virologists, it aims to stimulate ideas on how proteomics can further the

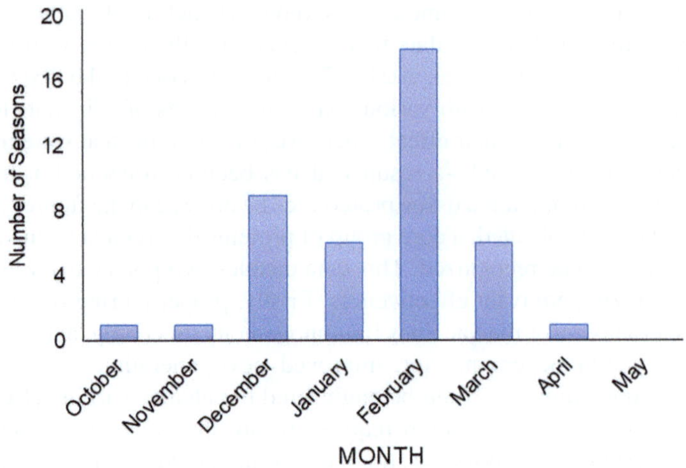


**Fig. 9.9** Influenza activity peaks between October and May in the USA. The graph shows that the highest influenza activity occurred more often in February than in any other month

understanding of how viruses impact cells at the molecular level, aid in the diagnosis of viral infection, and contribute to vaccine development. This chapter fulfils both aims. While the influenza virus is not unique in being comprised of a variety of different strains, the consistent circulation of different strains contributes to the need for yearly updates of the influenza vaccine. These variations, known as antigenic drift, can lead to a mismatch between the circulating strain and the vaccine strain, which reduces the efficacy of the yearly influenza vaccine. This chapter describes proteomic studies that accumulated data with the goal of developing enhanced methods to prevent or treat influenza infections across all seasons. In the first study the aim is to recognize common host machinery affected by different influenza strains with the hope that a host-directed therapy (HDT) could be developed (Haas et al. 2023). By targeting a commonly utilized host protein, the HDT would inhibit infection by all common influenza strains. An HDT could be consistently used year over year, eliminating the current mismatches that occur between vaccines and viral strains. If an infection occurred, another type of HDT could be used to prevent the spread of infection. By eliminating the strain-based uncertainty in preventing and treating influenza virus infection and transmission, these HDTs would have the added benefit of increasing vaccine participation rates.

Another study described in this chapter did not strictly focus on influenza or the action of any virus but rather on the mechanism of action of adjuvants (Li et al. 2022a). The roles of adjuvants in vaccine efficacy are often overlooked. Adjuvants can enhance the immune response, improve vaccine efficacy, reduce the antigen requirement, target the stimulation of specific immune components, and boost a vaccine's immunogenicity in vulnerable populations. Although they have been used in vaccines for over 100 years (Di Pasquale et al. 2015), the mechanism of action of adjuvants is still poorly understood. The study described in this chapter used a comprehensive proteomic approach to better understand host proteins that were critical to adjuvant vaccine enhancement. By understanding these mechanisms, it opens the possibilities of engineering molecules that could enhance adjuvant efficacy even further.

There is one very important quality in the studies discussed in this chapter that illustrates the progress made in proteomics since the term was popularized in 1994 (Wilkins et al. 1996). For many years, proteomic studies encompassed the collection and reporting of data, describing changes in proteins in different systems. The hope was that basic research labs would seize upon the data and further explore it and develop products that could impact human health (e.g., diagnostic biomarkers, therapeutic agents, prognostic indicators, etc.). The studies described here show that the bar has been raised for proteomics. Proteomic studies have gone from simply listing differentially expressed proteins to incorporating studies that use a variety of molecular biology techniques to validate important findings. While proteomic studies used to simply be the discovery-driven component that generated hypotheses that would be tested by others, they have now evolved into being both the discovery and hypothesis-driven arms of research projects.

## References


Abrams EM, Zafack JG, Jensen C, Bettinger JA, Top KA, Hildebrand KJ et al (2024) Vaccination and egg allergy: considerations for family physicians. Can Fam Physician 70:10–12

Bachmann MF, Jennings GT (2010) Vaccine delivery: a matter of size, geometry, kinetics and molecular patterns. Nat Rev Immunol 10:787–796

Baldo V, Bertoncello C, Cocchio S, Fonzo M, Pillon P, Buja A et al (2016) The new pandemic influenza A/(H1N1)pdm09 virus: is it really "new"? J Prev Med Hyg 57:E19–E22

Benaud C, Gentil BJ, Assard N, Court M, Garin J, Delphin C et al (2004) AHNAK interaction with the annexin 2/S100A10 complex regulates cell membrane cytoarchitecture. J Cell Biol 164:133–144

Betakova T, Nermut MV, Hay AJ (1996) The NB protein is an integral component of the membrane of influenza B virus. J Gen Virol 77:2689–2694

Bhadoria P, Gupta G, Agarwal A (2021) Viral pandemics in the past two decades: an overview. J Family Med Prim Care 10:2745–2750

Biehl JP, Vilter RW (1982) Proceedings of the society for experimental biology and medicine. Nutr Rev 40:183–186

Bouhaddou M, Memon D, Meyer B, White KM, Rezelj VV, Correa Marrero M et al (2020) The global phosphorylation landscape of SARS-CoV-2 infection. Cell 182:685–712.e19

Bouvier NM, Palese P (2008) The biology of influenza viruses. Vaccine 26(Suppl 4):D49–D53

Boytz R, Słabicki M, Ramaswamy S, Patten JJ, Zou C, Meng C et al (2023) Anti-SARS-CoV-2 activity of targeted kinase inhibitors: repurposing clinically available drugs for COVID-19 therapy. J Med Virol 95:e28157

Briedis DJ, Lamb RA (1982) Influenza B virus genome: sequences and structural organization of RNA segment 8 and the mRNAs coding for the NS1 and NS2 proteins. J Virol 42:186–193

Butler D (2009) How severe will the flu outbreak be? Epidemiologists race to pin numbers on the global H1N1 spread. Nature 459:14–15

Camfield S, Chakraborty S, Dwivedi SKD, Pramanik PK, Mukherjee P, Bhattacharya R (2024) Secrets of DNA-PKCs beyond DNA repair. NPJ Precis Oncol 8:154

Cao Y, Zhu X, Hossen MN, Kakar P, Zhao Y, Chen X (2018) Augmentation of vaccine-induced humoral and cellular immunity by a physical radiofrequency adjuvant. Nat Commun 9:3695
Castello A, Álvarez L, Kamel W, Iselin L, Hennig J (2024) Exploring the expanding universe of host-virus interactions mediated by viral RNA. Mol Cell 84:3706–3721
Chen X (2023) Emerging adjuvants for intradermal vaccination. Int J Pharm 632:122559
Clancy S (2008) Genetics of the influenza virus. Nat Educ 1:83
Cros JF, Palese P (2003) Trafficking of viral genomic RNA into and out of the nucleus: influenza, Thogoto and Borna disease viruses. Virus Res 95:3–12
Dangi T, Jain A (2012) Influenza virus: a brief overview. Proc Natl Acad Sci India Sect B Biol Sci 82:111–121
Dawood FS, Iuliano AD, Reed C, Meltzer MI, Shay DK, Cheng PY et al (2012) Estimated global mortality associated with the first 12 months of 2009 pandemic influenza A H1N1 virus circulation: a modelling study. Lancet Infect Dis 12:687–695
Deinhardt-Emmer S, Jäckel L, Häring C, Böttcher S, Wilden JJ, Glück B et al (2021) Inhibition of phosphatidylinositol 3-kinase by pictilisib blocks influenza virus propagation in cells and in lungs of infected Mice. Biomolecules 11:808
Di Pasquale A, Preiss S, Tavares Da Silva F, Garçon N (2015) Vaccine adjuvants: from 1920 to 2015 and beyond. Vaccines (Basel) 3:320–343
Dias RA (2024) Towards a comprehensive definition of pandemics and strategies for prevention: a historical review and future perspectives. Microorganisms 12:1802–1840
Dou D, Revol R, Östbye H, Wang H, Daniels R (2018) Influenza A virus cell entry, replication, virion assembly and movement. Front Immunol 9:1581
Drancourt M, Raoult D (2016) Molecular history of plague. Clin Microbiol Infect 22:911–915
Dubois J, Traversier A, Julien T, Padey B, Lina B, Bourdon JC et al (2019) The nonstructural NS1 protein of influenza viruses modulates *TP53* splicing through host factor CPSF4. J Virol 93:e02168–e02118
Eichberg J, Maiworm E, Oberpaul M, Czudai-Matwich V, Lüddecke T, Vilcinskas A et al (2022) Antiviral potential of natural resources against influenza virus infections. Viruses 14:2452
Els MC, Laver WG, Air GM (1989) Sialic acid is cleaved from glycoconjugates at the cell surface when influenza virus neuraminidases are expressed from recombinant vaccinia viruses. Virology 170:346–351
Facciolà A, Visalli G, Laganà A, Di Pietro A (2022) An overview of vaccine adjuvants: current evidence and future perspectives. Vaccines (Basel) 10:819
Feng JJ, Marston S (2009) Genotype-phenotype correlations in ACTA1 mutations that cause congenital myopathies. Neuromuscul Disord 19:6–16
Fiore M, Forli S, Manetti F (2016) Targeting mitogen-activated protein kinase-activated protein kinase 2 (MAPKAPK2, MK2): medicinal chemistry efforts to lead small molecule inhibitors to clinical Trials. J Med Chem 59:3609–3634
Gillet LC, Navarro P, Tate S, Röst H, Selevsek N, Reiter L et al (2012) Targeted data extraction of the MS/MS spectra generated by data-independent acquisition: a new concept for consistent and accurate proteome analysis. Mol Cell Proteomics 11(O111):016717
Glenny A, Pope C, Waddington H, Wallace U (1926) Immunological notes. xvii–xxiv. J Pathol Bacteriol 29:31–40
Guinea R, Carrasco L (1995) Requirement for vacuolar proton-ATPase activity during entry of influenza virus into cells. J Virol 69:2306–2312
Gygi SP, Rochon Y, Franza BR, Aebersold R (1999) Correlation between protein and mRNA abundance in yeast. Mol Cell Biol 19:1720–1730
Haas KM, McGregor MJ, Bouhaddou M, Polacco BJ, Kim EY, Nguyen TT et al (2023) Proteomic and genetic analyses of influenza A viruses identify pan-viral host targets. Nat Commun 14:6030
Hale BG, Steel J, Medina RA, Manicassamy B, Ye J, Hickman D et al (2012) Inefficient control of host gene expression by the 2009 pandemic H1N1 influenza A virus NS1 protein. J Virol 84:6909–6922

Hatta M, Goto H, Kawaoka Y (2004) Influenza B virus requires BM2 protein for replication. J Virol 78:5576–5583

https://study.com/learn/lesson/pandemics-in-history.html. Accessed Oct 2025

https://wonder.cdc.gov/controller/datarequest/D158;jsessionid=31F4CB2777713D3A244AC82A1995. Accessed Nov 2025

https://www.cdc.gov/fluvaxview/dashboard/vaccine-doses-distributed.html. Accessed Nov 2025

Hu Y, Gor V, Morikawa K, Nagata K, Kawaguchi A (2017) Cellular splicing factor UAP56 stimulates trimeric NP formation for assembly of functional influenza viral ribonucleoprotein complexes. Sci Rep 7:14053

Hussain M, Galvin HD, Haw TY, Nutsford AN, Husain M (2017) Drug resistance in influenza A virus: the epidemiology and management. Infect Drug Resist 10:121–134

Imai H, Dansako H, Ueda Y, Satoh S, Kato N (2018) Daunorubicin, a topoisomerase II poison, suppresses viral production of hepatitis B virus by inducing cGAS-dependent innate immune response. Biochem Biophys Res Commun 504:672–678

Israeli E, Agmon-Levin N, Blank M, Shoenfeld Y (2009) Adjuvants and autoimmunity. Lupus 18:1217–1225

Iwasaki A, Omer SB (2020) Why and how vaccines work. Cell 183:290–295

Janeway C (1989) Immunogenicity signals 1,2,3 … and 0. Immunol Today 10:283–286

Jolly C, Winfree S, Hansen B, Steele-Mortimer O (2014) The Annexin A2/p11 complex is required for efficient invasion of Salmonella Typhimurium in epithelial cells. Cell Microbiol 16:64–77

Kainov DE, Müller KH, Theisen LL, Anastasina M, Kaloinen M, Muller CP (2011) Differential effects of NS1 proteins of human pandemic H1N1/2009, avian highly pathogenic H5N1, and low pathogenic H5N2 influenza A viruses on cellular pre-mRNA polyadenylation and mRNA translation. J Biol Chem 286:7239–7247

Kampenusa I, Niedre-Otomere B, Trofimova J, Pole I, Pakarna G, Savicka O et al (2024) Circulation and codetections of influenza virus, SARS-CoV-2, respiratory syncytial virus, rhinovirus, adenovirus, bocavirus, and other respiratory viruses during 2022-2023 season in Latvia. Viruses 16:1650

Kang Y, Shi Y, Xu S (2023) Arbidol: the current demand, strategies, and antiviral mechanisms. Immun Inflamm Dis 11:e984

Kaur R, Batra J, Stuchlik O, Reed MS, Pohl J, Sambhara S et al (2022) Heterogeneous ribonucleoprotein A1 (hnRNPA1) interacts with the nucleoprotein of the influenza a virus and impedes virus replication. Viruses 14:199

Kędzierska H, Piekiełko-Witkowska A (2017) Splicing factors of SR and hnRNP families as regulators of apoptosis in cancer. Cancer Lett 396:53–65

Knowles JA, Golden B, Yan L, Carroll WR, Helman EE, Rosenthal EL (2011) Disruption of the AKT pathway inhibits metastasis in an orthotopic model of head and neck squamous cell carcinoma. Laryngoscope 121:2359–2365

Ko EJ, Kang SM (2018) Immunology and efficacy of MF59-adjuvanted vaccines. Hum Vaccin Immunother 14:3041–3045

Kościuczuk EM, Lisowski P, Jarczak J, Strzałkowska N, Jóźwik A, Horbańczuk J et al (2012) Cathelicidins: family of antimicrobial peptides. A review. Mol Biol Rep 39:10957–10970

Lamb RA, Choppin PW, Chanock RM, Lai CJ (1980) Mapping of the two overlapping genes for polypeptides NS1 and NS2 on RNA segment 8 of influenza virus genome. Proc Natl Acad Sci USA 77:1857–1861

Lamb RA, Lai CJ, Choppin PW (1981) Sequences of mRNAs derived from genome RNA segment 7 of influenza virus: colinear and interrupted mRNAs code for overlapping proteins. Proc Natl Acad Sci USA 78:4170–4174

Lee MF, Tham SK, Poh CL (2025) Antiviral strategies targeting enteroviruses: current advances and future directions. Viruses 17:1178

Li Y, Li Z, Zhao Y, Chen X (2021) Potentiation of recombinant NP and M1-induced cellular immune responses and protection by physical radiofrequency adjuvant. Vaccines (Basel) 9:1382

Li Y, Li Z, Chen X (2022a) Comparative tissue proteomics reveals unique action mechanisms of vaccine adjuvants. iScience 26:105800
Li Z, Kim KH, Bhatnagar N, Park BR, Jeeva S, Jung YJ et al (2022b) Physical radiofrequency adjuvant enhances immune responses to influenza H5N1 vaccination. FASEB J 36:e22182
Li S, Hu X, Ahn YM, Medina A, Ye L, Heffner A et al (2025) A quantitative high-throughput screening pipeline to identify small molecule inhibitors of Chikungunya nsP2 protease. Sci Rep 15:33479
Lin TH, Zhu X, Wang S, Zhang D, McBride R, Yu W et al (2024) A single mutation in bovine influenza H5N1 hemagglutinin switches specificity to human receptors. Science 386:1128–1134
Liu R, Sheng Z, Huang C, Wang D, Li F (2020) Influenza D virus. Curr Opin Virol 44:154–161
Lo U, Selvaraj V, Plane JM, Chechneva OV, Otsu K, Deng W (2014) p38α (MAPK14) critically regulates the immunological response and the production of specific cytokines and chemokines in astrocytes. Sci Rep 4:7405
Londino JD, Lazrak A, Noah JW, Aggarwal S, Bali V, Woodworth BA et al (2015) Influenza virus M2 targets cystic fibrosis transmembrane conductance regulator for lysosomal degradation during viral infection. FASEB J 29:2712–2725
Ludwig C, Gillet L, Rosenberger G, Amon S, Collins BC, Aebersold R (2018) Data-independent acquisition-based SWATH-MS for quantitative proteomics: a tutorial. Mol Syst Biol 14:e8126
Ma C, Li F, Musharrafieh RG, Wang J (2016) Discovery of cyclosporine A and its analogs as broad-spectrum anti-influenza drugs with a high in vitro genetic barrier of drug resistance. Antivir Res 133:62–72
Meany EL, Klich JH, Jons CK, Mao T, Chaudhary N, Utz A et al (2025) Generation of an inflammatory niche in a hydrogel depot through recruitment of key immune cells improves efficacy of mRNA vaccines. Sci Adv 11:eadr2631
Mohammadzadeh V, Atapour-Mashhad H, Shahvali S, Salehi B, Shaban M, Shirzad M et al (2025) Hydrogels as advanced drug delivery platforms for cancer immunotherapy: promising innovations and future outlook. J Nanobiotechnol 2:545
Mostafa A, Abdelwhab EM, Mettenleiter TC, Pleschka S (2018) Zoonotic potential of influenza A viruses: a comprehensive overview. Viruses 10:497–534
Muramoto Y, Noda T, Kawakami E, Akkina R, Kawaoka Y (2013) Identification of novel influenza A virus proteins translated from PA mRNA. J Virol 87:2455–2462
Mykkänen OM, Grönholm M, Rönty M, Lalowski M, Salmikangas P, Suila H et al (2001) Characterization of human palladin, a microfilament-associated protein. Mol Biol Cell 12:3060–3073
Nitiss JL (2009) Targeting DNA topoisomerase II in cancer chemotherapy. Nat Rev Cancer 9:338–350
Noda T (2021) Selective genome packaging mechanisms of influenza A viruses. Cold Spring Harb Perspect Med 11:a038497
O'Hagan DT, Valiante NM (2003) Recent advances in the discovery and delivery of vaccine adjuvants. Nat Rev Drug Discov 2:727–735
Olafsdottir T, Lindqvist M, Harandi AM (2015) Molecular signatures of vaccine adjuvants. Vaccine 33:5302–5307
Palese P, Shaw ML (2007) Orthomyxoviridae: the viruses and their replication. In: Knipe DM, Howley PM (eds) Fields virology. Lippincott Williams and Wilkins, Philadelphia
Paragas J, Talon J, O'Neill RE, Anderson DK, García-Sastre A, Palese P (2021) Influenza B and C virus NEP (NS2) proteins possess nuclear export activities. J Virol 75:7375–7383
Parra M, Stahl S, Hellmann H (2018) Vitamin $B_6$ and its role in cell metabolism and physiology. Cells 7:84
Pascale RM, Joseph C, Latte G, Evert M, Feo F, Calvisi DF (2016) DNA-PKcs: a promising therapeutic target in human hepatocellular carcinoma? DNA Repair (Amst) 47:12–20
Payne SH (2015) The utility of protein and mRNA correlation. Trends Biochem Sci 40:1–3
Pekosz A, Lamb RA (1998) Influenza C virus CM2 integral membrane glycoprotein is produced from a polypeptide precursor by cleavage of an internal signal sequence. Proc Natl Acad Sci USA 95:13233–13238

Perwitasari O, Yan X, O'Donnell J, Johnson S, Tripp RA (2015) Repurposing kinase inhibitors as antiviral agents to control influenza A virus replication. Assay Drug Dev Technol 13:638–649
Pinto RM, Lycett S, Gaunt E, Digard P (2021) Accessory gene products of influenza A virus. Cold Spring Harb Perspect Med 11:a038380
Pollard AJ, Bijker EM (2021) A guide to vaccinology: from basic principles to new developments. Nat Rev Immunol 21:83–100
Pradere JP, Dapito DH, Schwabe RF (2014) The Yin and Yang of Toll-like receptors in cancer. Oncogene 33:3485–3495
Pulendran B, Arunachalam PS, O'Hagan DT (2011) Emerging concepts in the science of vaccine adjuvants. Nat Rev Drug Discov 20:454–475
Pulendran B, Arunachalam PS, O'Hagan DT (2021) Emerging concepts in the science of vaccine adjuvants. Nat Rev Drug Discov 20:454–475
Reinhart K, Huang S, Kniss K, Reed C, Budd A (2025) Influenza-associated pediatric deaths – United States, 2024-25 influenza season. MMWR Morb Mortal Wkly Rep 74:565–569
Rupali P (2019) Introduction to tropical medicine. Infect Dis Clin N Am 33:1–15
Sampath S, Khedr A, Qamar S, Tekin A, Singh R, Green R et al (2021) Pandemics throughout the history. Cureus 13:e18136
Sautto GA, Kirchenbaum GA, Ross TM (2018) Towards a universal influenza vaccine: different approaches for one goal. Virol J 15:17
Sieczkarski SB, Whittaker GR (2005) Viral entry. Curr Top Microbiol Immunol 285:1–23
Sullivan LC, Berry R, Sosnin N, Widjaja JM, Deuss FA, Balaji GR et al (2016) Recognition of the major histocompatibility complex (MHC) class Ib molecule H2-Q10 by the natural killer cell receptor Ly49C. J Biol Chem 291:18740–18752
Sussman J, Stokoe D, Ossina N, Shtivelman E (2001) Protein kinase B phosphorylates AHNAK and regulates its subcellular localization. J Cell Biol 154:1019–1030
Takashita E (2021) Influenza polymerase inhibitors: mechanisms of action and resistance. Cold Spring Harb Perspect Med 11:a038687
Thiruvaiyaru A, Mattila S, Sadeghi M, Naumenko K, Merits A, Varjosalo M et al (2025) Proximity interactome of alphavirus replicase component nsP3 includes proviral host factors eIF4G and AHNAK. PLoS Pathog 21:e1013050
Trifonov V, Khiabanian H, Rabadan R (2009) Geographic dependence, surveillance, and origins of the 2009 influenza A (H1N1) virus. N Engl J Med 361:115–119
Türei D, Valdeolivas A, Gul L, Palacio-Escat N, Klein M, Ivanova O et al (2021) Integrated intra- and intercellular signaling knowledge for multicellular omics analysis. Mol Syst Biol 17:e9923
UniProt Consortium (2025) UniProt: the universal protein knowledgebase in 2025. Nucleic Acids Res 53:D609–D617
Veit M, Thaa B (2011) Association of influenza virus proteins with membrane rafts. Adv Virol 2011:370606
Wang M, Veit M (2016) Hemagglutinin-esterase-fusion (HEF) protein of influenza C virus. Protein Cell 7:28–45
Wang L, Jiang H, Shen SM, Wen CX, Xing Z, Shi Y (2018) Inhibition of autophagy and chemokine induction by sphingosine 1-phosphate receptor 1 through NF-κB signaling in human pulmonary endothelial cells infected with influenza A viruses. PLoS One 13:e0205344
Wilkins MR, Pasquali C, Appel RD, Ou K, Golaz O, Sanchez JC et al (1996) From proteins to proteomes: large scale protein identification by two-dimensional electrophoresis and amino acid analysis. Biotechnology (NY) 14:61–65
Wille M, Holmes EC (2020) The ecology and evolution of influenza viruses. Cold Spring Harb Perspect Med 10:a038489
Wilson BR, Bogdan AR, Miyazawa M, Hashimoto K, Tsuji Y (2016) Siderophores in iron metabolism: from mechanism to therapy potential. Trends Mol Med 22:1077–1090
Zhang L, Wang J, Muñoz-Moreno R, Kim M, Sakthivel R, Mo W et al (2018) Influenza virus NS1 protein-RNA interactome reveals intron targeting. J Virol 92:e01634–e01618

Zhang D, Yang Y, Li M, Lu Y, Liu Y, Jiang J et al (2022) Ecological barrier deterioration driven by human activities poses fatal threats to public health due to emerging infectious diseases. Engineering (Beijing) 10:155–166
Zhang S, Cai Z, Li H (2023a) AHNAKs roles in physiology and malignant tumors. Front Oncol 13:1258951
Zhang C, Wei B, Liu Z, Yao W, Li Y, Lu J et al (2023b) Bafilomycin A1 inhibits SARS-CoV-2 infection in a human lung xenograft mouse model. Virol J 20:18
Zhao T, Su Z, Li Y, Zhang X, You Q (2020) Chitinase-3 like-protein-1 function and its role in diseases. Signal Transduct Target Ther 5:201
Zhao T, Cai Y, Jiang Y, He X, Wei Y, Yu Y et al (2023) Vaccine adjuvants: mechanisms and platforms. Signal Transduct Target Ther 8:283
Zhao S, Liu M, Zhou H (2024) Identification of novel M2 macrophage-related molecule ATP6V1E1 and its biological role in hepatocellular carcinoma based on machine learning algorithms. J Cell Mol Med 28:e70072
Zhu Y, Cong Y, Sun Y, Sheng S, Liu C, Jiang J et al (2025) Molecular patterns of matrix protein 1 (M1): a strong predictor of adaptive evolution in H9N2 avian influenza viruses. Proc Natl Acad Sci USA 122:e2423983122

# Chapter 10
# Proteomic Applications in Vaccine Development

**Abstract** Controversy surrounding vaccines is not new; the first vaccine proven to be effective in preventing disease was subjected to wild rumors (https://www.who.int/news-room/spotlight/history-of-vaccination/history-of-smallpox-vaccination). In 1796 Dr. Edward Jenner inoculated James Phipps with pus from a cowpox sore taken from the hand of Sarah Nelmes (see Fig. 10.1). While 8-year-old James fell ill with a mild case of cowpox, he eventually recovered. Poor James was not yet free from his life as a "lab rat," as Dr. Jenner later inoculated him with pus taken from an individual suffering with smallpox. Fortunately for James, he was unaffected by this inoculation and shown to be resistant to human smallpox. While Jenner's vaccine proved to be effective in preventing smallpox infection, many people were opposed to this remedy. Some ill-informed Church members believed that suffering from smallpox was part of God's judgment for man and attempting to use a vaccine against it was subverting God's will (https://www.jenner.ac.uk/about/edward-jenner). Another controversy may have been self-inflicted as Jenner named this novel technique "vaccination," which originated from vacca, the Latin word for "cow." This word association caused some people to believe the vaccine worked by giving people an animal disease or even turned them into or gave them certain characteristics of a cow. In addition, Jenner could not completely explain how the smallpox vaccine worked as it would be another 50 years before Louis Pasteur developed the germ theory of disease.

**Keywords** Smallpox · Vaccine · Vaccine development · Antigen · Epitope mapping · Peptide array · On- and off-target binding · Immunopeptidomics · Immunoinformatics · Multi-epitope vaccine · Subtractive proteomics



T. D. Veenstra, *Proteomic Applications in Virology*, Advances in Experimental Medicine and Biology 1511, https://doi.org/10.1007/978-3-032-22340-1_10

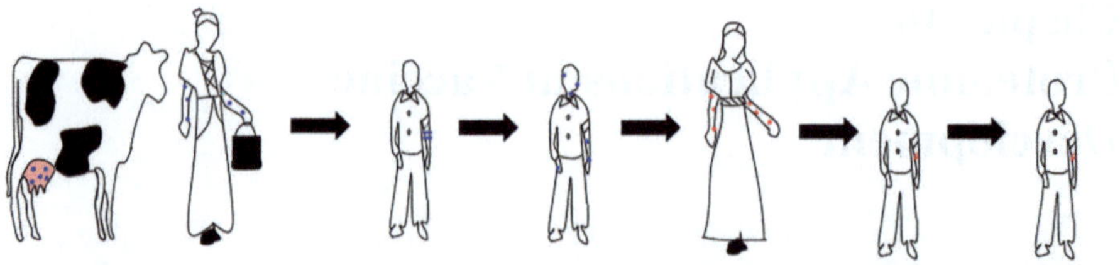


**Fig. 10.1** Dr. Edward Jenner's experiments leading to the development of the smallpox vaccine. Dr. Jenner inoculated James Phipps with cowpox, a virus similar to smallpox. Because of its similarity to cowpox, James exhibited immunity to smallpox when he was later inoculated with this virus. (Image was obtained from https://commons.wikimedia.org/wiki/File:Edward_Jenner-_Smallpox.svg#Licensing and licensed under the Creative Commons Attribution-Share Alike 4.0 International license)

## 10.1 Introduction

Regardless of the controversy, by 1801 the vaccine was shown to be effective against smallpox. By the mid-nineteenth century smallpox vaccination was mandatory in Britain as well as other parts of the world, including the USA (https://www.who.int/news-room/spotlight/history-of-vaccination/history-of-smallpox-vaccination). While smallpox was eradicated in some regions of Europe by 1900, over two million people still died from this disease in other parts of the world. With continued research and improvement in vaccine production, stability, and delivery, smallpox was eliminated in western Europe, North America, and Japan by the 1950s. In 1976, confirmed cases of smallpox were found only in sub-Saharan Africa. The last known case of naturally acquired smallpox was Ali Maow Maalin in Merca, Somalia, in 1977 (https://www.cdc.gov/smallpox/about/history.html#:~:text=Ali%20Maow%20Maalin%20was%20the,22%2C%20he%20developed%20a%20fever). The last person to die from smallpox was Janet Parker in 1978. She worked as a photographer at Birmingham University Medical School, one floor above the Medical Microbiology Department (https://www.cdc.gov/smallpox/about/history.html#:~:text=Ali%20Maow%20Maalin%20was%20the,22%2C%20he%20developed%20a%20fever). It is believed she contracted the virus via an airborne route through the building's duct system or by direct contact while visiting a smallpox research laboratory. On May 8, 1980, smallpox was considered eradicated by the World Health Assembly joining rinderpest, a fatal viral cattle disease, as the only other infectious disease to be considered eradicated (https://www.cdc.gov/smallpox/about/history.html#:~:text=Ali%20Maow%20Maalin%20was%20the,22%2C%20he%20developed%20a%20fever; King et al. 2020). While any remaining stocks of rinderpest were destroyed, stocks of smallpox are kept in the Centers for Disease Control in Atlanta, Georgia, USA, and the Vector Laboratory near Novosibirsk, Russia. Considering smallpox killed an estimated 300 million people in the twentieth century, it is no wonder that its eradication is considered the greatest achievement in international public health (Fig. 10.1).

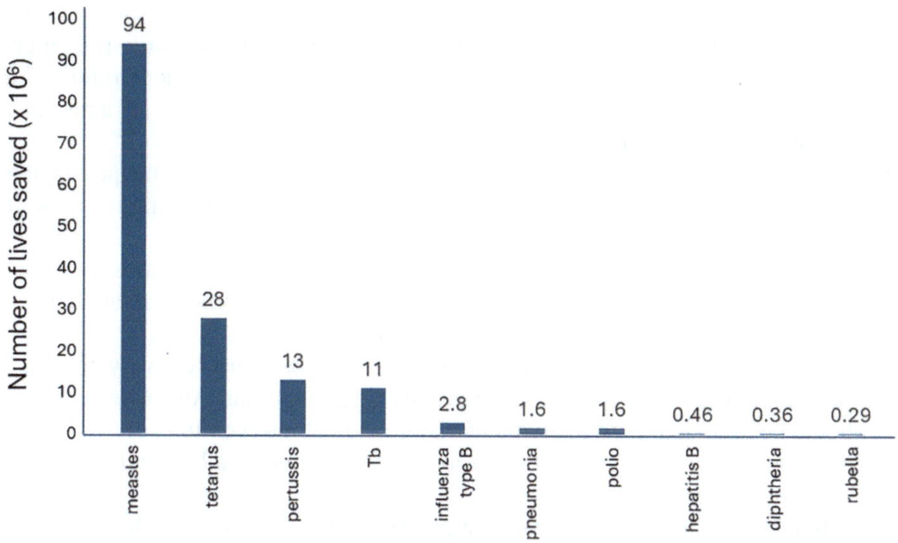


**Fig. 10.2** Bar graph showing the estimated number of lives saved via vaccination against a variety of different infectious diseases

Regardless of the controversies that continue to propagate, vaccinations continue to transform public health. Many diseases that accounted for most of the childhood death have almost disappeared, especially in countries with coordinated immunization programs. The number of lives saved by vaccination against a variety of infectious diseases is estimated to be greater than 150 million people (see Fig. 10.2) (Shattock et al. 2024). The World Health Organization (WHO) estimates that each year current immunization programs save between 2 and 3 million lives, contributing to the reduction in mortality of children less than 5 years of age globally from 93 deaths per 1000 live births in 1990 to 39 deaths per 1000 live births in 2018 (World Health Organization 2020). These numbers represent an astonishing 68% drop in mortality of children under 5 years of age in a period of less than two decades.

## 10.2 Characteristics of a Good Vaccine

Vaccine development has become increasingly refined since Jenner rubbed cowpox pus into James Phipps' open wound. In addition, the number of different vaccine forms (e.g., live attenuated, adenovirus, subunit, mRNA, etc.) has increased. Regardless of the type, there are still several key characteristics that an ideal vaccine typically possesses (Ada 1991).

1. Effectiveness: The most important factor is the body's immune response to the vaccine must be robust. The vaccine needs to produce a high quantity (i.e., titer) of antibodies and/or immune cells that recognize the target pathogen. Ideally a robust immune response would be generated without the need for (or a minimal amount of) an adjuvant.
2. Induction of immunological memory: A vaccine should stimulate the production of immunological memory cells that can generate a fast and strong response if the actual pathogen infects an individual.
3. Broad protection and cross-reactivity: With the propensity for viruses to mutate and give rise to additional strains, an ideal vaccine would recognize and induce immunity across various strains of the same pathogen.
4. Safety: Although it is impossible to eliminate adverse events for every person, a vaccine showed be well tolerated across diverse populations. Any adverse events should be mild and dissipate without major medical intervention.
5. Endurance: A vaccine should confer long-term immunity reducing the need for repeated booster shots.
6. Stability: Particularly in the case of a pandemic, an ideal vaccine should be stable under routine storage conditions to enable global distribution. A vaccine that possesses long-term stability at ambient temperature would be ideal, especially for developing nations in which refrigeration may be limited.
7. Affordability: Ideally the cost of the vaccine should make it affordable to all peoples, regardless of their socioeconomic status.

## 10.3 Identification of Optimal Antigens and Epitopes

Although there have been remarkable advances in technology, especially in the past 5–6 years, there is no generally accepted vaccine design strategy. Strategies include the production of live-attenuated or inactivated vaccines that utilize the entire pathogen. Subunit and conjugated vaccines only require a component (e.g. protein, oligosaccharide, etc.) of the pathogen, while DNA and RNA vaccines use genetically engineered materials whose design is based on the pathogen's structure. This genetic material is then injected into the host where its cells take up the vaccine and produce a proteinaceous antigen to which the immune system responds. Regardless of the selected strategy, the most critical characteristic of a vaccine is its ability to induce an immune response. Optimizing the vaccine requires determining the appropriate antigen and epitope that maximally stimulates the host's immune response.

To refresh the reader's memory, an antigen is any substance that induces an immune response, while an epitope is the specific region within the antigen to which antibodies and T-cell receptors bind. Since an epitope is located within an antigen, the two structures are intimately connected. The first epitope-based vaccine against infectious diseases was introduced in 1985 against *Vibrio cholerae* and *Escherichia coli* (Jacob et al. 1985). The basic strategy is to identify epitopes that will produce the strongest, highest antibody-titer immune responses. The three key steps in epitope-based vaccine design are prediction of the immunogenic regions using in silico epitope mapping, design and production of the immunogenic construct, and experimental evaluation of the efficiency of the vaccine produced from this construct.

Identifying a strategy that predetermines which epitope will produce the optimal immune response is a major challenge. It is a worthwhile endeavor, however, compared to the continued use of live-attenuated or weakened vaccines, as an epitope vaccine should help to focus the immune response (Tang and Zhang 2025; Parvizpour et al. 2020; Bhattacharya et al. 2025; Gershoni et al. 2007). Pathogens possess several potential antigens that contain large numbers of epitopes. Obviously not all these epitopes would stimulate the immune system to provide a strong, high-titer antibody response. Allowing the immune system to focus on a small collection of epitopes known to result in a strong immune response would provide stronger protection against infection. Unfortunately, some epitopes that generate the greatest response are not optimal for vaccine production as they prevent the immune system from responding to other epitopes that may more effectively neutralize the virus. Viruses are known to present "enticing" epitopes so that their weaknesses are obscured. For example, the HIV-1 gp120 variable loops are highly immunogenic and surface accessible but rapidly mutate. This mutation rate allows HIV to evade immune surveillance by constantly changing the affinity of the presently available anti-HIV antibodies (Wyatt et al. 1998; Shcherbakov et al. 2015).

Epitope-based vaccines should be selected for their ability to potently neutralize a pathogen rather than interact with an epitope based on its accessibility (i.e., on the surface of the pathogen). In addition, epitopes should be selected from regions of the virus that are highly conserved and cannot be readily modified without impacting the survival of the virus itself. The problem of producing immunological responses against "undesirable" epitopes could be obviated using epitope vaccines. The possibility that viruses contain epitope mimetics of native host proteins has been connected to the rise in autoimmunity diseases (Olson et al. 2001; Fallahi et al. 2023; Balz et al. 2020). The hypothesis is that the immune system creates antiviral antibodies against epitopes that cross-react with host epitopes and generate the undesirable effects observed in autoimmune disorders, such as rheumatoid arthritis. Finally, epitope-based vaccine production would be cheaper since viruses used to create live-attenuated or inactivated vaccines have their own specific production protocols. Using synthetically created epitopes would enable a standardized production platform making vaccine production cheaper (Jacob et al. 1985; Wei et al. 2024). It would also be safer since large quantities of infectious viruses would not need to be produced, as they presently are for live-attenuated and inactivated vaccines.

### 10.3.1 Epitope Mapping Using Peptide Arrays

The key to developing an effective epitope-based vaccine is to identify epitopes that will provide the most effective immune response. Different methods for identifying and/or mapping epitope-paratope pairs (i.e., the specific region of the antibody the epitope binds to) include X-ray crystallography, cryo-electron microscopy, nuclear magnetic resonance (NMR), hydrogen/deuterium exchange (HDX), and screening peptide arrays (Lin et al. 2024). X-ray crystallography, cryo-electron microscopy, NMR, and HDX provide high-resolution information about specific interacting residues between antibodies and antigens; however, they lack the throughput required to screen thousands of different epitopes. X-ray crystallography can be limited by an antibody's and antigen's stability, size, and solubility, which affect the ability to form an analyzable organized crystallize lattice (Grewal et al. 2024). The throughput of these methods also makes acquiring data on the impact mutations within an epitope make on the antigen/antibody interaction very laborious and time-consuming.

### 10.3.2 Construction of Synthetic Peptide Arrays for Broad Proteome Coverage

Peptide arrays, however, provide high throughput and allow thousands of different epitopes to be screened against a specific antibody (Kazim and Atassi 1981). While they provide throughput, peptide arrays lack the detailed structural resolution obtained using the techniques mentioned in the preceding paragraph. Peptide arrays consist of synthetically prepared overlapping peptides between 15 and 20 amino acids in length (Hudson et al. 2012). The sequences of the peptides are based on the possible proteins found within the pathogen. This information can be generated by translating data within the genome sequence of the organism. The peptides typically overlap by about 11–14 amino acids, ensuring that at least one antigen contains the complete epitope sequence. The peptides are individually presented to an antibody of interest to identify the peptides that bind with the highest affinity. The peptides are presented in the wells of an enzyme-linked immunosorbent assay (ELISA) plate, cellulose membrane, glass slide, or Luminex suspension beads (Suprun et al. 2019). The binding of an antibody to a peptide is detected using a labeled secondary antibody that binds to the primary antibody and emits a fluorescent signal.

Proteomic technology has made an enormous impact on the ability to multiplex the interrogation of peptide epitopes. In 2014, investigators reported on the ability to generate a miniaturized peptide array composed of 2.1 million overlapping peptides that covered the entirety of the human proteome (Forsström et al. 2014). The peptides were synthesized in parallel using an in situ photolithographic process. These peptides were 12 residues in length and overlapped by 6 amino acids.

To synthesize the peptide arrays, a UV light source is combined with digital micromirrors to selectively activate small squares of an array. The UV light source is required since the process uses amino acids with photolabile protective groups (Fodor et al. 1991). Once added to a specific position on the array, amino acids will only incorporate into those peptides that are activated by the UV beam. Via repeated cycles of activation and coupling, all the unique peptides can be synthesized in parallel. Once synthesized, the array is incubated with antibodies, and binding is detected using a secondary antibody conjugated with a fluorophore. The fluorescent intensity measured using a high-resolution microarray scanner is taken as a measure of the binding intensity between the primary antibody and peptide.

Focused-content arrays for more detailed epitope mapping and alanine scanning (described later) were designed using a mix of 12-mer and 15-mer peptides overlapping by 11 and 14 amino acids, respectively (Forsström et al. 2014). Peptides of both lengths were designed to cover the entire sequences of the protein fragments used for immunization. Additional peptides corresponding to the 15-mer peptides, but with the middle amino acid substituted by an alanine, were also included in the design, and all peptides were randomly distributed.

### *10.3.3 Polyclonal Antibody Testing Using a Protein Array*

To validate the arrays, three polyclonal antibodies against fragments of the human proteins 3-hydroxy-3-methylglutaryl-coenzyme A reductase (HMGCR), human epidermal growth factor receptor 2 (HER2), and hyaluronidase-1 (HYAL1) were evaluated. These combinations of antibodies and peptide antigens were useful test cases since the three-dimensional structure of these interactions had previously been determined. Technical replicates revealed two major epitopes for both HMGCR and HYAL1, while three major epitopes were found within HER2. All the major epitopes identified were known to be part of defined structural elements in the full-length proteins.

Based primarily on the reproducibility of the previous results, a 2.1-million peptide array covering all human proteins was synthesized. This array was used to measure the on- and off-target binding of three polyclonal antibodies targeting protein-L-isoaspartate (D-aspartate) O-methyltransferase (PCMT1), tyrosine kinase 2 (TYK2), and ribosomal RNA processing protein 1 homolog B (RRP1B). The anti-PCMT1 antibody showed high intensity binding to three distinct epitopes (GAAAP, GRLI, and VGSGS). The most intense binding to the GAAAP epitope was when it was located within a PCMT1-derived peptide. Intense binding to this epitope in peptides originating from non-PCMT1 proteins, however, was also observed. Off-target binding was observed to be more intense with the GRLI and VGSGS epitopes than that found for binding to PCMT1-derived peptides. Antibody binding to TYK2 was found at two major epitopes comprised of the amino acid sequences VTGT and APRF. While the intensity of binding to the VTGT epitope within a TYK2-derived peptide was most intense, off-target binding to a protocadherin gamma subfamily

A2 (PCDHGA2)-derived peptide was almost as intense. Although most of the bound peptides contained either of these two epitopes, two peptides with no apparent sequence similarity to these epitopes also bound anti-TYK2 antibody with high intensity. The anti-RRP1B antibody showed a significant degree of off-target binding, as most of the most intense binding peptides contained only a few key amino acids found in the RRP1B epitope. A common feature in the off-target peptides was a PF-K amino acid pattern in which these amino acids were separated by a variable number of residues. While the peptide that exhibited the second most intense binding to the anti-RRP1B antibody contained a TGPS-F pattern that is found in the RRP1B protein, neither of the RRP1B peptides that contained this motif (i.e., ILVSP**TGPS**RVA and **GPS**RVA**F**DPEQ) bound to the antibody. Overall, the off-target binding of this antibody was substantial, either suggesting the original epitope used to produce this antibody was poorly selected or the antibody was not carefully validated during its production phase.

Peptide arrays that contain human proteome-wide collections of epitopes can also be useful to validate antibody cross-reactivity. In a further experiment, the proteome-wide binding of a monoclonal antibody targeting RNA-binding motif protein 3 (RBM3) was evaluated. RBM3 is over-expressed in breast, ovarian, colorectal, and malignant melanoma cancers and is a positive prognostic marker for these conditions (Jogi et al. 2009; Ehlen et al. 2011; Jonsson et al. 2011; Hjelm et al. 2011). Evaluation of this antibody against the entire human proteome revealed the most intense binding to a RBM3 peptide containing the epitope GAHGR (Forsström et al. 2014). None of the other top 20 binding peptides originated from the RBM3 protein; however, they all had parts of the GAHGR epitope. While the antibody exhibited significant cross-reactivity, at least this cross-reactivity was driven via interactions with a similar epitope.

## 10.4 On- and Off-Target Binding

The significant amount of off-target binding is one of the trade-offs for the throughput gained when using peptide arrays for identifying antigen-antibody interactions. Part of the reason for this feature is that all the synthesized peptides contain linear epitopes that are outside the context of their native three-dimensional space. These linear peptides can adapt almost any conformation in the presence of an antibody. This flexibility would allow off-target peptides with homologous sequences to the on-target epitope to adapt their conformation to bind to the antibody (i.e., "induced fit"). While a peptide can adapt the conformation necessary to bind to the antibody being tested, it may not be able to display this conformation within its full-length protein. Therefore, positive off-target findings in the protein array may not be reflective of what occurs naturally within the host.

If the hypothesis that synthesized peptides exist in their native conformation for only small periods of time is true, the probability that a peptide antigen will generate an antibody that reacts with the native protein is small (Lerner et al. 1981). The implication of this hypothesis is that a peptide antigen-generated antibody may not recognize the native protein in plasma, cells, or tissues: the three locations in which pathogens thrive. Antibodies generated using synthetic peptide antigens may only interact with denatured proteins, which would limit their ability to neutralize viruses and generate immune responses such as complement activation. This hypothesis, along with the significant off-target binding observed in the antibody screening of peptide arrays described above, leads to two conclusions. The first conclusion is that synthetic peptide antigens should be designed to represent unstructured regions of the target protein (i.e., amino- and carboxy-terminal regions). The second conclusion is that peptide arrays for assessing proteome-wide antibody binding may only provide reliable results for antibodies generated against unstructured peptide antigens.

### *10.4.1 Physiological Effect of Off-Target Antibody Binding*

While the previously described peptide arrays were not used to directly identify epitopes that would make effective vaccine candidates, it does demonstrate the scale at which proteomics has enabled antibody validation. The ability to generate peptide microarrays at this scale allows both on- and off-target antibody binding to be identified across the entire human proteome. Is it important to identify off-target binding? Absolutely, as the recent COVID-19 pandemic demonstrated. Severe acute respiratory syndrome coronavirus 2 (SARS-CoV-2) pathogenesis is still not fully understood and causes effects on multiple organ systems. Many people reported having "long COVID" that persists long after the infection is resolved (Barizien et al. 2021). Frenzied development resulted in multiple efficacious vaccines against SARS-CoV-2; however, in some cases their administration resulted in off-target effects such as allergic reactions, myocarditis, immune-related thrombosis, and thrombocytopenia in healthy individuals (Shukla and Misra 2022). While some of these effects could be a result of undetected off-target binding by the antibody generated by the SARS-CoV-2 vaccine, another explanation could be an anti-idiotype immune response (see Fig. 10.3) (Nakamura et al. 2023). An anti-idiotype response occurs when an antibody binds to a unique structural feature of another antibody. In this scenario an antibody's response to an antigen can induce an antibody response to the original antigen-specific antibody. The initial antibody (termed "Ab1") has regions within its variable domain that contain immunogenic amino acid sequences (i.e., idiotypes) that can induce an immune response that generates an anti-Ab1 antibody (termed "Ab2"). Since it reacts to paratope of Ab1, the paratope of Ab2 would structurally resemble the original antigen. This mimicry results in two outcomes. Ab2 can bind to Ab1 resulting in the neutralization of Ab1 through the formation of immune complexes that are cleared by the body. Since its paratope structurally resembles the original antigen (i.e., the SARS-CoV-2 spike protein), Ab2 could bind to the ACE2 receptor, either blocking or stimulating function or

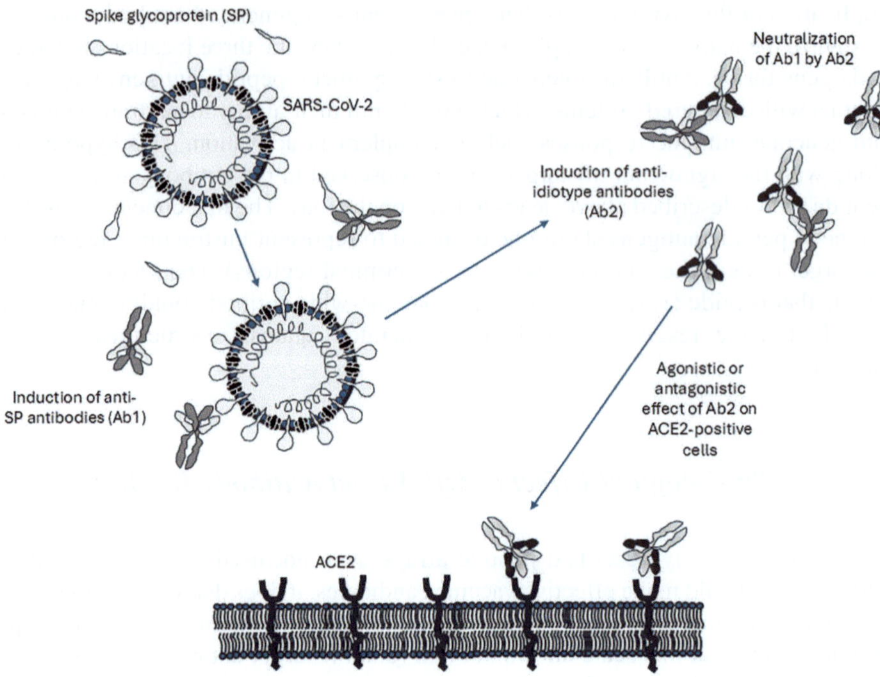


**Fig. 10.3** Anti-idiotype antibodies and SARS-CoV-2. Both SARS-CoV-2 infection and vaccination produce antibodies (Ab1) against the spike glycoprotein (SP), which prevent viral binding to angiotensin-converting enzyme 2 (ACE2) receptors. The presence of Ab1 can induce the production of anti-idiotype antibodies (Ab2) that neutralize Ab1 and bind to ACE2 receptors. The binding of Ab2 to cells containing ACE2 receptors can produce physiological effects that correlate to symptoms observed in patients with "long-COVID"

inducing a complement-mediated or immune-cell attack on ACE2 expressing cells (Murphy and Longo 2022).

Many physiological effects are mediated via ACE2 (Al-Saigh et al. 2024). While the spike glycoprotein of SARS-CoV-2 suppresses ACE2 signaling, it can also trigger toll-like receptors and induce inflammatory cytokines (Zhao et al. 2021). The anti-idiotype responses induced by Ab2 could mimic the effects of the SARS-CoV-2 spike glycoprotein. While clinical assessment of antibody responses to the SARS-CoV-2 vaccines has focused on its neutralization properties, little attention has been given to the impact of idiotype antibodies produced via this vaccine. Indeed, some of the off-target effects observed in the SARS-CoV-2 vaccines, such as myocarditis and thrombotic events, may be related to the induction of Ab2 antibodies (Paque and Miller 1991). Some of the neurological effects of SARS-CoV-2 infection or vaccination may be associated with Ab2 induction as ACE2 is expressed on neuronal tissue as well as lung tissue. It has also been observed that many of the neuropathological effects of SARS-CoV-2 infection are similar to Ab2-mediated effects seen in other viral models (Jahantigh et al. 2025).

## 10.5 Advantages and Disadvantages of Peptide Array Antibody Validation

Could the use of peptide arrays described above be useful in recognizing the presence of idiotypic antibodies that are generated postvaccination? The short answer is yes. The basis of this answer lies in their potential application in identifying autoantibodies in patients suffering from autoimmune diseases. In such a study, a human proteome-wide peptide array of overlapping epitopes could be interrogated with a patient's plasma and antigen-antibody interactions identified. This process could be customized to identifying idiopathic antibodies induced by primary vaccination. The process would require characterizing the primary antibody's paratope domain and generating an array containing peptides with random combinations of amino acids corresponding to this region. While the number of peptides becomes exponentially large, the required number could be minimized by in silico modeling prior to construction of the array to identify the peptides that would bind the paratope. The patient's serum could then be depleted of the primary antibody and tested for the presence of the anti-Ab1 antibody (i.e., Ab2). If the test was positive, the depleted serum would be tested against a human-wide peptide array to identify off-target effects of Ab2 binding.

As previously mentioned, one disadvantage of peptide arrays is that native epitopes are typically conformational, meaning that their three-dimensional structure within the antigen is a factor in their binding affinity. The high-throughput nature of peptide arrays limits them to identifying linear epitopes. Although this is a limitation, its throughput allows the determination of residues that are key to their antibody binding affinity. Through the process of alanine scanning mutagenesis (ASM; also known as "alanine walking"), every residue in an epitope can be systematically replaced with an alanine residue, while alanine residues are replaced with glycine (see Fig. 10.4) (Kokoszka and Kay 2015). The binding affinity of each mutated epitope is reanalyzed to identify residues that are critical for antibody or T-cell receptor recognition. While ASM works well for linear epitopes, it can lead to confusing results if the epitope is conformational. For example, 32 residues of the

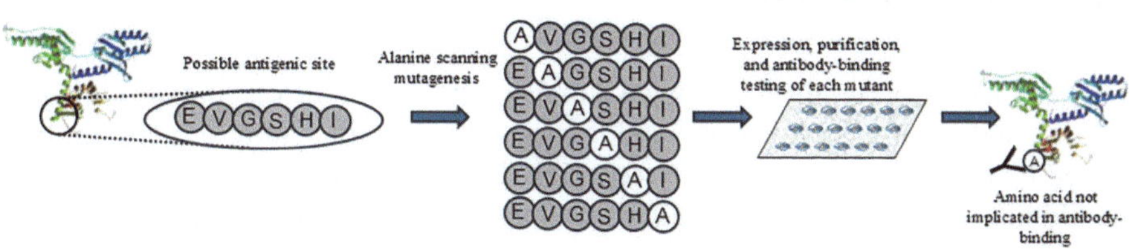


**Fig. 10.4** Alanine scanning mutagenesis workflow for epitope mapping of antibody-antigen interactions. Mutant versions of the antigen protein are expressed and purified with alanine residues sequentially substituted for the amino acids present within the antigen's epitope. The binding affinity between the purified mutant protein and its antibody is tested to determine the effect of alanine substitution at each specific location. If the substitution showed no effect, it is concluded that the amino acid in the wild-type protein did not contribute to the binding of the epitope to the antibody

severe acute respiratory syndrome (SARS) coronavirus glycoprotein receptor binding domain (RBD) were subjected to ASM, and the binding of the mutant peptides to ACE2 was evaluated (Forsström et al. 2014). Six of the residues were known to be part of the actual binding interface via co-crystallization data of the RBD/ACE2 complex. While 11 of the 32 ASM exchanges had a dramatic effect on binding of RBD to ACE2, only 3 of these residues were part of the 6 known to be in contact between the 2 proteins. Mutagenesis of the other three residues had no effect on binding. Replacement of the remaining eight residues has a dramatic effect on binding, although they were outside the binding surface boundary. The result demonstrates the importance of residues outside the direct binding interface in maintaining the conformation of the epitope enabling the correct contact between directly interacting RBD/ACE2 residues. This finding illustrates a fundamental principle of science; one result rarely tells the whole picture. While a peptide array combined with ASM may be useful in characterizing the binding between a linear epitope and antibody, without a three-dimensional structural of the complex, it is impossible to be sure of the role each residue plays in the interaction.

Like any in vitro technique, peptide arrays may not always provide accurate data that reflects the in vivo reality. They do, however, provide a high-throughput screen to evaluate on- and off-target antibody binding to the entire known human proteome. Like almost every scientific method, results from arrays of this magnitude require significant validation. These arrays, however, decrease the number of targets that require validation to a manageable number. Another often-overlooked value in these experimental platforms is that they "reveal" data that is not captured by more focused techniques. Most scientists are aware of the expression "the absence of evidence is not the evidence of absence." Proteome-wide peptide arrays can identify potential sites of off-target antibody binding that lower coverage and resolution methods such as immunoprecipitation, Western blotting, or ELISA do not sample. Discovery of off-target antibody binding is critical for vaccine development as they enable the recognition and explanation of potential adverse events caused by vaccination.

## 10.6 Immunopeptidomics

One area where proteomics has made a major contribution in vaccine development is immunopeptidomics. Immunopeptidomics is the specific study of peptides presented by major histocompatibility complex (MHC) molecules (class I and II) and should not be confused with immunoproteomics, which is the broader study of all proteins associated with the immune system (Morrison and Weiss 2001; Shapiro and Bassani-Sternberg 2023). These MHC molecules, called human leukocyte antigens (HLA) in humans, are present on most nucleated cells and present the large repertoire of peptides taken from the cell's interior and display them on its surface for recognition by cytotoxic T cells ($T_c$) or helper T cells ($T_h$). Cytotoxic T cells recognize class I MHC proteins, whereas $T_h$ cells recognize class II MHC proteins that are expressed mostly on antigen-presenting cells (APCs) of the immune system (Gomase et al. 2024). The peptides presented by MHC molecules are derived from the degradation of cellular and circulating proteins taken up from the outside of the cells in the case of APCs. While MHC-II molecules mainly present peptides originating from proteins phagocytosed by immune cells, MHC-I molecules normally present peptides derived from the cell's own housekeeping proteins but will also present peptides derived from pathogens if the cell becomes infected (Swain 1983). Both MHC-I and MHC-II molecules are highly polymorphic, so they can bind and present a large variety of peptides (Hewitt 2003). As of 2023, there were 25,228 and 10,592 known HLA class I and II known alleles, respectively, which encode for 14,461 and 9160 different HLA class I and II protein allomorphs, respectively (Hewitt 2003). Each HLA allomorph presents thousands of different peptides; therefore, the entire human immunopeptidome is likely to contain greater than a million unique peptides (Admon 2023)!

### *10.6.1 Using MHC-I as a Prefilter for Antigen Recognition*

So how can identifying peptides presented by MHC proteins assist in vaccine development? To illustrate its potential, let us discuss cancer. Tumor development and maintenance involve several abnormal cellular events often initiated by genetic mutations that result in the expression of genes involved in cancer transformation pathways. These mutations produce proteins with either a gain or loss of function that results in a lack of cell cycle regulation. There is active research trying to identify protein antigens involved in these cancer pathways that can be used to produce vaccines that will stimulate the body's own immune system to produce responses (i.e., antibodies, cytotoxic T-cells, etc.) that neutralize these aberrant pathways. Identifying these tumor antigens, however, remains a major challenge for two reasons. Many of the proteins that cause cancer are believed to be present at a very low copy number. Although it has made enormous strides, current proteomic technology is unable to characterize the very low abundant proteins within cells. A larger hurdle to overcome is the complexity and dynamic nature of the human proteome. The human proteome is in a constant state of flux; therefore, it is impossible to fully characterize. Identifying tumor antigens requires some method of pre-fractionating candidates that could potentially produce tumor vaccines. One strategy for recognizing tumor-associated proteins is to use the cell's own method of presenting these antigens to T cells via MHC-I molecules. MHC-I molecules provide a view into the ever-changing environment within the cell that is constantly being evaluated by CD8+ T cells. Identifying peptides presented by MHC-I molecules of tumor cells would provide unique T-cell targets that would generate an immune response targeting the malignant cell. Not only would these peptides serve as effective antigens for developing vaccines, but they would also be specific to the tumor cell since they would be differentially expressed or processed compared to their normal cellular counterparts. This feature would limit the risk of autoimmunity, ensuring that only tumorigenic cells are targeted by the immune system.

### 10.6.2 *Identification of Immunopeptides*

The first reported use of proteomics technology to physically identify large numbers of MHC-bound peptides was in 1992 by Dr. Donald Hunt (see Fig. 10.5) (Gfeller and Bassani-Sternberg 2018). His lab immunoprecipitated HLA-A2.1 molecules—one of the most widely distributed MHC-I molecule in humans—from a human B lymphoblastoid cell line. Peptides bound to HLA-A2.1 were released using acid extraction and isolated using filtration. These peptides were fractionated using reversed-phase liquid chromatography (LC) and identified using tandem mass spectrometry ($MS^2$). This study detected approximately 200 distinct signals with a signal-to-noise ratio greater than two; however, with this being the early days of LC/$MS^2$, only 8 of these signals could be unambiguously identified. The identified peptides shared a distinct structural motif and were nine residues in length. Assuming that the remaining signals were different peptides, the data suggested that well over 1000 different peptides were presented by HLA-A2.1 in this cell line. It was hypothesized that many more peptides from pathogen-infected and transformed cells, as well as antigens associated with autoimmune diseases, would be identified as the sensitivity of the LC/$MS^2$ technology improved.

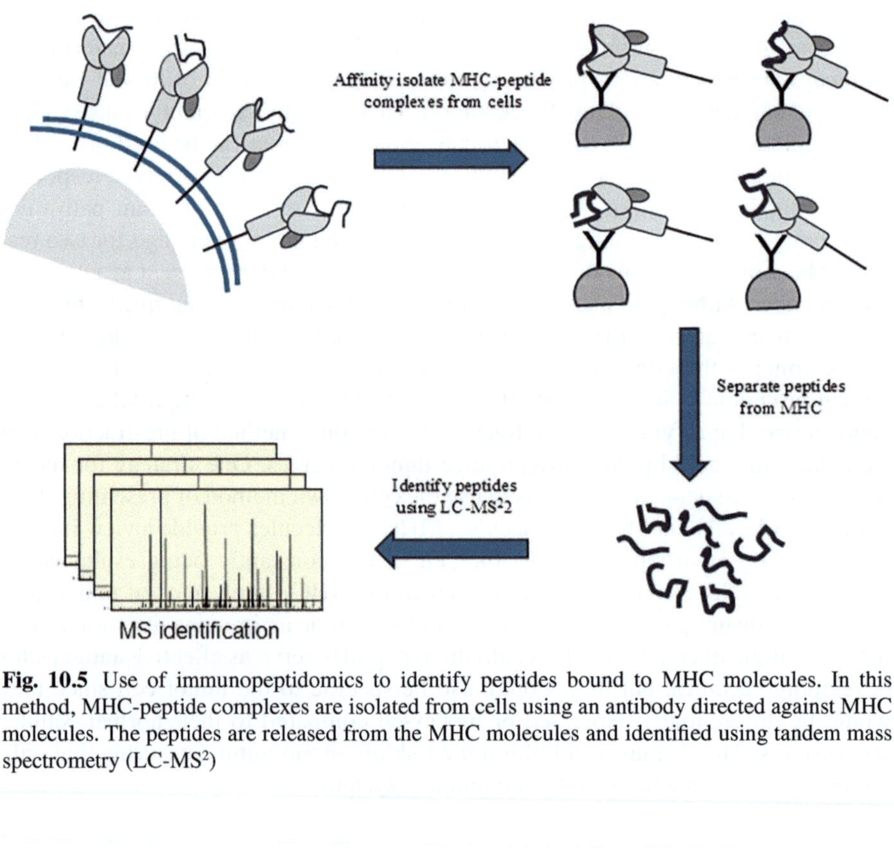


**Fig. 10.5** Use of immunopeptidomics to identify peptides bound to MHC molecules. In this method, MHC-peptide complexes are isolated from cells using an antibody directed against MHC molecules. The peptides are released from the MHC molecules and identified using tandem mass spectrometry (LC-$MS^2$)

Jumping forward 30 years, both instrument sensitivity and sample preparation techniques significantly improved, resulting in a much larger number of MHC-associated peptides being identified using LC/MS$^2$. Dr. Hunt's lab continued to spearhead immunopeptidomics by identifying over 2500 *phosphorylated* MHC-associated peptides across 15 different types of cancer cells, with about 40% of them being found in multiple cancer types (Hunt et al. 1992). About 80% of these phosphopeptides generate central memory T-cell responses in healthy donors, suggesting these peptides are antigenic and have circumvented central tolerance. The observation that these cancer cell-specific peptides generated an immune response in healthy individuals makes them promising vaccine candidates for immunotherapy against cancer.

### 10.6.3 Application of Immunopeptidomics to Develop Tuberculosis Vaccines

Advances in sample preparation, LC/MS$^2$ sensitivity, and data analysis have now made it possible to identify MHC-bound peptides with the aim of identifying effective epitope vaccines that will confer protection against pathogen infection (e.g., viral, bacterial, etc.) or deranged cell proliferation (e.g., cancer). This technology allowed epitopes presented by conventional HLA class-I molecules to be identified in ovarian cancer, influenza, hepatitis C, and HIV (Mahoney et al. 2021; Ramakrishna et al. 2003; Testa et al. 2012; Wölk et al. 2012). Peptides from *Leishmania* have also being identified bound to HLA class-II molecule using this approach. These peptides were subsequently used as antigens to develop vaccines (Ternette et al. 2016; Mou et al. 2015).

*Mycobacterium tuberculosis* (*M.tb*), the causative pathogen responsible for tuberculosis (TB), causes more deaths globally than any other pathogen. Every year, ten million people fall ill resulting in approximately 1.5 million deaths. Tuberculosis is also the leading cause of death of people infected with HIV (https://www.who.int/health-topics/tuberculosis#tab=tab_1; Sullivan and Nathavitharana 2022). The primary vaccine for TB is the Bacillus Calmette-Guérin (BCG) vaccine, a live-attenuated form of *M.tb* (Ari et al. 2024). This vaccine is used in areas where TB is most prevalent but is not typically recommended in the USA due to the low incidence rate of the infection in that country. While it is the primary vaccine used to fight TB, BCG has variable efficiency around the globe, making it urgent that a more effective vaccine be developed. Since vaccines induce antigen-specific CD4+ T cells presented by MHC class II molecules on infected cells, identifying *M.tb* antigens bound to MHC molecules could hypothetically reveal an effective anti-TB epitope vaccine.

To identify an effective epitope vaccine against TB, human monocyte leukemia THP-1 cells were differentiated into macrophages and infected with BCG (Bettencourt et al. 2023). This cell line was selected since its HLA genotype is well characterized, enabling an accurate prediction of the binding of the identified peptides to these complexes. The cells were stimulated with cytokines to inhibit the pathogen's ability to downregulate antigen presentation. Both MHC-I and MHC-II complexes were immunoprecipitated, and the bound peptides were extracted and identified using LC/$MS^2$. A total of 43 and 94 *M.tb* peptides presented by MHC-I and MHC-II molecules were identified, respectively. As expected, most MHC-I bound peptides were 9 amino acid residues in length, while most MHC-II bound peptides ranged between 12 and 18 residues in length. These peptides originated from a total of 117 different antigens. Most of these antigens were involved in biosynthesis and transport of lipids that were associated with the cell membrane.

After further testing, three peptide antigens were selected for evaluation as vaccine candidates based on their intracellular expression levels and ability to induce interferon-gamma (IFN-γ) secretion. The first antigen, LAASLLSRV, was present within the *M.tb* galactofuranosyl transferase (glfT2) protein. This protein plays an important role in the pathogen's cell wall galactan region. Three peptide sequences were identified from the second antigen selected, fatty acid synthase (fas), an enzyme involved in de novo synthesis of fatty acids from acetyl-CoA. These peptides were found to bind to either MHC-I or MHC-II complexes. The final antigen was identified via peptides originating from antigen 85A (Ag85A), a mycolyltransferase that plays a role in the synthesis of the bacteria's cell wall. When expressed in viral vectors, Ag85A has been shown to improve the protective efficacy of BCG (Sullivan and Nathavitharana 2022); therefore, this antigen was selected for further epitope vaccine study. Two additional antigens, isoniazid-induced protein B (iniB) and proline-proline-glutamic acid family protein 15 (PPE15), were selected based on them being previously identified as binding to MHC-I molecules (Lewinsohn et al. 2017).

To determine if these antigens could offer greater protection against *M.tb* when administered in combination with the BCG vaccine, each was cloned into a viral vector and inoculated in mice either alone or in combination. The typical *M.tb* vaccine regimen is comprised of a BCG inoculation followed by boost of replication-deficient chimpanzee adenovirus ChAdOx1 at 10 days and a second boost with a modified vaccinia virus Ankara (MVA) vector that expressed each experimental antigen being evaluated as a new vaccine candidate 14 days after the initial BCG vaccination. Both the BCG and MVA vaccines were given intradermally, while ChAdOx1 was administered intranasally. This regimen, designated as BCM, has been validated as the best method to induce protection in a mouse *M.tb* aerosol challenge model. Each mouse was subjected to a *M.tb* challenge 18 days after administration of the BCG vaccine, and the number of colony-forming units (CFU) in the lung and spleen was measured.

The BCM regimen that included PPE15 showed a significant lower median CFU compared to BCG vaccination alone in the spleen but not in the lung when the data was analyzed using a Mann-Whitney U test (MW), but not when analyzed using Kruskal-Wallis followed by Dunn's multiple comparison test (KW-DM). None of the other individual antigens conferred greater protection efficiency beyond BCG vaccination. When the antigens glfT2, fas, and iniB were combined and administered as part of the vaccination regimen, however, its efficacy was significantly higher in the lung (MW $p = 0.0095$, KW-DM $p = 0.0144$) and spleen (MW $p = 0.0148$, KW-DM $p = 0.0141$) compared to BCG alone. This efficacy was observed even when the dose per antigen was given at a level 2.5-fold lower than the normal dose. This study provided a proof of concept for using proteomics in the form of immunopeptidomics to identify new candidate antigens that can be administered in combination to improve the efficacy of current vaccination standards for a specific disease. The proteomics method is unbiased and relies on empirical data presented by MHC complexes themselves. While it required three new antigens given in combination to improve *M.tb* protection efficiency, discovery of these antigens was quick, enabling rapid progression into vaccine studies.

## 10.7 Identification of Potential Vaccine Candidates Using Existing Proteomic Datasets

The Human Genome Project (HGP) was a tremendous scientific achievement in several ways (Schuler et al. 1996). The genetic sequencing and software tools that were invented during this project continue to evolve and have elevated genome sequencing to a routine technology. Beyond the simplicity of identifying known disease-causing mutations across entire genomes, genome-wide association studies (GWAS) for identifying disease-associated variations across human populations are now possible (Abdellaoui et al. 2023). A combination of gene sequencing and epigenetics can reveal how gene expression is regulated and how it influences disease development and the inheritance of traits across generations. Even for scientists not involved in generating genomic data, the HGP has enabled a plethora of genomic databases that can be interrogated to answer important research questions. This information legacy, which continues to grow, will provide foundational information that can answer specific hypotheses for decades to come.

Although not at the magnitude of genomics, there are now large collections of proteomic informatic sites that provide algorithms for analyzing proteins. These algorithms analyze protein sequences to identify structures, functions, and interactions between proteins within biological systems. So how can these databases assist in vaccine development? Since the genome and/or proteome of many infectious organism is already known, all that is required is to classify the proteins based on functions such as host-pathogen interaction, immune system evasion, pathogenesis, etc. The key is to analyze the data to find protein sequences that meet the criteria for making them potentially effective targets for vaccine development.

Immunoinformatics is the term used to describe the analysis of proteomic data to discover vaccine targets and other immune system related information. As its name implies, immunoinformatics integrates the fields of immunology and bioinformatics to analyze immune-related data (Nag et al. 2025). The data can be analyzed to assist in the design of vaccines, therapeutics, and diagnostics or simply to evaluate molecular interactions between hosts and pathogens. Immunoinformatics offers several benefits for vaccine design including speed, efficiency, and targeted selection of antigen candidates. Besides prioritizing vaccine targets and identifying epitopes, immunoinformatics can predict immunogenicity and assess cross-reactivity with host proteins. Having access to the protein sequence data through translation of the information in genomic databases or the proteomic data itself accelerates the vaccine development process, increasing the chances of developing effective immunization strategies and enhancing safety. In the following sections, an application of using both experimental data and informatics to generate vaccine candidates against the filarial nematode *Wuchereria bancrofti* is described (Aarthy et al. 2024).

## 10.8 *Wuchereria bancrofti* Vaccine Development via Proteomics and Immunoinformatics

Lymphatic filariasis is a tropical disease that causes chronic conditions such as lymphedema, elephantiasis, and hydrocele (Drame et al. 2016; Nutman 2001). This disease leads to physical disability, economic hardship through reduced productivity, and social stigma because of the patient's physical deformities. It is most common within developing tropical countries that lack medical resources making it an excellent candidate for a vaccine to stop its transmission. More than 90% of cases of lymphatic filariasis are a result of transmission via the bite of an infected mosquito carrying the nematode *Wuchereria bancrofti* (*W. bancrofti*).

The first step to identifying potential vaccine candidates is a careful filtering of the proteins within an organism's known or predicted proteome. An overview of the filtering scheme used to identify possible vaccine candidates for this project is shown in Fig. 10.6. The 13,058 known *W. bancrofti* coding genes were translated into their protein sequence (https://parasite.wormbase.org/Wuchereria_bancrofti_prjeb536/Info/Index/). A BLAST analysis of these proteins was conducted against known human proteins to eliminate homologous proteins that could generate antibodies that cross-react with human proteins (Aarthy et al. 2024; Altschul et al. 1997). Since the aim of the vaccine is to produce antibodies that would prevent the pathogen from thriving, a BLAST analysis of the remaining proteins against the Database of Essential Genes (DEG) database was performed (Luo et al. 2021). This database identifies genes that are essential to an organism's survival. Fifty-one proteins that are considered essential to *W. bancrofti's* survival were identified.

Fifty-one potential vaccine targets are still too many to test, so these protein candidates were evaluated using WoLF PSORT. WoLF PSORT is an algorithm that determines the most probable subcellular location of a protein (Horton et al. 2007). Of the 51 proteins analyzed, 24 were associated with the plasma membrane, and 6 were predicted to be secreted. These 30 proteins were selected as potential vaccine targets since they exist in regions that are readily exposed to the human immune system. Since algorithms are not always accurate, the locations of these 30 proteins

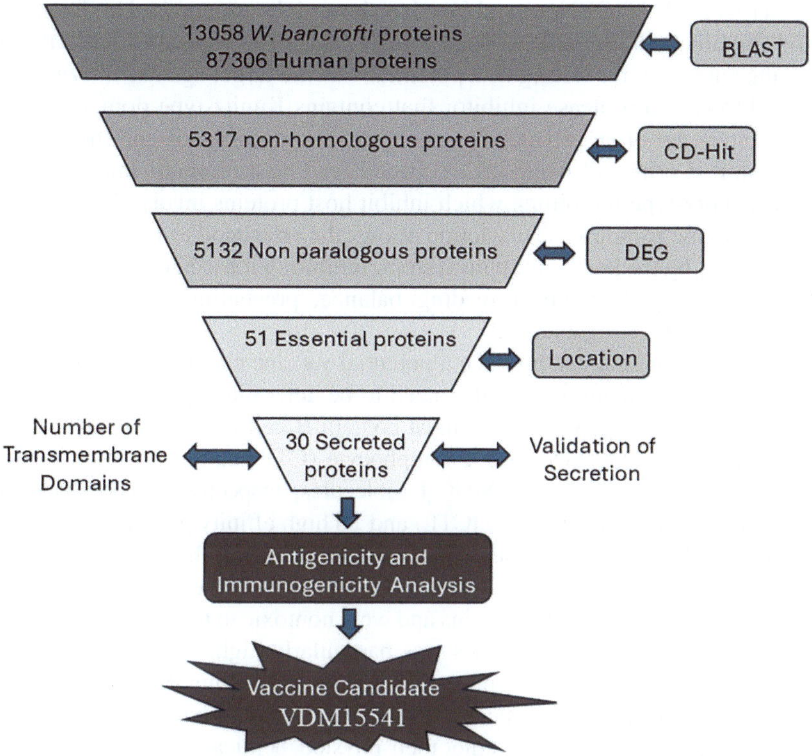


**Fig. 10.6** Workflow using subtractive proteomics and immunoinformatics to analyze the proteome of *Wuchereria bancrofti* to identify vaccine candidates and therapeutic targets

were further scrutinized for the presence of a signal peptide and searched against SecretomeP, a database of secreted mammalian proteins. Only 4 of the 30 proteins passed both filters. Since membrane-bound proteins typically contain hydrophobic α-helices, these four proteins were analyzed using the transmembrane hidden Markov model (TMHMM) algorithm, which calculates the probability of a protein containing hydrophobic helices based on its primary structure (Li and Gallin 2004). Since they are difficult to purify, membrane-bound proteins containing multiple transmembrane helices were eliminated from further consideration. This filtering left two remaining proteins, VDM10251 and VDM15541, that were either secreted or contained one or fewer transmembrane domains. While classified as a secreted protein, VDM10251 is not truly secreted as it is a collagen protein that is associated with the extracellular matrix. Fortunately, VDM15541 is a 165-amino-acid-long, secreted protein associated with molting and embryogenesis. It is homologous to the protease inhibitor protein MLT-11 of the nematode *Caenorhabditis elegans* (*C. elegans*). MLT-11 has been shown to be critical for the development of *C. elegan's* adult exoskeleton (Frand et al. 2005), suggesting that VD15541 plays a similar role in *W. bancrofti*. Analysis of the sequence and physicochemical properties of

VD15541 using ANTIGENpro and VaxiJen showed the protein had high antigenicity and immunogenicity values. Therefore, VDM15541 was considered the most promising candidate for testing as a potential vaccine target against *W. bancrofti*.

VDM15541 is a protease inhibitor that contains Kunitz-type domains. Kunitz-type inhibitors are often produced by blood-feeding arthropods and the pathogens they transmit (Černý and Arora 2024). Blood-feeding arthropods inject saliva that contains Kunitz-type inhibitors, which inhibit host proteins involved in blood clotting and immune responses. This action allows the arthropod to feed while remaining undetected by the host's immune system. Inhibiting these Kunitz-type inhibitors could upset the pathogen/host feeding balance, preventing the pathogen from spreading disease amongst hosts.

Having a protein that can produce a potential vaccine candidate is just the beginning. There are still many factors that need to be determined prior to actual experimental testing. An effective vaccine must contain B cell as well as CD4+ helper T cells (HTLs) and CD8+ cytotoxic T lymphocyte (CTL) epitopes, which recognize antigens bound to MHC-II and MHC-I molecules, respectively. The sequence of VDM15541 was assessed using NetCTL, and 26 high affinity potential B-, CD4 + -, and CD8 + -cell epitopes were found (Karlos et al. 2024). Of these, only six bound to MHC-I molecules while none to MHC-II. Fortunately, all these epitopes were conserved across various *Wolbachia* strains and were nontoxic to host cells. The predicted antigenicity of four of these epitopes was particularly high, so these were selected from continued analysis. B-cell epitopes were predicted using ABCPred, BepiPred, and BCePred, which employ artificial neural network (ANN)-based methods that analyze amino acid sequences and predict their physicochemical properties (Costa et al. 2013). The ABCPred algorithm predicted 12 B-cell epitopes, while 2 other algorithms BepiPred and BCePred predicted 7 and 6 possible epitopes, respectively. Six of these epitopes overlapped in at least two of the analyses and were further analyzed.

### *10.8.1 Design and Characterization of a Multi-epitope Vaccine Construct*

These ten potential B-cell and MHC-I epitopes were linked together to form a putative vaccine construct (see Fig. 10.7). MHC-I epitopes were sequentially linked using EAAAK linkers, while the six B-cell epitopes were connected to each other via GPGPG linkers (Aarthy et al. 2024). A 45-amino-acid-long β-defensin molecule was added to the N-terminal to serve as an adjuvant and connected to the MHC-I epitope via an EAAAK linker. The final vaccine candidate construct consisted of 182 amino acids and for the sake of this chapter is termed *Wb*10.1α. *Wb*10.1α had a calculated average antigenic probability score of 0.9425, which is considerably higher than the 0.8 threshold required for any vaccine candidate, suggesting it is likely to exhibit antigenic properties. The solubility probability of the vaccine candidate was 0.694 as determined using SolPro, suggesting it will be soluble within the human body (García de la Torre et al. 1997).

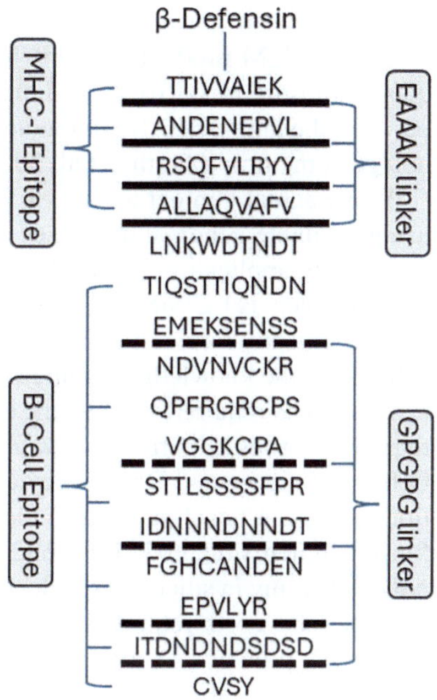


**Fig. 10.7** Design of a multi-epitope vaccine construct to induce an immune response against *Wuchereria bancrofti*

For a vaccine to work, it must be capable of interacting with various immune system factors. To evaluate how it may interact with other proteins, PSI blast-based secondary structure PREDiction (PSIPRED) was used to predict for the presence of secondary structure elements within *Wb*10.1α (Buchan et al. 2024). This algorithm predicted the candidate vaccine possesses a single α-helix, six β-strands, and several regions of random coils. Based on its primary and predicted secondary structure, the interaction of the candidate vaccine with toll-like receptors (TLR) 3 and 4 and human leukocytes antigens were determined using HDock (Luo et al. 2010). The HDock results predicted strong interactions between *Wb*10.1α and TLR3 and TLR4 human proteins and nine HLA alleles. These nine HLA alleles also showed binding with MHC-I epitopes. Modeling of *Wb*10.1α alone or in complex with TLR3, TRL4, and HLA alleles showed that the complexes had higher stability than the vaccine candidate by itself.

### *10.8.2 Immune Simulation of Predicted Vaccine Construct*

The natural human immune response to *Wb*10.1α was analyzed using C-language Immune Simulator (C-ImmSim), an algorithm that incorporates existing immune system knowledge and predicts how it responds to an antigen (Castiglione et al. 2007). The algorithm predicted that *Wb*10.1α stimulates the immune system leading to the production of mostly IgM antibodies, with a small amount of IgGs. The simulation predicted that initial administration of the vaccine would produce low antibody levels; however, three doses at 3-month intervals would result in an increase in IgM production. This increase was predicted to continue with subsequent exposure to *Wb*10.1α. C-ImmSim also predicted a stronger binding affinity between the anti-*Wb*10.1α antibodies and its target and the establishment of immunological memory (Aarthy et al. 2024).

While actual clinical testing of *Wb*10.1α has not been performed, its discovery using available genomic databases and informatics demonstrates another aspect of proteomics application to vaccine development. While the development of technologies such as LC/MS and protein arrays is most readily associated with the proteomics era, proteomics has also spurred the rise of informatic tools whose accuracy is based on the knowledge acquired through empirical genomic and proteomic studies. The steps required to show the efficacy of a vaccine candidate are complex and require rigorous experimental validation and clinical testing. A report from the Office of Science and Data Policy estimates that it costs an average of $886.8 million to bring a novel vaccine to the US market (https://aspe.hhs.gov/sites/default/files/documents/dddeb3748f486324a493a8a6d27f4338/aspe-vaccine-costs-brief.pdf). While any in silico approach is limited by its reliance on predictions, it reduces the time and costs required to identify potential vaccine targets. Even a simple organism's proteome is too complex to experimentally test the vaccine efficacy of every potential antigen. Identifying candidates using computational analysis is an effective starting point for discovering effective vaccines against filariasis and other conditions. In fact, several vaccine development studies have been initiated in the past few years using computational tools to identify useful starting candidates. Obviously, further research is needed to perform the experimental studies, immunogenicity assessments, vaccine formulation optimization, and long-term efficacy evaluations to establish how effective any vaccine performs. The results of these future research studies will ultimately determine the value of holistic, computational approaches for identifying vaccine candidates.

## 10.9 Vaccine Candidate Identification Using a Proteomics/Bioinformatics Strategy

As mentioned previously, the first step in producing a vaccine is identification of potential vaccine targets, whether they be in the form of an antigen or a specific epitope. Proteomics has contributed to identification of vaccine candidates in three major areas. First is the direct experimental identification of proteins present within the proteome of a pathogenic organism. Second is the availability of protein sequence information within proteomic databases and/or the ability to translate protein sequences from gene sequences within genomic databases. Third is the availability of bioinformatic tools that have been developed over the years to analyze large amounts of omics data generated via both proteomics and genomics.

### 10.9.1 Analysis of Biological Samples to Identify Vaccine Candidates

As described in Chap. 1, LC/MS has been the driving force enabling high-throughput proteome characterization. Analysis of a pathogen's entire proteome can identify proteins involved in host-pathogen interactions, immune evasion, and pathogenesis. One drawback of this approach, however, is the sheer number of proteins that are identified, which makes it challenging to select good vaccine candidates.

Bioinformatics, and more specifically immunoinformatic resources, have become a necessity for evaluating proteomic data for the recognition of vaccine candidates as described previously. Unfortunately, analyzing experimentally characterized or predicted proteins within a proteome does little to prefilter vaccine candidates prior to immunoinformatic analysis. A reasonable strategy would be to experimentally identify antigens that have the characteristics necessary to produce the most robust immunogenic response. So what would indicate if an antigen produced an immune response within an infected individual? The obvious answer is the presence of antigen-specific antibodies within their plasma or serum. A more direct approach would be to analyze samples from infected individuals and see what pathogenic proteins their native antibodies are binding to. This approach was taken in a recent study attempting to identify vaccine candidate proteins for *Acinetobacter baumannii* (*A. baumannii*) (Acar et al. 2022). This bacterium is one of the most life-threatening pathogens globally and has developed extensive drug resistance (Naveed et al. 2025). Owing to this resistance, the frequency of infections caused by these bacteria is increasing. As the number of antibiotic options available against *A. baumannii* continues to decrease, the importance of developing a vaccine increases.

### 10.9.2 Antigen Identification Using Subtractive Proteomics

To execute this strategy, blood samples were acquired from 29 patients diagnosed with *A. baumannii* and 13 noninfected individuals (Acar et al. 2022). Along with blood samples, secreted and outer membrane proteins (OMPs) were isolated from antibiotic resistant and nonresistant strains of *A. baumannii* cells. Only testing secreted and OMPs ensures the identification of antigens that are accessible to the immune system.

Proteins from *A. baumannii* were immunoprecipitated using serum obtained from infected and noninfected patients. *Acinetobacter baumannii* proteins identified using serum from noninfected patients were subtracted from the total pool of proteins identified in serum obtained from both noninfected and infected individuals. This subtraction left a total of 34 antigenic proteins that were identified from the secretome and outer membrane portion of the resistant strain in serum from infected patients. Of these proteins, 15 were OMPs, while 18 were part of the pathogen's secretome. Two separate secretome preparations were subjected to immunoprecipitation. The first preparation used phenol extraction to concentrate the secreted proteins and the second preparation analyzed the secretome without any type of enrichment. A single protein, BVO885, was isolated from the non-enriched secretome suggesting it was highly abundant and immunodominant.

Of the 34 identified proteins, they covered a range of different molecular functions including membrane transport, ion binding, antibiotic binding, outer membrane assembly, and protein secretion. Some of the proteins possessed enzymatic activity and functioned as either a hydrolase, oxidoreductase, or cis-trans isomerase. The sequences of these proteins were analyzed using the Immune Epitope Database (IEDB) to predict the presence of B-cell epitopes, which are critical for antibody production (Zhang et al. 2008). Two of these proteins, BOV885 and BOVAB5, scored highly for potential epitopic regions. BOV885 is an OMP that was previously identified within the unenriched *A. baumannii* secretome, and BOVAB5 is a bifunctional protein that has hydrolase activity. Since many of the identified proteins were associated with the outer membrane, BOCTOPUS was used to assess their solubility (Hayat et al. 2016). An effective vaccine needs to be soluble, so insoluble candidates were eliminated at this stage. BOCTUPUS is an algorithm that predicts the presence of transmembrane β-barrels within outer membrane proteins. BOCTOPUS predicted 8 of the 34 proteins met the necessary solubility threshold along with other structural constraints. These eight proteins, along with BOV885, were further scrutinized as potential vaccine candidates.

### *10.9.3 Bioinformatic Analysis of* Acinetobacter baumannii *Vaccine Candidates*

The sequences of these candidates were analyzed using AlphaFold2 to predict their three-dimensional structures (Tunyasuvunakool et al. 2021). The location of previously predicted epitopic regions was mapped onto these structures and found to be in accessible regions of each protein. In addition, all eight candidates had potential O-glycosylation sites that would increase their immunogenicity. Further scrutinizing showed that both AdeC, a protein that is part of a multidrug efflux transporter complex (Pannek et al. 2006), and BOVE52 contained several transmembrane domains making their success as recombinant vaccines unlikely. Of the remaining candidates, β-barrel assembly machinery component D (BamD) is part of a membrane protein assembly complex that inserts β-barrel proteins into the cell's outer membrane. Although BamD's immune potential is not known, immunization of a mouse model infected with *A. baumannii* with BamA, which is part of the BamD complex, increased survival rate by 60% (Singh et al. 2017). FK506-binding protein-like B (FklB), a peptidyl-prolyl cis-trans isomerase, was also identified in this study as having vaccine potential. This protein has been experimentally shown to be highly immunogenic (Piri-Gharaghie et al. 2022). An oxidoreductase, BOV9Z6, was also identified as having high immunogenic potential based on the presence of an accessible B-cell epitope region (Acar et al. 2022). There were no previous reports suggesting that this protein had been previously suggested or tested as a possible vaccine candidate.

As previously mentioned, B0VAB5 and BOV885 were identified in the secretome of *A. baumannii*. BOVAB5 was predicted to have several favorable qualities that would make it a good vaccine candidate. It plays a role in peptidoglycan metabolism and has multiple LysM repeats (a protein domain found in a wide variety of extracellular proteins and receptors) suggesting it is located on the bacteria's accessible surface (Buist et al. 2008). It was predicted to contain several epitopic regions within accessible regions of the protein. Since BOV885 was the only protein immunoprecipitated directly from media containing *A. baumannii* and had several epitopic regions, it was of particular interest. This protein contained a total of eight significant epitopes that fell within a structurally unique β-sheet region. While analysis of the entire sequence of the protein suggested it was insoluble, limiting analysis to the epitope-containing region predicted this domain as being soluble. What was particularly interesting is when the sequences of BVO885 were compared in the antibiotic resistant and nonresistant strains of *A. baumannii*, it was found that the protein in the antibiotic-resistant strain had an additional 17 amino acid sequence. While the importance of this sequence is not known, it may be related to this strain's drug resistance.

The above study demonstrates the use of both proteomic and bioinformatic analysis to identify potential vaccine candidates. Using immunoglobulins from infected patients, vaccine candidates that have already been shown to induce an immune response to *A. baumannii* can be isolated and identified. Since any immunoprecipitation extracts a significant number of nonspecific binding proteins, an important feature of the experimental methods in this study was the use of serum from noninfected patients and protein extracts from a nonresistant strain of *A. baumannii*. These steps enabled recognition of proteins that were specific to the antibiotic-resistant strain. Combined with further bioinformatic analysis, the number of candidates was filtered to numbers that could be reasonably tested in preclinical studies, saving time and money.

## 10.10 Conclusions

The research and development phase for producing a vaccine typically takes greater than 10 years. However, many vaccines still fail in clinical development because of unreliable animal models or lack of measurable factors indicating the likelihood of protecting from an infection or disease (i.e., correlates of protection). The history of vaccine development is rife with promising phase 1 and 2 results and failed phase 3 efficacy trials due to a lack of correlates of protection. A recent report analyzing vaccine trials within clinicaltrials.gov showed that the probability of a vaccine progressing from phase 2 to licensure within 10 years was approximately 10.0%. Most of the successful approvals represented H1N1 or H5N1 vaccines.

Their development will continue to grow as more uses for vaccines become imagined. While vaccines have historically been associated with pathogens, new applications in cancer prevention are continuously being investigated. Proteomics will have a major impact on increasing the rate of vaccine success by reducing the time to licensure and associated research and development costs. The tools provided by proteomics for enhancing vaccine discovery are not simply limited to global identification of proteins within pathogens. Peptide arrays can be used to evaluate the specificity of antibody-epitope interactions, MHC and autoantibody enrichment methods can enrich potential vaccine targets, and bioinformatic tools can analyze protein characteristics to increase the chances of a potential vaccine candidate gaining licensure. It is an impressive repertoire of technologies considering proteomics started with a two-dimensional gel and amino acid analysis to identify a few stained protein spots.

## References

Aarthy M, Pandiyan GN, Paramasivan R, Kumar A, Gupta B (2024) Identification and prioritisation of potential vaccine candidates using subtractive proteomics and designing of a multi-epitope vaccine against Wuchereria bancrofti. Sci Rep 14:1970–1988

Abdellaoui A, Yengo L, Verweij KJH, Visscher PM (2023) 15 years of GWAS discovery: realizing the promise. Am J Hum Genet 110:179–194

Acar MB, Ayaz-Güner Ş, Güner H, Dinç G, Ulu Kılıç A, Doğanay M et al (2022) A subtractive proteomics approach for the identification of immunodominant *Acinetobacter baumannii* vaccine candidate proteins. Front Immunol 13:1001633–1001645

Ada GL (1991) The ideal vaccine. World J Microbiol Biotechnol 7:105–109

Admon A (2023) The biogenesis of the immunopeptidome. Semin Immunol 67:101766–101783

Al-Saigh NN, Harb AA, Abdalla S (2024) Receptors involved in COVID-19-related anosmia: an update on the pathophysiology and the mechanistic aspects. Int J Mol Sci 25:8527–8547

Altschul SF, Madden TL, Schäffer AA, Zhang J, Zhang Z, Miller W et al (1997) Gapped BLAST and PSI-BLAST: a new generation of protein database search programs. Nucleic Acids Res 25:3389–3402

Ari MM, Beig M, Sholeh M, Khoshmirsafa M (2024) The BCG vaccine, advantages, and disadvantages of introducing new generation vaccines against *Mycobacterium tuberculosis*. Clin Exp Vaccine Res 13:184–201

Balz K, Trassl L, Härtel V, Nelson PP, Skevaki C (2020) Virus-induced T cell-mediated heterologous immunity and vaccine development. Front Immunol 11:513

Barizien N, Le Guen M, Russel S, Touche P, Huang F, Vallée A (2021) Clinical characterization of dysautonomia in long COVID-19 patients. Sci Rep 11:14042–14042

Bettencourt P, Müller J, Nicastri A, Cantillon D, Madhavan M, Charles PD et al (2023) Identification of antigens presented by MHC for vaccines against tuberculosis. NPJ Vaccines 5:2

Bhattacharya M, Chatterjee S, Das A, Lee SS, Chakraborty C (2025) Machine learning (ML) and deep learning (DL) in vaccine target selection, design, development and characterization. Vaccine 71:128051

Buchan DWA, Moffat L, Lau A, Kandathil SM, Jones DT (2024) Deep learning for the PSIPRED protein analysis workbench. Nucleic Acids Res 52:W287–W293

Buist G, Steen A, Kok J, Kuipers OP (2008) LysM, a widely distributed protein motif for binding to (peptido)glycans. Mol Microbiol 68:838–847

Castiglione F, Duca K, Jarrah A, Laubenbacher R, Hochberg D, Thorley-Lawson D (2007) Simulating Epstein-Barr virus infection with C-ImmSim. Bioinformatics 23:1371–1377

Černý J, Arora G (2024) Proteases and protease inhibitors in saliva of hard ticks: biological role and pharmacological potential. Adv Parasitol 126:229–251

Costa JG, Faccendini PL, Sferco SJ, Lagier CM, Marcipar IS (2013) Evaluation and comparison of the ability of online available prediction programs to predict true linear B-cell epitopes. Protein Pept Lett 20:724–730

Drame PM, Meng Z, Bennuru S, Herrick JA, Veenstra TD, Nutman TB (2016) Identification and validation of Loa loa microfilaria-specific biomarkers: a rational design approach using proteomics and novel immunoassays. MBio 7:e02132–e02115

Ehlen O, Nodin B, Rexhepaj E, Brandstedt J, Uhlen M, Alvarado-Kristensson M et al (2011) RBM3-regulated genes promote DNA integrity and affect clinical outcome in epithelial ovarian cancer. Transl Oncol 4:212–221

Fallahi P, Elia G, Ragusa F, Paparo SR, Patrizio A, Balestri E et al (2023) Thyroid autoimmunity and SARS-CoV-2 infection. J Clin Med 12:6365

Fodor SP, Read JL, Pirrung MC, Stryer L, Lu AT, Solas D (1991) Light-directed, spatially addressable parallel chemical synthesis. Science 251:767–773

Forsström B, Axnäs BB, Stengele KP, Bühler J, Albert TJ, Richmond TA et al (2014) Proteome-wide epitope mapping of antibodies using ultra-dense peptide arrays. Mol Cell Proteomics 13:1585–1597

Frand AR, Russel S, Ruvkun G (2005) Functional genomic analysis of *C. elegans* molting. PLoS Biol 3:e312–e326

García de la Torre J, Carrasco B, Harding SE (1997) SOLPRO: theory and computer program for the prediction of SOLution PROperties of rigid macromolecules and bioparticles. Eur Biophys J 25:361–372

Gershoni JM, Roitburd-Berman A, Siman-Tov DD, Tarnovitski Freund N, Weiss Y (2007) Epitope mapping: the first step in developing epitope-based vaccines. BioDrugs 21:145–156

Gfeller D, Bassani-Sternberg M (2018) Predicting antigen presentation-what could we learn from a million peptides? Front Immunol 9:1716–1732

Gomase VS, Dhamane SP, Kakade PG (2024) Immunoproteomics: a review of techniques, applications, and advancements. Protein Pept Lett 31:827–849

Grewal S, Hegde N, Yanow SK (2024) Integrating machine learning to advance epitope mapping. Front Immunol 15:1463931–1463944

Hayat S, Peters C, Shu N, Tsirigos KD, Elofsson A (2016) Inclusion of dyad-repeat pattern improves topology prediction of transmembrane β-barrel proteins. Bioinformatics 32:1571–1573

Hewitt EW (2003) The MHC class I antigen presentation pathway: strategies for viral immune evasion. Immunology 110:163–169

Hjelm B, Brennan DJ, Zendehrokh N, Eberhard J, Nodin B, Gaber A et al (2011) High nuclear RBM3 expression is associated with an improved prognosis in colorectal cancer. Proteomics Clin Appl 5:624–635

Horton P, Park KJ, Obayashi T, Fujita N, Harada H, Adams-Collier CJ et al (2007) (2007) WoLF PSORT: protein localization predictor. Nucleic Acids Res 35:W585–W587

https://aspe.hhs.gov/sites/default/files/documents/dddeb3748f486324a493a8a6d27f4338/aspe-vaccine-costs-brief.pdf. Accessed 7 Apr 2025

https://parasite.wormbase.org/Wuchereria_bancrofti_prjeb536/Info/Index/. Accessed 7 Apr 2025

https://www.cdc.gov/smallpox/about/history.html#:~:text=Ali%20Maow%20Maalin%20was%20the,22%2C%20he%20developed%20a%20fever. Accessed June 2025

https://www.jenner.ac.uk/about/edward-jenner. Accessed June 2025

https://www.who.int/health-topics/tuberculosis#tab=tab_1. Accessed Mar 2025

https://www.who.int/news-room/spotlight/history-of-vaccination/history-of-smallpox-vaccination. Accessed Nov 2025

Hudson E, Uhlen M, Rockberg J (2012) Multiplex epitope mapping using bacterial surface display reveals both linear and conformational epitopes. Sci Rep 2:706–714

Hunt DF, Henderson RA, Shabanowitz J, Sakaguchi K, Michel H, Sevilir N et al (1992) Characterization of peptides bound to the class I MHC molecule HLA-A2.1 by mass spectrometry. Science 255:1261–1263

Jacob CO, Leitner M, Zamir A, Salomon D, Arnon R (1985) Priming immunization against cholera toxin and *E.* coli heat-labile toxin by a cholera toxin short peptide-beta-galactosidase hybrid synthesized in E. coli. EMBO J 4:3339–3343

Jahantigh HR, Elsharkawy A, Guglani A, Arora K, Patterson LD, Kumar M (2025) Neurobiological alterations induced by SARS-CoV-2: insights from variant-specific host gene expression patterns in hACE2-expressing mice. Viruses 17:329–341

Jogi A, Brennan DJ, Ryden L, Magnusson K, Ferno M, Stal O et al (2009) Nuclear expression of the RNA-binding protein RBM3 is associated with an improved clinical outcome in breast cancer. Mod Pathol 22:1564–1574

Jonsson L, Bergman J, Nodin B, Manjer J, Ponten F, Uhlen M et al (2011) Low RBM3 protein expression correlates with tumour progression and poor prognosis in malignant melanoma: an analysis of 215 cases from the Malmo Diet and Cancer Study. J Transl Med 9:114

Karlos GG, Bal H, Wang L (2024) NetCL: a unified programming framework for in-network computing. In: SC24: international conference for high performance computing, networking, storage and analysis, Atlanta, GA, USA, pp 1–20

Kazim A, Atassi M (1981) Haemoglobin binding with haptoglobin. Biochem J 197:507–510

King S, Rajko-Nenow P, Ropiak HM, Ribeca P, Batten C, Baron MD (2020) Full genome sequencing of archived wild type and vaccine rinderpest virus isolates prior to their destruction. Sci Rep 10:6563–6576

Kokoszka ME, Kay BK (2015) Mapping protein-protein interactions with phage-displayed combinatorial peptide libraries and alanine scanning. Methods Mol Biol 1248:173–188

Lerner RA, Green N, Alexander H, Liu FT, Sutcliffe JG, Shinnick TM (1981) Chemically synthesized peptides predicted from the nucleotide sequence of the hepatitis B virus genome elicit antibodies reactive with the native envelope protein of Dane particles. Proc Natl Acad Sci USA 78:3403–3407

Lewinsohn DA, Swarbrick GM, Park B, Cansler ME, Null MD, Toren KG et al (2017) Comprehensive definition of human immunodominant CD8 antigens in tuberculosis. NPJ Vaccines 2:8

Li B, Gallin WJ (2004) VKCDB: voltage-gated potassium channel database. BMC Bioinformatics 5:3–7

Lin N, Miyamoto K, Ogawara T, Sakurai S, Kizaka-Kondoh S, Kadonosono T (2024) Epitope binning for multiple antibodies simultaneously using mammalian cell display and DNA sequencing. Commun Biol 7:652–663

Luo W, Pei J, Zhu Y (2010) A fast protein-ligand docking algorithm based on hydrogen bond matching and surface shape complementarity. J Mol Model 16:903–913

Luo H, Lin Y, Liu T, Lai FL, Zhang CT, Gao F et al (2021) DEG 15, an update of the database of essential genes that includes built-in analysis tools. Nucleic Acids Res 49:D677–D686

Mahoney KE, Shabanowitz J, Hunt DF (2021) MHC phosphopeptides: promising targets for immunotherapy of cancer and other chronic diseases. Mol Cell Proteomics 20:100112–100122

Morrison KL, Weiss GA (2001) Combinatorial alanine-scanning. Curr Opin Chem Biol 5:302–307

Mou Z, Li J, Boussoffara T, Kishi H, Hamana H, Ezzati P, Hu C et al (2015) Identification of broadly conserved cross-species protective Leishmania antigen and its responding CD4+ T cells. Sci Transl Med 7:310ra167

Murphy WJ, Longo DL (2022) A possible role for anti-idiotype antibodies in SARS-CoV-2 infection and vaccination. N Engl J Med 386:394–396

Nag R, Srivastava S, Rizvi S, Ahmed S, Raza ST (2025) Innovations in vaccine design: computational tools and techniques. Adv Pharmacol 103:375–391

Nakamura Y, Moi ML, Shiina T, Shin-I T, Suzuki R (2023) Idiotope-driven T-cell/B-cell collaboration-based T-cell epitope prediction using B-cell receptor repertoire sequences in infectious diseases. Viruses 15:1186

Naveed M, Abid A, Aziz T, Saleem A, Arshad A, Javed K et al (2025) Fragment optimized chalcone derivatives targeting OmpA protein as a therapeutic approach against multidrug resistant Acinetobacter baumannii. Sci Rep 15:3917–3935

Nutman TB (2001) Lymphatic filariasis: new insights and prospects for control. Curr Opin Infect Dis 14:539–546

Olson JK, Croxford JL, Calenoff MA, Dal Canto MC, Miller SD (2001) A virus-induced molecular mimicry model of multiple sclerosis. J Clin Invest 108:311–318

Pannek S, Higgins PG, Steinke P, Jonas D, Akova M, Bohnert JA et al (2006) Multidrug efflux inhibition in Acinetobacter baumannii: comparison between 1-(1-naphthylmethyl)-piperazine and phenyl-arginine-beta-naphthylamide. J Antimicrob Chemother 57:970–974

Paque RE, Miller R (1991) Autoanti-idiotypes exhibit mimicry of myocyte antigens in virus-induced myocarditis. J Virol 65:16–22

Parvizpour S, Mohammad MP, Razmara J, Mohammad Rafi MA, Omidi Y (2020) Epitope-based vaccine design: a comprehensive overview of bioinformatics approaches. Drug Discov Today 25:1034–1042

Piri-Gharaghie T, Doosti A, Mirzaei SA (2022) Identification of antigenic properties of acinetobacter baumannii proteins as novel putative vaccine candidates using reverse vaccinology approach. Appl Biochem Biotechnol 194:4892–4914

Ramakrishna V, Ross MM, Petersson M, Gatlin CC, Lyons CE, Miller CL et al (2003) Naturally occurring peptides associated with HLA-A2 in ovarian cancer cell lines identified by mass spectrometry are targets of HLA-A2-restricted cytotoxic T cells. Int Immunol 15:751–763

Schuler GD, Boguski MS, Stewart EA, Stein LD, Gyapay G, Rice K et al (1996) A gene map of the human genome. Science 274:540–546

Shapiro IE, Bassani-Sternberg M (2023) The impact of immunopeptidomics: from basic research to clinical implementation. Semin Immunol 66:101727–101237

Shattock AJ, Johnson HC, Sim SY, Carter A, Lambach P, Hutubessy RCW et al (2024) Contribution of vaccination to improved survival and health: modelling 50 years of the Expanded Programme on Immunization. Lancet 403:2307–2316

Shcherbakov DN, Bakulina AY, Karpenko LI, Ilyichev AA (2015) Broadly neutralizing antibodies against HIV-1 as a novel aspect of the immune response. Acta Nat 7:11–21

Shukla AK, Misra S (2022) Clinical implications of anti-idiotype antibodies in COVID-19. J Basic Clin Physiol Pharmacol 33:727–733

Singh R, Capalash N, Sharma P (2017) Immunoprotective potential of BamA, the outer membrane protein assembly factor, against MDR *Acinetobacter baumannii*. Sci Rep 7:1–11

Sullivan A, Nathavitharana RR (2022) Addressing TB-related mortality in adults living with HIV: a review of the challenges and potential solutions. Ther Adv Infect Dis 9:1–19

Suprun M, Getts R, Raghunathan R, Grishina G, Witmer M, Gimenez G et al (2019) Novel Bead-Based Epitope Assay is a sensitive and reliable tool for profiling epitope-specific antibody repertoire in food allergy. Sci Rep 9:18425–18438

Swain SL (1983) T cell subsets and the recognition of MHC class. Immunol Rev 74:129–142

Tang X, Zhang W, Zhang Z (2025) Developing T cell epitope-based vaccines against infection: challenging but worthwhile. Vaccines (Basel) 13:135–154

Ternette N, Yang H, Partridge T, Llano A, Cedeño S, Fischer R et al (2016) Defining the HLA class I-associated viral antigen repertoire from HIV-1-infected human cells. Eur J Immunol 46:60–69

Testa JS, Shetty V, Hafner J, Nickens Z, Kamal S, Sinnathamby G et al (2012) MHC class I-presented T cell epitopes identified by immunoproteomics analysis are targets for a cross reactive influenza-specific T cell response. PLoS One 7:e48484

Tunyasuvunakool K, Adler J, Wu Z, Green T, Zielinski M, Žídek A et al (2021) Highly accurate protein structure prediction for the human proteome. Nature 596:590–596

Wei Y, Qiu T, Ai Y, Zhang Y, Xie J, Zhang D et al (2024) Advances of computational methods enhance the development of multi-epitope vaccines. Brief Bioinform 26:bbaf055

Wölk B, Trautwein C, Büchele B, Kersting N, Blum HE, Rammensee HG et al (2012) Identification of naturally processed hepatitis C virus-derived major histocompatibility complex class I ligands. PLoS One 7:e29286

World Health Organization (2020) Child mortality and causes of death. WHO. https://www.who.int/gho/child_health/mortality/mortality_under_five_text/en/

Wyatt R, Kwong PD, Desjardins E, Sweet RW, Robinson J, Hendrickson WA et al (1998) The antigenic structure of the HIV gp120 envelope glycoprotein. Nature 393:705–711

Zhang Q, Wang P, Kim Y, Haste-Andersen P, Beaver J, Bourne PE et al (2008) Immune epitope database analysis resource (IEDB-AR). Nucleic Acids Res 36:W513–W518

Zhao Y, Kuang M, Li J et al (2021) SARS-CoV-2 spike protein interacts with and activates TLR41. Cell Res 31:818–820

# Index

**A**
*Acinetobacter baumannii*, 281
Acquired immunodeficiency syndrome, 76
Acute respiratory distress syndrome, 83, 168, 224
AddaVax, 244, 246, 247
Adenoviruses, 32, 56, 106, 138, 141, 232, 261, 274
Adjuvants, 241–244, 246–250, 252
Affinity purification mass spectrometry (AP-MS), 116, 176–178, 181, 184, 232, 241
Alanine scanning mutagenesis, 269
Ali Maow Maalin, 260
Amantadine, 231
Angiotensin-converting enzyme 2 (ACE2), 81, 131, 133–135, 161, 164, 167–169, 174, 175, 179, 183, 267, 268, 270
Antibiotic, 60, 79, 121, 237, 281–283
Antibodies, 14, 22–24, 53–56, 62, 65, 78, 92, 131, 133, 135, 141, 152, 168, 183, 196, 208, 209, 241–243, 262–265, 267–269, 271, 276, 279, 281
Antibody arrays, 22, 23
Antibody-titer, 263
Antigen presentation, 211, 241–243, 274
Antigen presenting cells, 95, 241, 271
Antigens, 55, 64, 243, 249, 263–265, 267, 271–274, 278, 279, 281
Anti-idiotype response, 267
Antiretroviral therapy, 197, 206
Anti-viral therapies, 52
Apoptosis, 41, 93, 178, 199, 200, 211, 234, 237
Aptamers, 168, 170

**B**
Bacteriophages, 40, 128
Baloxavir marboxil, 231
Baltimore Classification scheme, 37
Biotyper CA®, 61
B lymphocytes, 78
*Bombyx mori* nucleopolyhedrovirus (BmNPV), 109
Borg CR10 scale, 95
Bottom-up proteomics, 8, 9, 13

**C**
Capsid, 33–35, 40, 42, 43, 105, 132, 138, 152, 161, 162, 178, 191–193, 195, 196
Cardiovascular disease, 207, 211, 213
Carvedilol, 180
C-C chemokine receptor type 5, 42
CD4 receptors, 42, 105
CD4+ T cells, 76, 133, 194, 206, 273
CD40, 95–97, 242
CDC, 58, 225, 232
Cell-surface receptor, 129
Chikungunya virus, 109
Cholesterol metabolism, 86, 87, 89
Chromatin immunoprecipitation, 115
Clinical virology, 52, 59–62
*Clostridia bacteria*, 79
Colony-forming units, 274
Community transmission, 129
Complement system, 84, 87
COVID-19, 52, 64, 72, 81–94, 97, 98, 159–184, 225, 243, 267
COVID-19 Meta-Analysis of Plasma Proteins (CoViMAPP), 183


T. D. Veenstra, *Proteomic Applications in Virology*, Advances in Experimental Medicine and Biology 1511, https://doi.org/10.1007/978-3-032-22340-1

COVIDpro, 182–183
CoVProt, 181–182
Cytokines, 77, 81, 84, 111, 145, 168, 170, 202, 203, 242, 243, 268, 274

**D**

Data-dependent analysis, 9
Data-independent analysis, 9
Dendritic, 145, 194, 211, 243
Dengue virus, 106
Desmoykin, 234, 236
d'Hérelle, F., 105
Diagnosis, 52, 53, 56, 57, 59, 60, 62, 64, 66, 68, 80, 92, 97, 148, 154, 202, 206, 216, 252
Discovery-driven, 4, 5, 12, 24, 59, 109, 112, 113, 116, 118, 122, 133, 137, 152, 174, 214, 241, 253
Disease progression, 2, 83, 98, 203, 204, 216
DNA methyltransferase 1, 121
DNA viruses, 31, 39, 40
Double-stranded DNA, 36, 209
Doxorubicin, 121

**E**

Ebola, 32, 33, 35, 76, 119, 224
Electrospray ionization (ESI), 7, 11
Endocytosis, 34, 35, 40, 45, 130, 161, 164, 175, 229
Enveloped viruses, 34, 148
Enzyme-linked immunosorbent assays (ELISAs), 54, 56, 90, 93, 196, 264, 270
Epidemic, 32, 58, 59, 71, 190, 224, 225
Epidemiology, 52
Epitope, 263, 264, 282
Epstein-Barr virus, 32, 78
ESCRT-III, 151, 152
Extracellular matrix, 42, 86, 87, 169, 200, 203, 211, 277
Extracorporeal membrane oxygenation, 83

**F**

FDA, 61, 66, 98, 121, 180, 181
FLAG affinity tags, 177
Fractionation, 2, 6, 10, 12–17, 61, 108, 110, 168
Framingham Heart Study, 212

**G**

Gel electrophoresis, 2, 10, 12, 106, 131, 196
Gene ontology, 109
Genome-wide association studies, 81, 181, 275
Gilteritinib, 239
Glenny, A., 241
Golgi apparatus, 46, 146, 150, 230
gp41, 42, 106, 191–193, 195, 196
gp120, 41–43, 105, 132, 133, 191–196, 263
Gut mucosal, 210, 212

**H**

H1N1, 119, 145–148, 224, 227, 228, 231, 244, 283
H3N2, 225, 231–236, 238, 239
H5N1, 145–148, 227, 231–239, 244, 283
Hemagglutinin, 15, 44–46, 57, 64, 115, 225–227, 229, 244, 250
Hepatitis C, 117, 119, 121, 273
Hepatitis E virus, 153
Herpes simplex virus, 76, 137
Host-directed therapy, 233, 240, 252
Host receptors, 81
HSP70, 109, 179, 247, 249, 250
Human Genome Project, 5, 275
Human immunodeficiency virus (HIV), 32, 33, 41–45, 55, 76, 78, 97, 104, 105, 119, 133, 136, 163, 165, 189–217, 263, 273
Human leukocyte antigens, 78, 270
Human Protein Atlas, 171
Hunt, D., 272
Hydrocephaly, 78
Hydrochlorothiazide, 180
Hypothesis-driven, 2, 4, 11, 12, 24, 58, 59, 106, 112, 113, 133, 152, 253

**I**

Immunogenicity, 241–243, 250, 252, 276, 278, 280, 282
Immunoinformatics, 275–280
Immunoprecipitation, 14, 131–133, 152
Inflammation, 212
Influenza, 32, 41, 44–46, 55, 56, 64, 69, 76, 78, 79, 97, 105, 133, 136, 144, 145, 147, 163, 165, 175, 224–241, 244, 250–252, 273
Influenza A, 44, 45, 69, 105, 133, 144, 145, 147, 224–228, 230, 231, 233, 235–240

Influenza B, 226
Influenza C, 227
Innate immune system, 204, 211
Integrase, 40, 42, 43, 141, 191, 192
Interferon-gamma (IFN-γ), 95, 173, 274
International Committee on Taxonomy of Viruses, 31
Intestinal fatty acid binding protein, 209
Ion-mobility MS, 167
Isobaric tags, 144
Isobaric tags for relative and absolute quantitation (iTRAQ), 144, 145, 147

**J**
Jenner, E., 260

**K**
Kaposi's sarcoma, 132
Kaposi's sarcoma-associated herpesvirus, 132
Kyoto Encyclopedia of Genes and Genomes (KEGG), 109, 147, 183

**L**
Label-free, 140, 144
Lateral flow immunoassays, 54
Lestaurtinib, 238
Liquid chromatography, 7, 61, 134, 145, 166, 174, 244–245, 272
Liquid chromatography combined with tandem mass spectrometry (LC/MS), 11, 13, 14, 84, 108, 115, 122, 123, 134, 144, 174, 177, 280, 281
Long COVID, 93–97, 267
Lymphatic filariasis, 276

**M**
Macrophages, 76, 168, 194, 202, 215, 243, 249, 273
Major histocompatibility complexes, 241
Marburg virus, 32, 148
Mass spectrometry, 8, 12, 13, 15, 16, 21, 60, 62, 65, 67, 70, 114, 131, 134, 145, 165, 174, 176, 177, 196, 232, 233, 272
Matrix-assisted laser desorption/ionization (MALDI), 7, 10, 60–62, 65, 66
Matrix proteins, 42, 45, 46, 231
Metapneumovirus, 62
MF59, 244, 251
Microbiome, 77–79, 93, 127, 212
Microglia, 249
Middle Eastern Respiratory Syndrome, 32
Mitogen-activated protein kinase, 235
Moderna, 163
MPL/Alum adjuvant, 246
Multidimensional protein identification technology (MudPIT), 13
Multiple-reaction monitoring-MS, 62
Multivesicular bodies, 149, 150
Mycobacteria, 33
*Mycobacterium tuberculosis*, 273
Mycophenolic acid, 121

**N**
N-acetylneuraminic acid, 228
National Center for Advancing Translational Sciences, 180
Negative regulatory factor, 191
Negative-sense single-stranded segmented RNAs, 227
Neuraminidase, 45, 46, 64, 105, 225, 226, 229
Neutrophil chemotaxis, 200
Neutrophil extracellular traps, 84, 87
Neutrophils, 84, 145, 249
Neutrophil to lymphocyte ratio, 81
Next generation sequencing, 59
Non-enveloped virus, 35, 138, 152
Non-structural protein 1, 111, 113, 116
Nuclear export protein, 226, 236
Nucleic acid amplification tests, 57
Nucleic acid-based, 56
Nucleocapsid, 44, 45, 57, 64, 70, 160–162, 166, 170, 183, 191, 195

**O**
Olink®, 207–209, 212
Open reading frames, 111, 160, 191
Oseltamivir, 231
Ovalbumin, 244, 247, 249

**P**
Pandemic, 52, 58, 59, 64, 66, 71, 72, 76, 83, 91, 93, 97, 136, 163, 164, 172, 178, 184, 190, 211, 216, 223–225, 231, 232, 244, 251, 262, 267
Pan-specific drug, 232
Papanicolaou, 54
Parker, J., 260
Pathogenicity, 52, 128, 232
Peptide arrays, 264–267, 269, 270
Peramivir, 231

Pfizer-BioNTech, 163
pH1N1, 231–236, 238, 239
Phipps, J., 260, 261
Phosphoproteomics, 92, 235, 241
Pimodivir, 237
Plasma cells, 211
Plasma proteins, 84, 92, 95, 198
Plasma proteome, 6, 84, 87, 93, 197, 203, 208, 212
Point-of-care, 165
Polyadenylated mRNA, 230
Polymorphisms, 81, 183
Polymorphonuclear leukocytes, 249
Positive sense single stranded RNA, 160
Post-acute sequelae COVID-19, 93
Post-translational modifications, 5, 6, 9, 166
Pr55Gag, 191–193, 195
Pr160GagPol, 191–193, 195
Principal-component analysis, 140
Prognosis, 80, 81, 83, 84, 89, 92, 97, 98, 202
Prophylactic, 52, 197
Protein abundance, 13, 90, 138, 140, 144, 146, 173, 182, 208, 215
Protein arrays, 3, 20, 22–24, 207, 280
Protein-protein interactions, 11, 12, 16, 18, 45, 104, 118, 174, 195, 229, 232
Protein quantitation, 199
Proteomics, 2–6, 8, 10–14, 16, 17, 19, 20, 22, 24, 25, 29, 46, 61, 62, 66, 68, 69, 71, 72, 83, 84, 91, 92, 98, 106, 110, 112, 116, 117, 122, 123, 131, 133, 135–137, 143, 149, 151, 152, 154, 160, 163, 165, 167, 173, 180, 181, 184, 190, 196, 206, 207, 212, 214, 217, 239–241, 250, 251, 253, 267, 270, 272, 275, 277, 280, 284
Proviral DNA, 191, 192, 216
Proximity extension assay, 93, 207–209
Pull-down assays, 15, 114–117

**R**

Radiofrequency adjuvants, 243
Real time quantitative PCR (RT-qPCR), 57, 58
Renin-angiotensin-aldosterone system, 167
Replication, 29, 37–39, 41, 42, 44, 45, 54, 104, 105, 119, 129, 130, 137, 139, 141, 143, 148, 149, 153, 154, 160–162, 164, 168, 170, 179, 180, 190–194, 203–205, 207, 208, 210, 211, 216, 217, 226, 229, 231, 232, 237, 274
Respiratory disease, 84
Respiratory syncytial virus, 54, 111, 113, 116, 232
Retroviruses, 37, 40, 57, 79, 104
Reversed-phase chromatography, 11
Reverse phase protein arrays, 22
Reverse transcriptase, 37, 38, 40, 42, 43, 57, 191
Reverse transcription polymerase chain reaction (RT-PCR), 64–67, 70, 165–167
Ribavirin, 121
Rimantadine, 231
RNA polymerase, 37, 43, 45, 46, 57, 113, 160, 162, 192, 195, 226, 230, 231
Rubella, 78

**S**

Secretome, 281–283
Sequential window acquisition of all theoretical mass spectra (SWATH-MS), 197, 198, 244, 246, 247
Severe acute respiratory syndrome coronavirus 2 (SARS-CoV-2), 54, 55, 57, 58, 62, 64–70, 76, 81, 83, 85, 88–91, 93, 94, 97, 98, 119–121, 131, 133–136, 160–181, 183, 184, 232, 237, 239, 240, 251, 267, 268
Sialic acid, 44–46, 105, 133, 228, 229, 231
Single stranded DNA, 36
Smallpox, 52, 260
Spike protein, 70, 131–135, 174, 175, 183, 267
Stable isotope labeling of amino acids in cell culture (SILAC), 144, 147, 149, 151
Strong cation exchange, 12, 13
Surfactant B protein, 172

**T**

Targeted MS, 165, 166
Targeted proteomics, 62, 66
Tissue arrays, 22, 24
TMT, 144
Top-down proteomics, 8, 13
Translational machinery, 129, 147
Tumor necrosis factor, 168, 200, 207
2D-PAGE, 10, 11, 147, 196

**V**
Vaccination, 93, 250, 251, 260, 261, 268–270, 274
Vaccine, 2, 52, 64, 183, 190, 216, 217, 224, 232, 241–244, 250–252, 260–264, 267, 268, 270, 271, 273–284
Ventilators, 91, 97
Viral budding, 151
Viral entry, 40, 41, 79, 104, 105, 109, 130, 131, 133, 134, 148, 153, 164, 175, 179, 196, 199, 205, 206, 228, 231, 235
Viral infectivity factor, 104, 191
Virome, 77, 79, 128
Virulence, 128
Viruses.STRING, 117, 118
Virus overlay protein blot assays, 106
Virus protein r, 191
Virus protein unique, 191
VITEK® MS Plus, 61

**W**
Western blotting, 6, 11, 106, 135, 143, 216, 240, 249, 270
White blood cell, 81
WHO Clinical Progression Scale, 82
World Health Organization, 32, 71, 82–83, 261
*Wuchereria bancrofti*, 276–280

**Y**
Yeast two-hybrid, 17, 111–113, 176

**Z**
Zanamivir, 231
Zika, 40, 76, 78, 119, 224

MIX
Papier aus verantwortungsvollen Quellen
Paper from responsible sources
FSC® C105338

If you have any concerns about our products,
you can contact us on
**ProductSafety@springernature.com**

In case Publisher is established outside the EU,
the EU authorized representative is:
**Springer Nature Customer Service Center GmbH**
**Europaplatz 3, 69115 Heidelberg, Germany**

Printed by Libri Plureos GmbH
in Hamburg, Germany